Earned Benefit Program Management

Aligning, Realizing, and Sustaining Strategy

Best Practices and Advances in Program Management Series

Series Editor
Ginger Levin

RECENTLY PUBLISHED TITLES

Earned Benefit Program Management: Aligning, Realizing, and Sustaining Strategy
Crispin Piney

PMI-PBA® Exam Practice Test and Study Guide
Brian Williamson

The Entrepreneurial Project Manager
Chris Cook

Leading and Motivating Global Teams: Integrating Offshore Centers and the Head Office
Vimal Kumar Khanna

Project and Program Turnaround
Thomas Pavelko

Project Portfolio Management in Theory and Practice: Thirty Case Studies from around the World
Jamal Moustafaev

Project Management in Extreme Situations: Lessons from Polar Expeditions, Military and Rescue Operations, and Wilderness Exploration
Monique Aubry and Pascal Lievre

Benefits Realization Management: Strategic Value from Portfolios, Programs, and Projects
Carlos Eduardo Martins Serra

IT Project Management: A Geek's Guide to Leadership
Byron A. Love

Situational Project Management: The Dynamics of Success and Failure
Oliver F. Lehmann

Ethics and Governance in Project Management: Small Sins Allowed and the Line of Impunity
Eduardo Victor Lopez and Alicia Medina

Becoming a Sustainable Organization: A Project and Portfolio Management Approach
Kristina Kohl

Earned Benefit Program Management

Aligning, Realizing, and Sustaining Strategy

Crispin ("Kik") Piney, PfMP, PgMP

CRC Press is an imprint of the
Taylor & Francis Group, an **informa** business
AN AUERBACH BOOK

CRC Press
Taylor & Francis Group
6000 Broken Sound Parkway NW, Suite 300
Boca Raton, FL 33487-2742

First issued in paperback 2022

ISBN 13: 978-1-03-247659-9 (pbk)
ISBN 13: 978-1-138-03312-2 (hbk)

DOI: 10.1201/9781315314297

Publisher's Note
The publisher has gone to great lengths to ensure the quality of this reprint but points out that some imperfections in the original copies may be apparent.

Visit the Taylor & Francis Web site at
http://www.taylorandfrancis.com

and the CRC Press Web site at
http://www.crcpress.com

Dedication

To all my friends and colleagues for their support and advice over the years.

Contents

Preface

The integrated project management family is composed of projects, programs, portfolios, and, for completeness, a close relationship with operations. Whereas there is general agreement that the overall goals of integrated project management are to generate beneficial change, there is still a large amount of discussion as to the scope of each family member and the actual techniques for achieving success.

This book addresses all of the stages that compose the life cycle of a business-related undertaking—whether this undertaking is tagged as a project, a program, or a portfolio. Rather than simply reformulating existing concepts, the book aims to break new ground and present a novel and consistent insight and approach at each stage. This new approach starts from the definition of the concepts and continues through modeling the solution, scheduling, risk management, resourcing, procurement, and progress tracking, right up to the handover of the new capabilities for ongoing operations and maintenance. The challenge of understanding stakeholders, analyzing their relationship to the program, and evaluating how to communicate effectively is also integrated consistently with the unique set of overarching tools and concepts that form the backbone of this book.

P.1 What Started It All

For many years now, I have been working on ideas and concepts of what I would like to see as integrated project management (IPM). IPM is characterized by a consistent model that combines and aligns the scope and methodology of each of projects, programs, and portfolios. This integrated view is sometimes called *Organizational Project Management* (Project Management Institute 2014), but I use the term IPM to differentiate my approach from that of the Project Management Institute (PMI®).

As part of my IPM analysis, I realized that earned value management, despite being a valuable tool for tracking completion of project scope, is ill-suited to and insufficient for tracking program and portfolio progress, since these need to be measured in terms of value rather than content. This led me to develop a new method that I named the earned benefit method (Piney 2011). As explained later in this book, this method made use of a structure called the *component–benefit matrix* to link each component of the program or portfolio quantitatively to the corresponding benefits. Although the overall concept seemed to have considerable value, I was faced with a new puzzle: how to develop this matrix from the approved business case for the program.

By luck, and with the help of social media, I found out that Fernando Santiago was developing a set of concepts and a tool for representing and analyzing business cases for programs and portfolios. The tool, called the benefits realization map (BRM), is being developed as a commercial software-as-a-service offering. Discussions with Fernando have contributed to expanding my initial concept into a complete methodology and set of tools for effective program benefits management.

As you will discover in this book, there is a considerable amount of ground to cover before arriving back where I started—that is, tracking progress with earned benefit. However, that is not all; in writing this book, once I arrived back at earned benefit, I realized that the initial ideas that I had outlined in the 2011 paper, although still valid, were simplistic, and that the concepts that developed along with writing the initial chapters of this book could be applied to provide a more complete and realistic method. Paradoxically, therefore, the original trigger for the book did not survive the book that it generated. However, the original earned benefit ideas have left behind a rich legacy.

P.2 Some of the Original Ideas

I have done my best not to reproduce ideas that are described elsewhere in the literature. My aim has been to build up an original set of concepts that integrate project, programs, and portfolios in such a way as to provide a new, consistent framework of concepts and tools for business management. Some highlights are given here.

P.2.1 A Consistent and Integrated Model

Chapter 1 addresses the potential confusion as to the individual scope of *project*, *program*, and *portfolio*. It considers the distinction from different points of view, including the novel approach of identifying the different mindsets required by the managers in each of these *domains*.

P.2.2 Typical Problem and Analysis

Nothing focuses the mind as much as the description of a typical problem to be solved. Chapter 2 starts from the conventional wisdom that an unacceptable number of projects or programs fail. The chapter builds on a classic example to introduce some of the original analysis tools for improving the chance of program success, which will be expanded in the subsequent chapters.

P.2.3 Integrating the Frameworks

Having understood the separate scope of the domains, the next step, in Chapter 3, is to model clearly how these members of the project family work together. Because of the business dimension of IPM, the role of the business analyst is evaluated and included in the model.

The starting position for this integration was to understand the mutual interdependence between the three domains: that programs and portfolios depend on projects, and that programs contain their own portfolio of projects and programs, and that programs and portfolios are intimately linked with benefits management. Therefore, given all of these mutual dependencies, a unique, consistent, and complete life cycle and set of concepts is required and has been developed.

This approach provides a sound basis for taking a major step toward the future of project management: a future in which each of the three domains of project, program, and portfolio fits together with the other two and contributes to the overall success of the business (program management) in achieving its strategy (portfolio management) while controlling the creation of each key deliverable in the most efficient manner (project management). The differentiation between the domains is the key to success, as it allows the project, program, or portfolio manager to focus on his/her specific responsibilities, techniques, skills and objectives while relying on the other domain managers and subject matter experts, such as business analysts, to deliver their parts of the solution The objectives, the mindsets, the tools, and the techniques are different and complementary in each domain.

This framework is a prerequisite to success but depends critically on the existence of a clear and complete description of the problem to be solved.

P.2.4 An Honest Cost and Benefit Model for Programs

In Chapter 4, an original model (the *results chain model*, illustrated by a BRM) with its analysis algorithm for defining the whole strategic solution is explained and applied to a representative case study.

The model is built up over three chapters (4, 5, and 6) in such a way that the costs of the initiatives can (and should) be estimated independently of the predicted value of the results. This makes the model impervious to any political pressure aimed at forcing the numbers to support a potentially invalid proposal.

Once baselined, this model is then preserved with its original content throughout all subsequent stages of the program. The model is robust and maintains its basic integrity through changes due to optimization or adaptation to changing conditions.

Intermediate Goals

The model is based on decomposing the overall objectives into intermediate goals successively until this logical chain leads to the definition of the required solution. In this way, the logical chain is built up between the actions to be undertaken and the strategic objectives to be achieved, allowing the definition of intermediate progress indicators. The assumptions on which these intermediate indicators are based are also part of this logical chain; they can be used to identify conceptual risks within the program as well as to allow ongoing assessment of the validity of the model during program execution.

Sharing of Benefits

The model allows *enabling* initiatives to be justified—that is, component projects or programs that, on their own, cannot be validated directly because the value of their deliverables is spread across a number of areas.

P.2.5 Calculating the Model

The basis for a program proposal is composed of two sets of numbers and a solution framework. The first set of numbers incorporates the benefit expected from the program. The other set specifies the cost of achieving these benefits. The solution framework is the link between the costs and the benefits. The business case compares the benefits with the corresponding costs.

A major contributor to program failure is addressed in Chapter 5. In current program management practice, the details of the solution framework are not used to provide a quantified validation of the business case. The solution framework only serves to justify the proposed technical architecture of the solution on an overall, undetailed basis. This lack of a focused, quantified approach leaves the door open to unvalidated assumptions that can undermine the viability of the program from the outset or lead to endless discussions with no definite conclusion. This situation can arise for accidental or deliberate reasons and becomes more likely the higher the political pressure and the more complex the program.

Chapter 5 develops and applies a consistent numerical algorithm that links the costs to the benefits, and the benefits to the costs, taking into account all of the features of the solution model. The approach encourages objective analysis and ensures an honest and largely unbiased analysis and forecast. The algorithm then provides a consistent basis for change management, scheduling, procurement, progress tracking and stakeholder relationship management.

P.2.6 Disbenefits and Essential Links

Including Disbenefits

Every change, however beneficial, can have side effects that negatively affect the final outcome. Chapter 6 explains how the new model allows these negative effects to be included in the complete analysis and taken into account in the evaluation of the business case. The case study also explains how to evaluate options for mitigating the disbenefits.

The Concept of Essential Links

The underlying synergy of the program model shows how several nodes can contribute jointly to the value of another (successor) node. However, in some circumstances, one of these contributing nodes may be essential before the successor node can gain any value at all. A worked example demonstrates how overlooking this constraint can lead to faulty assumptions being made in the forecasting process, in optimization, in management of change, and with respect to the final outcome of a program.

P.2.7 A Case Study

The model and concepts from the previous chapters are applied in Chapter 7 to a representative case study. This analysis provides practical insights into the following:

- Calculating the cost and the value of every entity in the program model
- Developing a business case for the program as well as for component projects
- Identifying and potentially removing nonviable components
- Determining the effect of essential links
- Formally defining subprograms as well as the value of the concept of subprograms
- Incorporating disbenefits and the analysis of mitigation strategies
- Evaluating the potential costs of false assumptions

P.2.8 Scheduling

A novel approach to scheduling is presented in Chapter 8. Although this scheduling method is not specific to the results chain model, it fits in with the overall approach of this book to integrate existing ideas in the field of project management: It encompasses the ideas and capabilities of all of the existing scheduling algorithms (CPM, critical chain, PERT, etc.) and resolves the limitations and contradictions found in some of these classical methods.

This scheduling technique is then used on the model in the case study to determine the delivery schedule, along with a cash-flow and break-even forecast.

P.2.9 Risk Management

In line with the aim of providing original, consistent and integrated concepts, a *total risk and issue management* (TRIM) process is developed and explained in Chapter 9. The interaction between TRIM and the program model that has been developed enables a program-level analysis and treatment of potential problems and advantages arising from the structure of the program. The analysis addresses the whole program model and takes into account the relative importance of the components, the dependencies, the disbenefits, and the underlying assumptions. TRIM provides a coherent process determining the actions to address advantages and problems as well as any potential side effects of these actions.

P.2.10 Capacity Planning

One major headache in most programs comes from being compelled to share resources with other parts of the organization, such as the operations group. However realistically the solution modeling and scheduling have been carried out, unless this problem is solved, the entire plan will be compromised. An innovative organization-planning model for addressing this problem is explained in detail in Chapter 10.

P.2.11 Procurement

The viability of a business model depends on being able to acquire the associated goods and services in a manner compatible with the assumptions and constraints of the model. The link between the results chain model and a procurement approach that has been applied by an award-winning team is explained in Chapter 11. The procurement approach allows the program to access and incorporate value-added ideas from the potential suppliers into the original proposal in order to produce a better solution from a technical and financial point of view.

P.2.12 Implementation Tracking

As mentioned in the introduction, and expanded on in Chapter 12, the new concept of earned benefit is developed and applied in parallel with earned value to provide a more complete means of tracking progress at multiple levels during the entire delivery phase of the program. The earned benefit parameters and indicators enable the program manager to translate progress, as reported by project managers, into concepts of interest to the program's business-focused stakeholders. Benefits-related reporting of this type can enhance the program's chances of maintaining the support of senior managers or financial sponsors and preserving its budgets across multiple review cycles.

P.2.13 Progress Tracking and Review

Once the program implementation phase is complete, earned benefit indicates that the *potential* benefit has been earned. However, actual benefits take time to arrive once the corresponding beneficial conditions have been achieved. A corresponding set of indicators to be handed over at the end of implementation for subsequent tracking is developed in Chapter 13. Some of these indicators are based on the earlier cash-flow forecasts and therefore provide a direct link to the BRM, thereby ensuring consistency with all of the concepts included in the business case. This link ensures that progress and the final assessment are carried out in an unbiased manner that is traceable back to the initial model and case study.

Due to this end-to-end consistency, tracking the accrual of benefits can be used to review the model on which the program was based. This analysis can help both to adapt the current program as well as to improve planning for future programs of a similar type.

P.2.14 Stakeholder Relationship Management

Modeling and planning provide a good technical basis for a program. However, ultimate success depends critically on ensuring the correct level of support from the stakeholders of the program. One key step in gaining support is to understand which elements of the program are of most interest to each stakeholder. An analysis of this type can allow the program manager to enhance the attractiveness of the program or to reduce resistance in areas where specific stakeholders feel negatively affected by the program.

Chapter 14 shows how the original program model, indicating links and synergies, can be used to expand an initial, superficial stakeholder analysis to include additional areas of interest or concern.

P.2.15 Communications

Chapter 13 explained how to obtain relevant program information. Chapter 14 provided a model for analyzing stakeholders. Together these chapters allow the program manager to determine the correct information, as well as who should be interested in it. The final chapter, Chapter 15, provides the missing link: What is the role of information and how can it be communicated in a meaningful manner that is acceptable to the corresponding stakeholder?

P.3 References

Piney, C. *The Earned Benefit Method for Controlling Program Performance.* Proceedings of the PMI® EMEA Congress, 2011.

Project Management Institute. *Implementing Organizational Project Management: A Practice Guide.* Newtown Square, PA: Project Management Institute, 2014.

Acknowledgments

The ideas in this book have evolved and developed thanks to discussions with friends, colleagues, and the participants I have met and trained in conferences and courses. To them, I offer my thanks and recognition.

However, I would never have translated these ideas into a book without the contributions and interest of a number of key people. I do not know which order to list them in, as they all contributed greatly and in different ways. In my mind, they are all top of the list.

Ginger Levin can provide more valuable knowledge and helpful contacts than anyone else I know. She has been generous in her encouragement, help, and advice. Her confidence and enthusiasm for the book have often been greater than mine and have provided me with energy and inspiration. She has read all of my draft chapters and done her best to tighten up my style and my explanations.

Fernando Santiago has shown generosity and patience in comparing and contrasting his approach to mine. His model served as the trigger that allowed me to expand my initial, limited Earned Benefit concept into a total program management method.

Simon Addyman was kind enough to share his innovative approach for contractor engagement with me and to allow me to quote extensively from his report on the success in applying it.

David Hillson ("the Risk Doctor") has taken time and effort to expand my understanding of risk management and has provided a valuable sounding-board for my ideas, even when we were not in full agreement.

Over several years, Max Wideman has allowed me to benefit from his extensive experience and common sense in a number of areas of project and program management. He kindly agreed to co-author the final chapter in this book and improved both the structure and the content considerably.

John Wyzalek has shown belief in my concept and, as Senior Editor, has acted as champion for the book within Taylor & Francis.

The team from DerryField Publishing Services who looked after me deserve recognition for the tactful and professional way they have done their best to control my frequently idiosyncratic or inconsistent use of punctuation and layout. My thanks go therefore to Theron R. Shreve, Director; Darice Moore, my copy editor; and Marje Pollack, who looked after the typesetting.

The list of acknowledgements would be incomplete if I did not include my wife. She supported my effort throughout and has always had more confidence than I not only that I could, but also that I would, finish the book successfully. The fact that you are reading this proves that, as usual, she was right.

About the Author

Crispin Piney (known as "Kik") has been involved in the project world since joining the IT Group at CERN, the European Laboratory for Particle Physics, in the 1970s, working on cutting-edge development projects. He later moved to the Digital Equipment Corporation (DEC), initially in England, before relocating to DEC's European technical centre in the south of France. While he was with DEC, Kik gained a deeper understanding of the importance of methodology, business alignment, and stakeholder relationship management.

After the acquisition of DEC by Compaq, Kik left in 2000 and set up as an independent consultant and trainer in project management. He invested enthusiasm, time, and effort working as a volunteer with the Project Management Institute (PMI®) on the majority of their standardization efforts, from the Organizational Project Development Management Maturity Model (OPM3®) through each the standards for projects, for programs, and for portfolio management (he was listed as a "significant contributor" for the first edition of both the program and the portfolio standards). He also contributed actively to the PMI competency development framework, the lexicon, and the *Practice Guide for Scheduling*. He is one of the three authors of PMI®'s *Practice Guide for Project Risk Management*.

Kik gained his Project Management Professional (PMP®) certification with PMI prior to leaving DEC. He later became the first person in France to acquire the corresponding certifications for programs (PgMP®) and, when it became available, for portfolios (PfMP®).

He has published a number of articles ranging across all aspects of projects, programs, and portfolios. He has presented at most PMI® EMEA congresses since 2002, as well as at regional events.

Additionally, Kik provides training on a wide range of topics in the project management space. He has developed and delivered courses across a broad range of industries including construction, perfumery, information technology, steelworks, aviation, packaging, electronics, and banking.

His approach to understanding and expanding the field of knowledge in the area of project, program, and portfolio management is based on the following principles:

- The need for alignment between all domains in order to ensure consistency
- The need for clarity and precision in order to avoid misunderstandings
- The willingness to generalize and expand valid concepts in order to increase their generality and potential value
- The willingness to reject ideas and beliefs, however well established, that are based on false or doubtful premises, and to discover creative, verifiable alternatives with which to replace them
- And, finally, the wish to share his enjoyment of the subject and, wherever possible, his sense of fun

Chapter 1

Defining the Domains

Projects, programs, portfolios: there seem to be too many P-words, but too little understanding of the domains that compose the project management universe.

"When I use a word," Humpty Dumpty said, in rather a scornful tone, "it means just what I choose it to mean—neither more nor less."
"The question is," said Alice, "whether you can make words mean so many different things."
"The question is," said Humpty Dumpty, "which is to be master—that's all."

– Lewis Carroll (Charles L. Dodgson)
Through the Looking-Glass, Chapter 6, p. 205 (1934)
First published in 1872

1.1 In the Beginning . . .

In the beginning, when the concepts of project management were being formalized, although people were managing programs and portfolios of projects, the main focus was on *projects*. This focus was clear from the name of the Project Management Institute® (PMI®), founded in 1969—by a happy coincidence, the same year of the first manned moon landing. The precursor to *A Guide to the Project Management Body of Knowledge (PMBOK® Guide)* (PMI 2013a), the "Ethics, Standards, Accreditation" [ESA] report, was published in *Project Management Quarterly* in 1983. This ESA report led to the first edition of *The Project Management Body of Knowledge* in 1987. It took almost another 20 years for the PMI to break free from this generic approach and issue the first editions of *The Standard for Program Management* (2006b) and *The Standard for Portfolio Management* (2006a). Thanks to these publications, two areas of the total project management space were effectively separated into different but affiliated domains. It is hardly surprising, therefore, that confusion and resistance persist among all of the relevant stakeholders—management, practitioners, and potential clients—who could only see everything as projects until fairly recently.

This chapter explains the differences and why the distinctions add value to the participants and to the customers of projects, programs, and portfolios.

1.2 Projects, Programs, and Portfolios

There are a number of different definitions for the terms *project*, *program,* and *portfolio* (the P-domains), proposed both by individuals and by standardization bodies. In addition, there are also competent project managers who state that much of the differentiation is not only unnecessary but is effectively counterproductive. My experience is that there is some truth in each of the arguments, but none of the current explanations captures what the profession really needs from a practical point of view in order to make use of these concepts.

1.2.1 Why Does This Matter?

It is valid to ask the question "Is there any benefit in differentiating between the P-domains, or is the distinction artificial and actually damaging?" As with most ideas, differentiation has both benefits and disadvantages, and it is important to recognize both of them.

There is a benefit in distinguishing the activity of creating a deliverable (i.e., a project) from that of coordinating the creation and integration of multiple, related deliverables in a manner aligned to a common goal (a program), since each requires its own different level of focus and has specific types of success criteria. Similarly, the managers of projects or programs are not well placed to determine the relative priority of their endeavours with respect to the other work carried out to achieve the organization's strategy—the domain of portfolio management. However, there is the risk that the separation introduces an arbitrary distinction that limits the project manager's role and reduces the end-to-end flow from strategy to execution.

Overall, as will be explained, the distinction is useful if its implementation is managed correctly. To achieve this goal, the terminology may need to be adapted.

The following section addresses the need for adapted terminology by considering the differences from two points of view: the mindset of the population that would apply them and the corresponding tools required. This dual approach may in fact go some way to explaining the different points of view of the practitioners, of the definers, and of the standardization bodies, since these points of view all start from the assumption that each domain can be considered to be self-sufficient with no overlap between the models. This assumption is too simplistic, though; for example, many project managers use program management concepts in order to scope their projects effectively, and best practice program management requires using portfolio management tools in order to prioritize, select, and coordinate the program components.

The similarities and differences are explained in more detail in the following section, but first, some thought needs to be given to a non-P-domain—one that benefits (and suffers) from the Ps but is ultimately the arbiter of success or failure of projects, programs or portfolios—that is, the operational arm of the organization.

1.3 Projects and Operations

Most of the roles and responsibilities in projects are well understood. These are reviewed in a later section. However, operations is often ignored, or at least passed over briefly, in discussions on project management. If the role of the operational units is overlooked, the project deliverables may turn out to be impossible to integrate into the organizational environment.

The interworking of projects, programs, and portfolios aims to support the business development chain composed of: new deliverables creating additional capabilities that are then integrated into "business as usual" to deliver beneficial outcomes and ongoing business investment to support the long-term strategy. What is often overlooked in discussions on project, strategy, and the like is that *operations* is the part of the business that actually delivers wealth! operations will become more effective in delivering wealth if the "business as usual" (b-a-u) environment is upgraded to contain new capabilities, linked to strategic goals, that operations can apply. This statement has three important parts:

- *new capabilities*: These are the integrated deliverables from supplier-level projects.
- *linked to strategic goals*: This is the responsibility of benefits management linked to the investor level.
- *that operations can apply*: This generally entails changing the operational processes to allow the deliverables mentioned above to be exploited effectively—the classical role of program management.

The next section applies this concept in analyzing the roles of the main categories of stakeholders.

1.4 Understanding the Roles

There are two major categories of stakeholders to be considered:

- The *investor*, whose interest is in creating a worthwhile return and who agrees to apply money, time, and resources to obtain this objective.
- The *supplier*, whose role is to create a product or service that can satisfy the investor and who expects to receive money or equivalent remuneration for this work.

Two additional roles are required to ensure that these two categories of stakeholders work effectively together:

- The *business* ensures that the supplier produces the correct product as required by the investor.
- *operations* provides ongoing, value-added services (often known as "business-as-usual") by exploiting what the supplier provides.

These ideas are explained in more detail later.

Stakeholder category (domain client)	Focus of the work	Objective of the effort	Role of the manager	Control	Role of operations
Investor	**Strategic** advantage	*Return*	**Portfolio** management	*Vision*	**Exploit** the service
		Creates		Directs	
Business	Beneficial **outcome**	*Benefit*	**Program** Management	*Mandate*	**Transition** to business as usual
		Creates		Directs	
Supplier	New **deliverable**	*Capability*	**Project** Management	*Charter*	**Support** development
		Creates		Directs	
Implementer	Specified **task**	*Element*	**Technical** Development	*Brief*	

Figure 1.1 The Categories of Stakeholders with the Corresponding Project and Operational Implications

(Text introducing Figure 1.1 can be found on the following page.)

These groups of stakeholders work together, each in relationship with a P-domain (shown in Figure 1.1). Note that although it is not usually considered explicitly as a domain in PMI's standards, the task level has been included, since this is where the building blocks of the solution are created. Further, there are both top-down and bottom-up interactions between the domains.

1.4.1 Investor Focus

The investor has a financial interest in an enterprise that can or should provide a valid return on his or her investment. If the enterprise already exists, operations is responsible for ensuring effective b-a-u (shown in "role of operations" in Figure 1.2). To change the way in which the enterprise operates, the investor will have a strategic goal divided into a set of strategic criteria that define areas in which the investor sees a need to progress ("focus of the work"). A number of possible initiatives for achieving the strategic goal need to be defined for each of these criteria; these initiatives should be specified and planned, and managed as an integrated portfolio ("role of the manager"). The responsibility for delivering them falls to the business.

Stakeholder category (domain client)	Focus of the work	Objective of the effort	Role of the manager	Control	Role of operations
Investor	**Strategic** advantage	*Return*	**Portfolio** management	*Vision*	**Exploit** the service
		Creates		Directs	
Business	Beneficial **outcome**	*Benefit*	**Program** Management	*Mandate*	**Transition** to business as usual

Figure 1.2 Investor Focus

From Investor to Business

The portfolio manager determines the scope of the corresponding initiatives and the sequence in which these steps should be executed, and then he or she "directs" the business to carry out the corresponding work in order to create the capability assumed to create a beneficial outcome and integrate it into the operational environment. To accomplish this, a document called the (program) mandate is used for definition and control based on the overall strategic direction in the "vision document" (see the Control column).

1.4.2 Business Focus

The business is responsible for applying its resources in line with the overall business objectives. The amount of resources that should be used to make changes to the existing operational environment is defined from the investor layer, described above. The focus of the work (see Figure 1.3) is therefore to enhance the benefits capability of the organization by applying program management techniques; that is why program management is often defined in terms of "benefits delivery" despite the fact that most

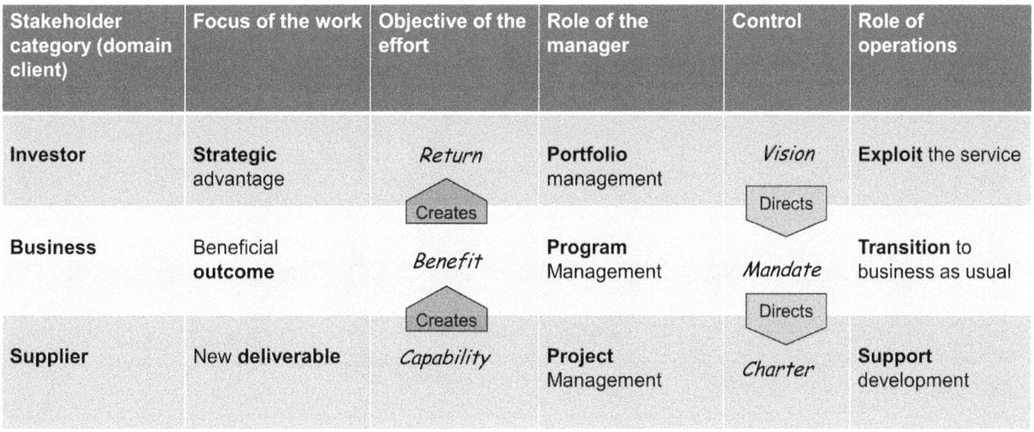

Stakeholder category (domain client)	Focus of the work	Objective of the effort	Role of the manager	Control	Role of operations
Investor	**Strategic** advantage	*Return*	**Portfolio** management	*Vision*	**Exploit** the service
		Creates		Directs	
Business	Beneficial **outcome**	*Benefit*	**Program** Management	*Mandate*	**Transition** to business as usual
		Creates		Directs	
Supplier	New **deliverable**	*Capability*	**Project** Management	*Charter*	**Support** development

Figure 1.3 Business Focus

programs are considered to be complete once the beneficial outcome has been created and before the benefits themselves have accrued.

From Business to Investor

The program should therefore create a capability from which the business can achieve a benefit and integrate that capability into the operational environment in a way it contributes to building an operational environment to deliver the required strategic advantage. Creation of the capability is the responsibility of the program manager and includes the following steps:

- defining the capability
- creating a working environment for developing it
- planning the major components of the capability
- authorizing the work to implement the components
- integrating the capabilities delivered by the components to ensure that they provide the required outcomes
- transitioning the result into operations; this step is carried out in collaboration between the program manager and the operations group

The implementation of the components entails involving the supplier layer.

From Business to Supplier

Once the required capability is clearly specified, an agreement is developed with the supplier. If the supplier is external to the organization, the choice of supplier and the resulting agreement must follow normal commercial rules (request for proposal, contractual agreements, etc.). For an internal supplier, the approach may follow the PMI approach of a mandate followed by a charter.

1.4.3 Supplier Focus

The role of the manager in the supplier layer is "project management" because the business layer requires a deliverable from the supplier (see Figure 1.4). The involvement of operations is normally greater when

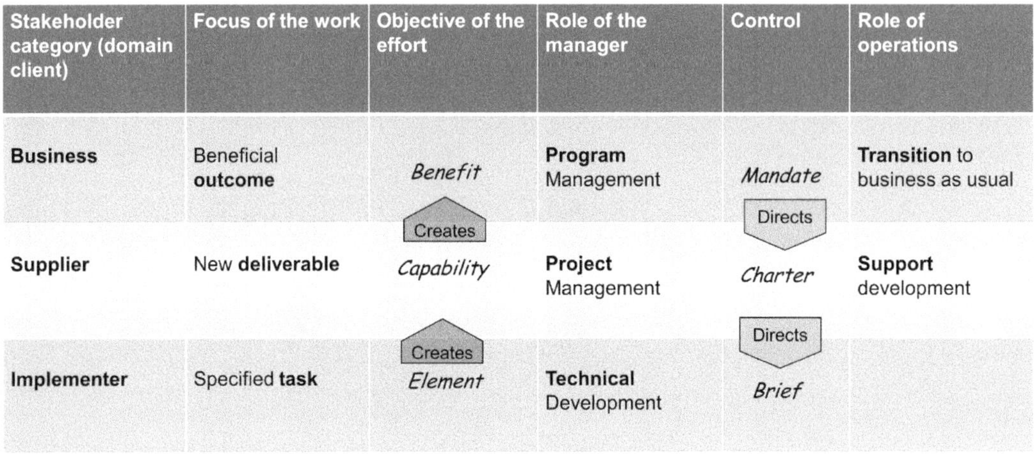

Stakeholder category (domain client)	Focus of the work	Objective of the effort	Role of the manager	Control	Role of operations
Business	Beneficial **outcome**	*Benefit*	**Program** Management	*Mandate*	**Transition** to business as usual
		Creates		Directs	
Supplier	New **deliverable**	*Capability*	**Project** Management	*Charter*	**Support** development
		Creates		Directs	
Implementer	Specified **task**	*Element*	**Technical** Development	*Brief*	

Figure 1.4 Supplier Focus

the supplier is internal to the organization, but that involvement can also be arranged in agreement between the project sponsor and the supplier to increase the chances of a smooth transition of the deliverable into the operational environment. The way in which the supplier interacts with the two neighboring domains is explained below.

From Supplier to Business

The supplier is expected to deliver regular status reports to the business as defined in the specification for the work and to provide the deliverable or deliverables in accordance with the formal agreement (e.g., a contract or charter). Final acceptance testing is normally carried out jointly between the business and the supplier. Acceptance testing may be handled in two steps: on reception, based on specific validation rests, and after integration, with the other program components prior to final transition to operations. This second validation step should be covered in the initial agreement for work between the business and the supplier.

From Supplier to Implementer

The supplier uses project or program management techniques to define the work packages and activities that compose the corresponding project or program. Each of these is described in a document similar to a simplified charter, called the brief. A specific implementation person or group is then directed and authorized, chartered, or contracted to carry out the work (titles such as "stream leader" are used for the person accountable for the work). The responsibility of the implementer is described in the following section.

1.4.4 Implementer Focus

Implementers will use whichever technique they think fit to manage the work at their level (e.g., project management tools, Excel activity lists, etc.; see Figure 1.5). At this level, the operations group is unlikely to be involved in their role of managing the service, although some of the operations staff may use their technical skills to implement some of the work. The challenges that are raised for capacity planning by the involvement of operational resources are analyzed in Chapter 10. The interface between this layer and the supplier layer will, however, be defined by the supplier, as explained next.

Stakeholder category (domain client)	Focus of the work	Objective of the effort	Role of the manager	Control	Role of operations
Supplier	New **deliverable**	*Capability*	**Project** Management	*Charter*	**Support** development
		Creates		Directs	
Implementer	Specified **task**	*Element*	**Technical** Development	*Brief*	

Figure 1.5 Implementer Focus

From Implementer to Supplier

The implementer will deliver estimates, status reports, and the tested element of the specified work to the supplier. In the same way as between the supplier and business layer (see Section 1.4.3 for specifics), acceptance testing will be carried out jointly on delivery, but full acceptance may be delayed until integration with the other related components is complete.

1.4.5 Review

Although the details given above are all relevant for the analysis, it is also useful to be able to provide a simple and succinct characterization of the domains (see Figure 1.6):

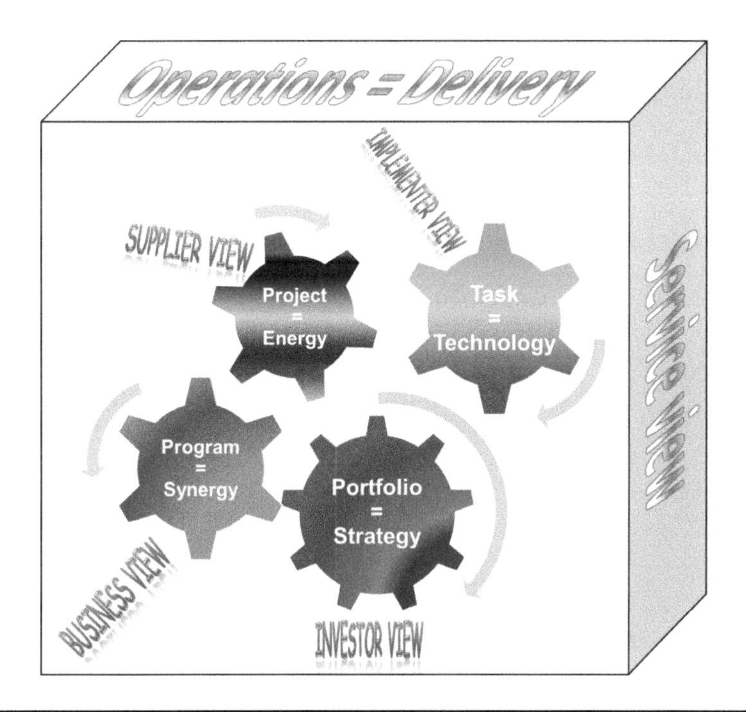

Figure 1.6 The Set of Domains

- A portfolio supports *strategy.*
- A program brings *synergy.*
- A project focuses on *delivery.*
- A task involves *technology.*
- Operations expends *energy.*

The roles and responsibilities are complementary, with commonalities and differences in the way in which they are carried out.

1.5 Commonalities

The commonalities among the P-domains relate to the how. However, these commonalities have to be adapted depending on the management focus to which they relate.

All of the domains have the following artifacts in common:

- Definition of the reason for the endeavor
  - The term *mandate* will be used to cover all domains
- Definition of the products, services or results (*statement of work*, or *SOW*) and the corresponding features and functions to be provided (the *scope statement*)
- The authorization to apply organizational resources to accomplish the endeavor (the *charter*)
- A hierarchical description (breakdown) of the work to be carried out to satisfy the SOW (project or program WBS or a portfolio component list)
  - For simplicity, this will be termed *activity register*
- Details of the components of the activity register (stored in the activity register dictionary):
  - Description of the logical dependencies between them
  - Estimates of time or effort for each component
  - Estimates and assignment of resources.
- Proposed sequence of components to provide the optimal result based on the charter (the *schedule* or *roadmap*)
- The need to manage uncertainties that could impact the achievement of the objectives (*risk management*)
- All other knowledge areas defined in the PMI standards

While the preceding list provides a brief overview of the commonalities, the major differences are in how to apply the corresponding tools in each of the domains. The differences are conditioned by the definition of the endeavor from the start (i.e., the very first document: the *mandate*), as explained next.

1.6 Differences

The difference is in the detail! The types of tools or the way in which they are used can be different in each domain. One other area of difference, due to the focus of the specific domain, is the way in which the work is controlled. Tools and control are analyzed separately in the following sections.

1.6.1 Tools

The tools are similar in intent but applied differently in each domain. The initial tool, the mandate, initiates this chain.

The Mandate

The mandate conditions the way in which the tools are applied within the domains.

- At the investor (portfolio) level, the mandate explains the overarching organizational strategy to be supported and the *benefit* that the strategy should deliver in the long term.
- At the business (program) level, the document describes the change to the operational environment that is needed to allow the portfolio goals to be achieved—the *outcome* to be created.
- At the supplier (project) level, the document provides a functional description of the *deliverable* or *capability* required as a component of the program.
- Although the task is not considered a domain, it is useful to examine it as well; at the implementer (task) level, the document provides a technical description of the *work* to be completed.

Other Tools

Some ways in which these differences in focus can affect the tools are explained below. More details are given in the corresponding PMI standards.

- Definition of the products, services, or results (SOW) and the corresponding features and functions to be provided (the Scope Statement)
  - No conceptual change
- The authorization to apply organizational resources to accomplish the endeavor (the Charter)
  - No conceptual change
  - At the implementer level, this will be included in the SOW
- A breakdown of the work to be carried out to satisfy the SOW (project or program WBS or the portfolio component list)
  - At portfolio level, this is the *project portfolio*
    - Managed as outlined in *The Standard for Portfolio Management*—Third Edition (PMI 2013b)
  - At program level, this is the *program component list*
    - Can (and should) also be managed as outlined in *The Standard for Portfolio Management* (PMI 2013b) because the set of components of the program can be managed as a portfolio of actions—although this approach is not mentioned in either *The Standard for Program Management*—Third Edition (PMI 2013c) or *The Standard for Portfolio Management*—Third Edition (PMI 2013b).
  - At project level, this is the *task list*
    - Managed using the scheduling tools
- Details of the components of the activity register
  - Description of the logical dependencies between them
    - Frequently overlooked for portfolios!
  - Estimates of time or effort for each component: resource estimates and assignment
    - No conceptual change
- The need to manage uncertainties that could impact the achievement of the objectives (risk management)
  - No change at the conceptual level
    - The practical differences are explained in the different standards
- All of the other knowledge areas
  - No change at the conceptual level
    - The practical differences are explained in the different standards.

1.6.2 Control

The contrast in the focus of control between the domains is the key differentiator. Although, as has been shown, the tactical management in the domains overlaps, there is a need for clear management roles and responsibilities. From this point of view, the type of control required by the domain client will condition requirements on the corresponding reporting. Since reporting depends on the tools used for tracking, and tracking—in turn—relies on the results of the planning, the specific tools used for each of these will depend on the category of the stakeholders and their domain focus. The different areas of interest of the stakeholders in the different domains are explained in more detail next with respect to the scheduling knowledge area. Similar differences can be identified in an analogous way in the other knowledge areas.

Developing the Schedule(s)

Once the dependencies, durations, resource requirements, and resource assignments are known, the schedule can be developed in line with the focus of the domain's client:

- At the investor level, the components need to be sequenced so as to provide the path to strategic advantage constrained by resources and investment, and optimized with respect to benefits—using the tools described in *The Standard for Portfolio Management* (PMI 2013b)—to create what should be known as the "critical route" linking the strategic initiatives.
  - Progress should then by tracked using key performance indicators (see, for example, Hynuk and Benoît 2010)
- At the business level, the program management approach should be used, to link deliverables from the supplier level to the needs of the investor level, and the roadmap is created using the "critical route" approach described in the previous bullet point
  - Tracking should then use the earned benefit method described in Chapter 12.
- At the supplier level, the project management approach should be applied as described in the *PMBOK® Guide*—Fifth Edition (PMI 2013a) applying critical path or critical chain techniques
  - Tracking should use tools similar to the earned value management (Fleming and Koppelman 2000)
- At the implementer level, the best practice approach for the corresponding technical area (e.g., information technology (IT) development techniques, building practices, etc.) should be applied. Reporting should comply with the rules specified in the corresponding SOW.

1.6.3 Easing the Transition

The previous sections have shown the value of the distinction between the project management domains. However, the challenge remains to make it easy for people to communicate unambiguously without having to go through a major mental shift in their use of language. One potential solution is presented below.

1.7 Compromise Terminology

Now that the integrated model has been developed, and the commonalities and differences between the domains have been investigated, these ideas will be used to develop a revised set of terms to take into account the fact that all of the domains are part of the project family.

At the most basic level, we have the *development project*: the classic view of a project as described in the *PMBOK® Guide*. The triple constraint of time, cost, and scope is a valid measure of success for this type of project. That is to say delivering what is requested, as requested, is exactly what is required. In a way, it can be compared with buying an article in a shop: You decide when you need it, whether it can do what you want, and how much you are allowed to spend on it.

The next level is the *business project* in which the whole is greater than the sum of its parts. This holistic view matches the concept of a *program*. For this type of project, the project manager has to understand the measure of success in the marketplace in which the outcome of the project will be used, whether business-focused or nonprofit-related benefits. The project manager in this case is sometimes referred to as the "business-savvy project manager." Business change projects, for example, are part of this domain. To continue the example, the business-savvy approach would entail buying the full set of equipment, training, maps, and the like *and* using them to go on a trip.

For the third level, the *enterprise project*, you need to take into account that, as ever, the main restriction on achieving your ambition is the level of available resources (e.g., money, people, time, etc.). Planning and delivery in this area require the use of project *portfolio* management tools. The term *portfolio management* is open to confusion because the everyday use of the term portfolio is generally associated with a financial investment portfolio. Although project portfolio management does not normally have a defined end-date (i.e., it does not fit the formal PMI definition of a project), portfolios are generally planned and replanned periodically in line with a specified strategic goal and current environmental factors. To follow on from the previous example, your strategic goal in the travel area might be to visit all countries whose name starts with the letter *B*. Your *enterprise project* for next year will have to take into account previous trips, your resources, and agreed-upon priority criteria.

That means that each of what is classically described as project, program, or portfolio is to be considered as a specific *domain* within the practice of project management. The advantage of this approach—which could be called *integrated project management*—is explained below.

1.8 Desired Result

The advantage of the Integrated Project Management approach is as follows. Based on comments on the Web and elsewhere, existing project managers can actually be managing any of these three domains—and most believe that they have, or would like to have, involvement at the business or investment level. The new naming structure supports this broader involvement, and the definition of a continuous career path, much more easily than the existing state, where somewhat artificial barriers are created by using totally different names for each of these project domains.

Another advantage is that this approach aligns with the real-life environment in which business managers who have no direct experience of managing projects take on the role of program managers. Lack of hands-on project-management experience is a potentially valuable feature because program management requires first and foremost solid business knowledge and experience, rather than a detailed project management understanding.

Integrated Project Management also avoids the other current confusion in which everything is called a "project" without clarifying the degree of business responsibility; for example, is an "IT project" responsible only for developing/acquiring a specific IT tool (an *output* project), or will it be measured on the benefit of the change it brings to the processes and operational efficiency of the organization (a *business* project), or is it responsible for ensuring that IT provides the capabilities required by the business on an evolving basis (an *enterprise* project)? To use the ubiquitous project management expression: "It depends!" What is, however, important is to ensure that the question is asked and answered clearly before the project is launched to ensure the expectations of all interested parties are aligned.

This is not only a matter of definition but also of technique, as explained previously.

1.9 Conclusion on Ps, O, and T

The important point to bear in mind is that each project domain shown in Figure 1.7 requires the use of different management techniques, as described for example in the three corresponding PMI standards (project, program, or portfolio). Using the example of an IT project, defining the domain of the project will determine how it should be managed from the initial definition right through to closure. Clear definition of the domain of responsibility should ensure that the correct expectations are set from the very start since, for example, the scope for a *development* project will not provide any promise of result from the creation of the deliverable as would be the case for a *business* project.

Once the project is well-defined, the authorization and subsequent steps should then follow the guidelines relative to the specified domain (project, program or portfolio) thereby ensuring that it is managed in accordance with the best practices in the corresponding domain.

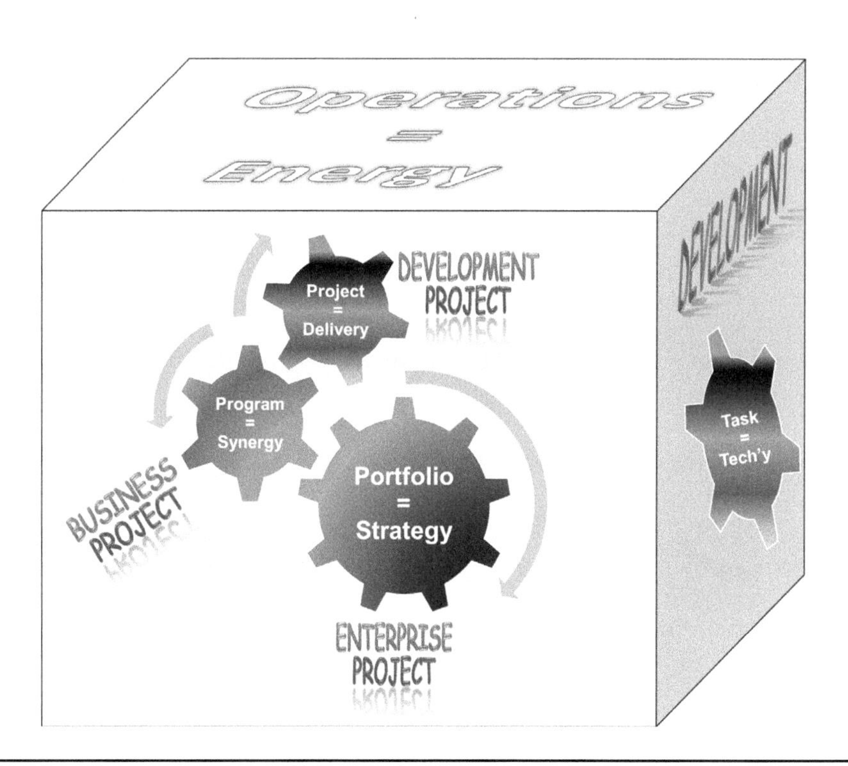

Figure 1.7 The Domains with Their Key Characteristics

1.10 Loose Ends

Although all of this may appear to be neat at first sight, the distinction between a classical (development) project and a program (a business project) is not really clear-cut because there are many supplier projects that would profit from the use of program management mechanisms. Two such examples are (1) evolutionary projects, and (2) integration projects. But then, these should really be assimilated with a business project to achieve lasting benefits.

1.10.1 Evolutionary Projects

A number of deliverables can or should be delivered in an evolutionary way, because, for example, of the need for flexibility and adaptation either to lack of long-term information, or to the need for repeated adjustments based on intermediate results—this is the domain of *agile* techniques. Although the agile techniques would be principally applied at the task (technology) work, they will have an impact at least on the development domain and possibly also beyond it.

1.10.2 Integration Projects

Projects for which there is a clear deliverable, rather than an outcome or benefit, but which rely on interdependent "subprograms" for their execution should also be implemented using program management tools for the definition, control, and integration of the subprograms. An example would be the development of a new aircraft, a project that seems to be midway between a deliverable (supplier view: the aircraft) and a benefit (business view: the capability to create and sell more such aircraft). The real synergy gained by managing all of the component subprojects in a coordinated way also argues strongly for adopting the program paradigm.

1.11 Towards Total Program Management

The preceding analysis shows that program management holds a central position in the set of project management domains. For the following reasons, successful program management calls on skills from all of the domains.

Program management requires the coordination of a number of interdependent components. Simple programs contain only projects. However, in larger or more complex programs, the component projects may themselves be organized as programs. In addition, some programs deliver their benefits incrementally and will manage the operation of the new services from within the program—this is described as "non-project work"—until the program is complete. Finally, coordination and control of all of the program components requires the use of portfolio management skills.

1.12 Summary

This chapter has presented a number of ways of understanding the differences and common features of projects, programs, and portfolios, and the role of operations in delivering value to an organization. The importance of understanding the differences resides in the fact that each of these project domains requires different, specialized ways of thinking and working that together build towards a complete, beneficial strategic result.

1.13 Now to Focus on Programs

Now that a clear definition of programs has been established and the differentiation with respect to the other related domains has been defined, we will be able to analyze the current challenges within the domain of program management, with a focus on benefits management and to develop a coherent approach for achieving success in this domain.

1.14 References

Fleming, Q. W., and Koppelman, J. M. *Earned Value Project Management.* Newtown Square, PA: Project Management Institute, 2000.

Hynuk, S., and Benoît, R. "Measuring Portfolio Strategic Performance Using Key Performance Indicators." *Project Management Journal* 41, no. 5 (2010): 64–73.

Project Management Institute. "Ethics, Standards, Accreditation: Special Report." In *Project Management Quarterly.* Newtown Square, PA: Project Management Institute, 1983.

Project Management Institute. *The Project Management Body of Knowledge.* Newtown Square, PA: Project Management Institute, 1987.

Project Management Institute. *A Guide to the Project Management Body of Knowledge (PMBOK® Guide)—Fifth Edition.* Newtown Square, PA: Project Management Institute, 2013a.

Project Management Institute. *The Standard for Portfolio Management.* Newtown Square, PA: Project Management Institute, 2006a.

Project Management Institute. *The Standard for Portfolio Management—Third Edition.* Newtown Square, PA: Project Management Institute, 2013b.

Project Management Institute. *The Standard for Program Management.* Newtown Square, PA: Project Management Institute, 2006b.

Project Management Institute. *The Standard for Program Management—Third Edition.* Newtown Square, PA: Project Management Institute, 2013c.

Chapter 2

Understanding the Problem

This chapter has been written in collaboration with Fernando Santiago of P3M Solutions, Inc.

Every few years, we are shocked by the Chaos report, which states that two thirds of projects are failing. After 15 years and the introduction of new approaches, including the critical chain and agile, project delivery has hardly improved dramatically. There are still fundamental problems that have not yet been addressed. This chapter proposes an approach and modeling technique which, if correctly applied, will eliminate most of the causes of these failures.

2.1 Overview

Despite significant investments in technology and organizational change, companies fail to make the connection between strategy formulation and execution. Information technology (IT) projects are a key component of implementing strategy, and the results are dismal:

- Kaplan and Norton (2015) tell us that nine out of ten companies fail to implement strategy.
- Program failures and cost overruns plague three-quarters of large federal IT programs.
- Estimates suggest that as much as $20 billion of U.S. taxpayers' money is wasted each year (Full House Committee on Oversight and Government Reform 2013).
- The Standish Group (2015) estimates that $81 billion worldwide will be spent on software projects that are abandoned.

If organizations are failing to implement strategy and wasting fortunes in IT, there is clearly a problem in portfolio management. The mainstream practice uses a bottom-up approach:

- Put ideas through a funnel process, based on a business case that gets progressively elaborated.
- Rank opportunities for investment based on their value, risk, and alignment with strategy.
    - This sometime involves complex multivariate analytical tools that few people understand or trust.
- Select the top of the list for execution, in a fixed cycle—monthly, quarterly, or annually, for example.
- Confirm the benefits in the business case after delivery—or not at all!

While this approach is logical and sound, in many occasions it leads to unsatisfactory results:

- Business cases may be based on fiction or be otherwise inappropriate for objective decision making.
- Executives who bring a last-minute list of projects "on a napkin" that jump the queue and get approved.
- The verification of benefits becomes an elusive task and simply does not get done.
- Millions are invested, and many times there is no clear idea of what the organization is trying to achieve.

It is not surprising that this form of program and portfolio management has not succeeded in helping companies implement strategy. The term *portfolio* comes from finance and leads to the root of the problem: the assumption that opportunities for investment are independent and that each one generates measurable financial returns. Finance departments have dictated the need to come up with "hard numbers" and a rate of return for every project. This approach works well when projects represent incremental improvement of an existing system that is not changing significantly. As an example in manufacturing, a new press will cut scrap by 20 percent, so you put the new press in and measure scrap—end of story. The link between the *apparently* independent projects in a portfolio should be provided by an effective program management approach to define and build on the synergies between the components. There will of course always be some projects that are truly independent and can only be managed at the portfolio level domain, but these do not tend to be the major loss-making projects.

2.2 Synergy

2.2.1 The Need for Synergy

Synergy can be seen, for example, when there is a strategy for transforming the business. As Porter states, "While operational effectiveness is about achieving excellence in individual activities, or functions, strategy is about *combining* activities" (1996, p. 10). The key word in the quotation is *combining*, as it goes against the concept of independent opportunities for investment. In a transformational strategy, activities are represented by projects that deliver capabilities, which then interact to generate results and sustainable competitive advantage. In his article, Porter presents Southwest Airlines (SW) as an example of a strategy that relies on combining activities. In a nutshell, to offer lower fares, SW does not provide meals, baggage checking, or seat assignments, and operates short flights from small airports. This set of changes enables SW to reduce gate turnaround time and increase plane utilization. In addition, it uses only one model of aircraft to reduce maintenance costs. As a result, SW offers lower fares than its competitors and point-to-point flights; after more than 20 years, it is still in the market and thriving. Talking about SW, Porter affirms (p. 11): "Its competitive advantage comes from the way its activities fit and reinforce one another."

In the case of Southwest, would it make sense to create a separate business case for not checking luggage? How much of the reduction in turnaround time at the gate can be attributed to that? Of course one can come up with a number, but the real benefit depends on all of the components working together. For the strategy to work, all of the key components need to be in place, so creating unconnected business cases and prioritizing then separately does not work in this scenario. However, since the initial business case was developed, changes in customer expectations and in software booking systems have led SW to provide seat assignments without damaging the overall strategy. This change now allows the passengers to be boarded more rapidly, thereby supporting the objective of reducing turnaround time at the gate.

The alternative approach to bottom-up portfolio management is the basis for this book: a top-down business case approach using a results chain, as shown in Figure 2.1, to capture the interactions

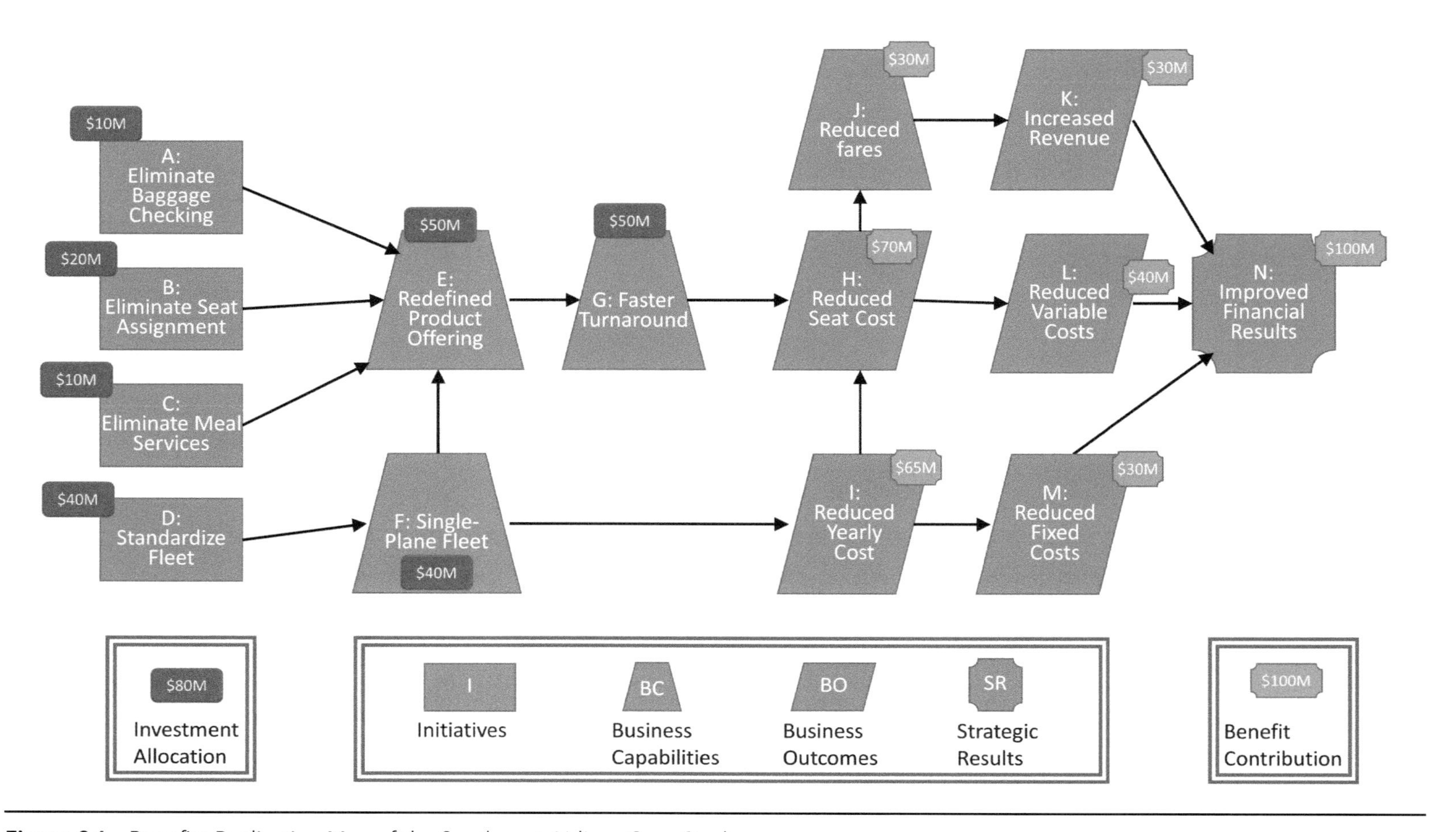

Figure 2.1 Benefits Realization Map of the Southwest Airlines Case Study

between financial outcomes, business outcomes, capabilities and initiatives. This results chain can then be expanded to include costs and benefits in order to generate a comprehensive benefits realization map. This approach is relevant to organizations that are planning a transformation, as opposed to incremental improvement. The top-down business case estimates the benefits the organization will realize from the total transformation with respect to the results it would obtain using the current business model. The difference between the transformed and the current scenarios generates a stream of benefits for the overall business case for transformation. These benefits are then propagated through the results chain based on the relative contribution of each component, to create the benefits realization map. Sounds simple? It is—in theory. The challenge is to understand the steps required to achieve the desired end state as well as the assumptions taken as to the value of the changes.

Continuing with Porter, Figure 2.1 is a revised interpretation of the SW case study. The overall benefits of implementing the strategy are estimated at $100 million, which are propagated from right to left, based on the relative contribution of each element to these benefits. In the first steps, 30 percent of the $100 million is contributed by K = Increased Revenue, a financial outcome. The source of this $30 million is then attributable to J = Reduced Fares, a business capability; this reduction in fares is enabled by the Reduced Seat Cost (node H). A second sub-chain starts from the right with a $40 million cost reduction (benefit) for L = Reduced Variable Costs, attributed to H = Reduced Seat Cost. These two sub-chains link together at H, giving a total of contribution by H of $40 million + $30 million = $70 million. What this is telling us is that 70 percent of the benefits are attributable to this business outcome and, as such, it should be a focal point for the transformation.

In a similar way, investments for the initiatives are propagated from left to right. The allocation of $40 million to D = Standardize Fleet enables the fleet to be standardized (node F), which then contributes to redefining the product offering (node E) with the corresponding cost allocations of $40 million for F and $50 million for E. Having a single-plane fleet (node E) delivers the yearly cost reduction (node I), which contributes $65 million to the total benefit.

2.2.2 The Problem with Synergy

Unless a diagram such as shown in Figure 2.1 is developed, there can be a number of issues and errors with the business justification. The three main errors are: double counting of benefits, massaging of figures, and loss of focus.

Double Counting

As Count Galeazzo Ciano (1947) wrote in his diaries: “success has many fathers.” In the business environment, each manager tends to attribute the main value of a program to the contribution from his or her part of the organization. In the SW example, it would be easy to imagine that the various managers for baggage checking, seat assignment, and meal services would each set their own (elevated) estimate on the contribution of their operational improvements. The sum of these might well exceed the total potential benefit of the program. This biased financial modeling is frequently a source of inflated promises leading to impossible targets and canceled or “failed” programs.

Massaged Figures

Once a strategic idea gains management support, the implicit pressure to prove its value tends to rise. This pressure can induce staff to reassess estimates of cost and effort in order to ensure that the business justification will be accepted.

Loss of Focus

Loss of focus on the details of the solution approach can occur because programs run for a long time. Over time, people and circumstances change. A financial justification that is based only on estimates and promises but does not have a solid underlying logical justification may well be ignored when later decisions have to be made. As time goes by, the interdependencies between all of the components can be overlooked or mismanaged.

2.2.3 Characteristics of the Benefits Realization Map

To mitigate the problems outlined earlier in the chapter, the key for this approach is the identification of the proper business outcomes, which in essence translate the strategy into action-oriented and measurable results. Once the links between actions and outcomes have been defined, the ambiguity is removed from the strategy. The what-contributes-to-what structure is clear so a reliable value mapping can be developed.

In a similar way, the estimated cost of each initiative can then be evaluated based on its specification, ignoring the value of the program as a whole. Insisting on an independent cost estimation in this way is key to ensuring the development of an honest business case for the program as a whole. The organization can then focus on achieving those key business results while the project managers can focus on delivering the component projects according to specification.

Another advantage of this approach is that it removes the need to come up with an independent business case for every project. Without this integrated approach to justifying the initiatives, the all-too-frequent approach is to create business cases based on fiction for those key projects that will transform the company.

Finally, for consultants and business analysts, this business realization map technique represents an opportunity to provide their customers with a fresh approach to implementing strategy; when nine out of ten companies are failing at this, and one trillion dollars is wasted every year just in IT, the business opportunities for everyone are significant. This book can be used as a basis for developing a consistent and reliable professional consulting service that would guarantee an improved level of consistency in place of the currently unreliable and wasteful practices. Even a 1 percent saving on the wasted trillion dollars can definitely be characterized as significant!

2.3 Business and Benefits Management: The Core Concepts

The concept of benefits management focuses on the basic objective for effective business management: to get more out of the business than you put into it, and to ensure the results can be maintained. In more formal terms, the outcome of any endeavor should be potentially more valuable than the investment made to achieve that outcome. In the map that will be developed, every component of the solution contributes toward the *value* of the endeavor, and each component requires an allocation of the overall *cost* of the investment in the corresponding implementation. These terms and a number of others are expanded in later chapters and are formally defined in the Glossary.

The difficulty in designing a strategy to achieve the business goal is created by the distance that exists between the initial investment and the final outcome of the endeavor. This distance can be measured in terms of many criteria: level of detail, mindset, time, visibility, accountability, authority, scope of control, and impact. The challenge, therefore, is to create a reliable and verifiable chain to connect the two extremes—the (technical) investments to the (strategic) outcomes—while linking these to the value contributions and the cost allocations.

As explained by Fernando Santiago (2012), top-down and bottom-up views of the opportunity contribute to the benefits realization map. First, top-down: A benefit comes from exploiting a new or existing situation; the opportunity is created by the outcome resulting from new capabilities; the capabilities are provided by outputs from specific activities.

Seen bottom-up (and chronologically), this chain of activities starts with projects, managed in a coordinated manner, then integrates the projects' outputs to create capabilities, which are then used to change the internal or external environment in a potentially beneficial way (a transition phase toward an outcome). Only once this change is in place can one really hope to benefit; the transition phase can be disruptive and initially counterproductive due, for example, to the need for people to retrain and gain new competencies, and for new processes to become fully accepted.

If the planning misses any of these steps, we can fall into a situation in which the costs exceed the benefits.

In fact, the real benefits only accrue once the transition is complete, and the new situation is operational. The change is the enabler, but operational stability is the creator of enhanced value and the resulting benefit. SW, for example, only saw its profits rise after all of the changes were in place. In many programs, however, the benefits can start to accrue before the program is complete, due to a phased implementation and successive transitions of the capabilities. A multi-country program, for example, with an initial implementation in a single country would start to generate financial benefits before the full roll-out is complete.

As mentioned above, a number of factors need to be taken into account to map out the route to achieving the strategic objectives.

2.3.1 Levels of Detail

The hierarchical structure of the benefits realization map is the first concept that needs to be understood, as it underpins the *top-down/bottom-up* terminology.

The top level is defined as the strategic overall view: where we wish to be. The desirability of this destination is measured in terms of *value* to the organization. The initial definition at the strategic level cannot and should not identify and define all of the details of the journey toward this valued destination.

The bottom level corresponds to the creation of the building blocks (initiatives) from which the final strategic outcomes will be built. This bottom level is where the *cost* of the work can be evaluated independently of other considerations. The focus for the lowest level can be considered as the technical level of the work. The people with day-to-day responsibility for one initiative at this level cannot simultaneously be concerned with the details of all of the other initiatives and even less take responsibility for the strategic details of the outcome. The range of their work affects the way in which the various actors need to envisage their responsibilities.

2.3.2 Mindset

The contrast between (bottom-level) technical requirements and (top-level) strategic vision explains the difference in the required mindsets. At the lowest level, the person responsible has to focus on providing the best possible building block consistent with the constraints defined from a higher level. At this lowest level, the building blocks can be termed *deliverables*. Deliverables are defined with the intention of achieving a given purpose, and it is successively the responsibility of higher levels to define and achieve these purposes in order to integrate the results with the aim of satisfying the ultimate strategic objectives. Successively higher levels, therefore, have successively broader and, by necessity, less minutely detailed, views of the path ahead. The higher levels integrate and apply. The lower levels concentrate and comply. The different levels therefore view time in different ways.

2.3.3 Time

The top level will define a window of strategic opportunity, whereas the lowest level needs to evaluate viable delivery dates. In accordance with the mindsets just defined, the lowest level considers time with respect to the current date (e.g., "tomorrow is the deadline"), whereas, at the highest level, the strategic vision is based on a projecting scenarios into the future and assessing them as if the future has arrived (e.g., "tomorrow is the new today").

2.3.4 Visibility

Because of the different level of focus required, each level also has a different degree of visibility of the final goal. To enhance the consistency between deliverables and to increase motivation, it is useful for the people at the lowest level to have an overall understanding of the final objective; however, their everyday focus must be on their current responsibility. In a similar way, the people responsible for the highest level need to understand the overall architecture of the solution, but they cannot spend their energy on the minutiae of each component.

2.3.5 Accountability

Accountability is an important consideration, because it happens only too often that people at the lower level are made accountable for areas over which they have no control, and in which they are not expected—or even allowed—to get involved. The benefits realization map is designed to make the areas of accountability clear by defining a hierarchy from deliverables to capabilities, from capabilities to outcomes, and from outcomes to benefits. In a well-regulated organization, the degree of accountability should be directly aligned to the level of authority.

2.3.6 Authority

In the resulting map, the concept of levels corresponds roughly to the levels of authority in the organization; both the development of the map and the assignment of authority are carried out in a top-down manner. The goal is to be able to align the scope, the accountability, the authority, the assigned costs, and the contributed value consistently at each level of the map so as to allow the solution to reflect the capabilities and structure of the delivery organization. The different levels of authority also imply different scope of control.

The diagrams in Chapter 1 on defining the domains provide a clear guideline to the required assignment of responsibility and accountability at each level:

- The portfolio manager should have overall budget level authority and be accountable for achieving the strategy.
- The program manager should have authority for defining the overall execution plan and be accountable for delivering the outcomes on which the strategic benefits depend.

- The project manager should have the authority to define and manage the creation of the required deliverables and be accountable for delivering according to the specifications provided by the program manager.

2.3.7 Scope of Control

Higher levels of the map are aligned to higher levels of authority and carry the need and the ability for wider-reaching control over the work to be managed. To a large extent, the higher levels have the responsibility to ensure the effective integration of the work from the lower levels as well as its application into the part of the environment over which the accountable manager has the corresponding authority. The concept of integration is a major consideration throughout the analysis and the map, in which the impact of each component on the others can be assessed and evaluated.

2.3.8 Impact

The result of the implementation of a component at the lowest level of the map will affect the final outcome of the strategy in that the component could make the strategy fail, but the component cannot ensure that the strategy succeeds: The strategy may fail at each level of the map, due to the inability to achieve the objectives defined at that level. However, none of the lower levels can ensure that the strategy succeeds. Only if the overall strategic plan is viable and correctly implemented will the strategy succeed. Top managers are therefore ultimately accountable for designing a successful strategy, but the responsibility for making it succeed depends on every one of the steps toward the overall outcome.

It is important to bear this mixture of shared but individual accountabilities in mind because it underlines the major contribution provided by an integrated methodology for analysis, design, implementation, and performance management, especially if supported by a tool that supports the methodology in a consistent manner.

2.4 Structure of the Benefits Realization Map

The map is designed to support top-down portfolio management but is closely linked as well to structuring a program. The goal is to develop the optimal architecture of initiatives through which to achieve a strategic or financial result; this architecture is represented by the benefits realization map.

The basic concept is that a portfolio is built up from the following *entities*:

- A set of *initiatives*—these are the *components* of the program portfolio.
  - Initiatives can be projects or programs.
  - Component programs will have their own benefits realization maps, thereby providing a hierarchical breakdown of the overall strategic approach.
- Each initiative creates either one or more *technical capabilities* and/or one or more *business capabilities.*
  - A technical capability is a means of performing a task, such as a tool.
  - A business capability is a means of acting on or within the current environment.
- Singly or together, business capabilities create *business outcomes*.
  - Business outcomes are changes to conditions or situations in the organization or the broader business environment.

- The business outcomes finally lead to *financial* or *strategic* outcomes.
  - A financial outcome is a change in the financial situation such as increased profitability.

Note: For this concept to be applicable also to organizations whose overall target (value) is not directly based on financial criteria, it would be more general to refer to this final step simply as *strategic outcomes*—to be defined as changes with respect to organizational goals such as reputation, market share, turnover, profits, happiness.

It is important to remember that outcomes can have negative as well as positive results. The positive ones can be referred to as *benefits*, with the corresponding qualifier (business, financial, strategic); the negative ones are *disbenefits*. These concepts match those of program management, since the characteristic of a program (e.g., in the definition provided in *The Standard for Program Management*—Third Edition [PMI 2013c]) is that it provides *benefits* that could not be achieved by a simple project approach. This mapping technique has has therefore been shown to include the whole range of capabilities from strategic planning right through to effective delivery of the relevant solution. It complements the program and benefits realization approach described in "Managing Successful Programmes" (see Figure 7.5 in Office of Government Commerce, 2011).

The benefits realization map ensures strategic consistency between all of the stages of the program, such as developing the implementation plan, managing procurements, and tracking and communicating progress.

The various phases and the corresponding life cycles are described in Chapter 3. The components and the linkages between them are defined in more detail in Chapters 4–7. Then Chapters 8–10 explain how to build a consistent implementation plan (scheduling, risk assessment, and capacity planning). The contribution of the benefits realization map in procurement is demonstrated in Chapter 11. The earned benefit technique for progress tracking is then explained in detail in Chapter 12, providing a benefits-focused approach and based on the same benefits realization map. Communicating this information and validating the initial map are addressed in the final two chapters.

2.5 Recap on the Benefits Realization Mapping Technique

The following set of definitions provides a top-down structure to the map:

- Business Objective (success criterion)
  - A measure of success of benefits delivery.
- Program Benefit
  - The achievement of one or more business objectives.
  - Contributes to achieving the organization's strategic vision.
- Outcome
  - The result of one or more actions, such as a potentially beneficial situation or scenario.
  - Can also have potentially damaging side effects (called *disbenefits*).
- Capability
  - A means that can be applied for achieving a result. There are two categories of capabilities: *technical* and *financial.*
  - Capabilities should contribute to outcomes.
- Deliverable
  - An output or a result of a project or subprogram.
  - Deliverables should contribute to capabilities.

- Initiative
  - A subprogram, a project or other work within a program
    - Projects create deliverables; subprograms create deliverables, capabilities, and outcomes.

2.6 Summary

This chapter has revisited the challenge of understanding why so many project-based endeavors are seen to fail. One core reason is the fact that a large number of such endeavors need to be structured, planned and managed as networks of mutually interdependent projects. Some of the difficulties of building an honest and verifiable business justification have been explained. The structure of the benefits realization map that is required to address this challenge has been developed and the way that this can address some of the main causes of failure has been presented.

2.7 The Need for End-to-End Control

The concepts and algorithms for building a complete benefits realization map will be explained in greater detail in Chapters 4–7, but, first we need to define a way of ensuring effective governance across the entire lifetime of the program.

2.8 References

Ciano, G. *The Ciano Diaries, 1939–1943,* Vol. 2. Portsmouth, NH: Heinemann, 1947.

Full House Committee on Oversight and Government Reform. "Wasting Information Technology Dollars: How Can the Federal Government Reform Its Investment Strategy?" Full House Committee on Oversight and Government Reform Hearing (2013). https://oversight.house.gov/hearing/wasting-information-technology-dollars-how-can-the-federal-government-reform-its-it-investment-strategy/ downloaded on February 9, 2016.

Kaplan, R. S., and Norton, D. P. "The Office of Strategy Management." *Harvard Business Review,* (October 2015). Downloaded on February 8, 2016, from https://hbr.org/2005/10/the-office-of-strategy-management

Office of Government Commerce. *Managing Successful Programmes (MSP®).* Norwich, England: The Stationery Office, 2011.

Porter, M. E. "What is Strategy?" *Harvard Business Review* (November–December 1996). Available from https://hbr.org/1996/11/what-is-strategy

Project Management Institute. *The Standard for Program Management—Third Edition.* Newtown Square, PA, USA: Project Management Institute, 2013c.

Santiago, F. *The BRM Tool,* 2012. See: https://www.youtube.com/watch?v=W-liDQt_WkE

The Standish Group. *Chaos Report,* 2015. Available from https://www.standishgroup.com/store/services/chaos-report-2015-blue-pm2go-membership.html

Chapter 3

A Life Cycle for Program Management, Benefits Management, and Business Analysis

All of the project management domains, as well as benefits management, have recognized the value of controlling progress by means of a life cycle. However, each life cycle seems to have been designed without sufficient consideration as to alignment with the other life cycles. This chapter provides a solution to this deficiency and adds a business analysis life cycle. It then introduces a case study and shows how the initial life cycle phases would be applied to it.

3.1 Introduction

The Standard for Program Management—Third Edition (Project Management Institute, 2013c) refers to benefits management and provides a diagram of the benefits management life cycle. However, it does not provide an explanation of how to apply benefits management in an integrated way across the program management life cycle. This chapter presents one possible approach to consider. The starting point therefore is a review of the program management phases as defined in *The Standard for Program Management*.

3.2 Current Life Cycles

3.2.1 The Program Management Phases

Figure 3.1 is adapted from various editions of *The Standard for Program Management* and is explained in detail in the following sections.

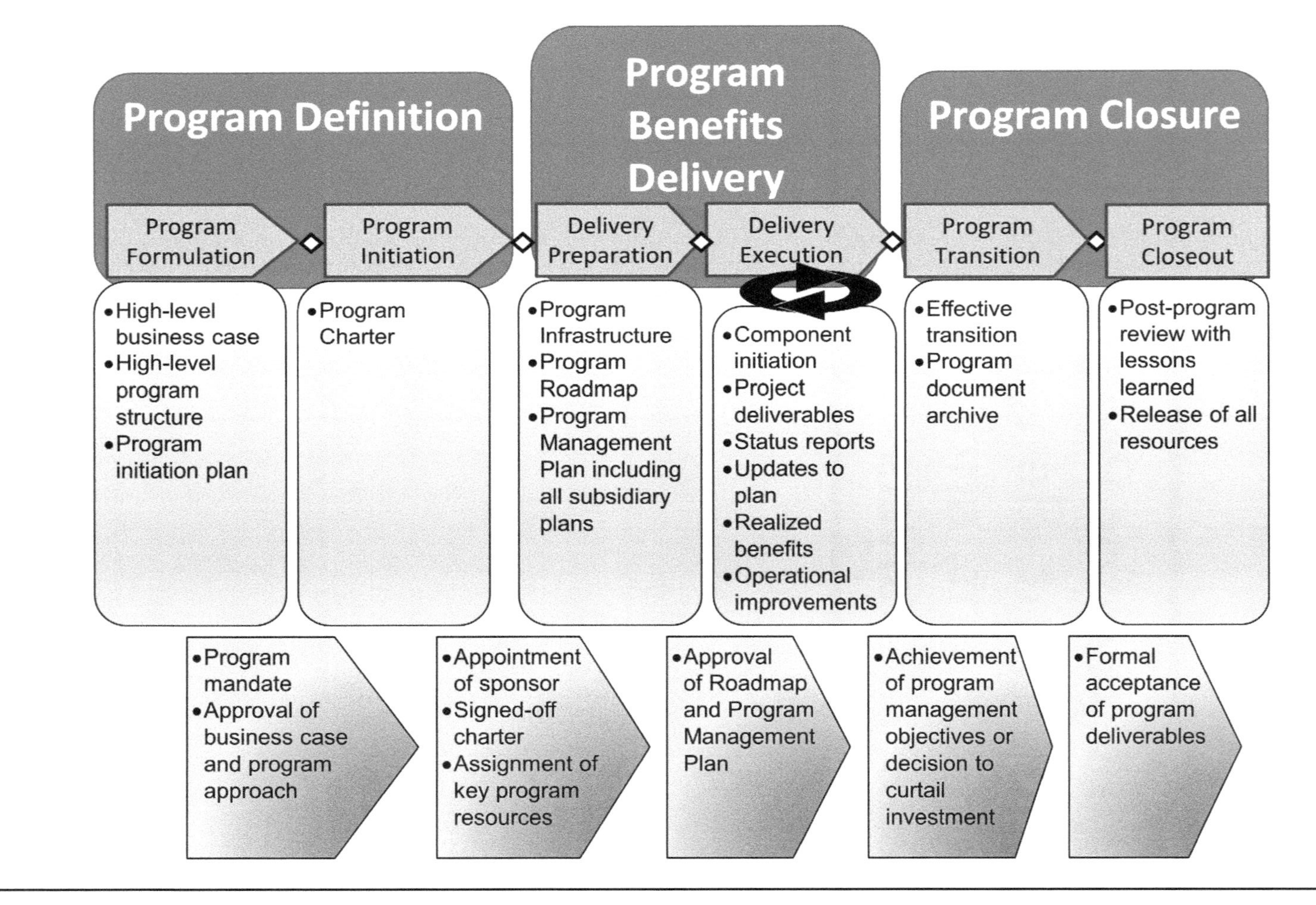

Figure 3.1 Program Management Life Cycle

3.2.2 Program Definition

The program definition phase comprises two subphases. The goal of this phase is to establish a firm foundation of support and approval for the program.

Program Formulation

Program-level deliverables include the following:

- The high-level business case
- The high-level program structure (checkpoints)
- The program initiation plan
- The program mandate
- The appointment of a program sponsor

Phase closure conditions include:

- Approval and publication of program mandate
- Approval of the business case and the program approach

High-Level Business Case

The high-level business case defines the conditions under which the proposed program would be viable. It lists:

- The need that the program addresses
- The areas of strategic alignment
- The assumptions on the value of the benefits
- An alternatives assessment
- The key stakeholders
- Investment assumptions in terms of finance, personnel, and internal resources
- The limit on the acceptable investment under which the program would be considered viable
- The level of confidence in the potential success of the program
- Unanswered questions

High-Level Program Structure

The high-level program structure outlines the initial concept on how the program should be organized. It includes ideas on:

- Governance, including committees, management office requirements, and reporting
- Timescales and milestones
- Funding
- Key resource categories
- Geographical considerations.

Program Initiation Plan

The objective of the program initiation plan is to present the actions, costs, and deliverables of the next stage of program definition for approval by the strategic decision-making body. It provides information on:

- Program manager selection, either explicitly (naming the designated person) or as a set of selection criteria
- Actions required to complete the initiation stage, plus the investment needed for that stage

Program Mandate

The program mandate provides the guidelines from the strategic decision-taking body to the sponsor and the future program manager on alignment, scope, and control of the program. It will be the basis for the charter in the next stage. It contains:

- The program vision
- Background information on the need for the program plus the level of importance to the organization
- The strategic objectives of the program
- The scope of the program
- The results required from the program
- The name of the sponsor for the program

The mandate designates and empowers the sponsor and launches the initiation stage.

Program Initiation

The aim of program initiation is to develop in greater detail how the program will be structured and managed.

The program-level deliverable for the initiation phase is:

- The program charter

The phase closure conditions for the initiation phase are:

- Appointment of program manager
- Signed-off charter
- Assignment of key program resources

Program Charter

The program charter designates the program manager and provides the authority to apply organizational resources to achieving the objectives of the program. Whereas the mandate indicates what the sponsor is appointed to accomplish, the charter is an agreement by the program manager on what he or she will take responsibility for delivering. The mandate can be compared to a request for proposal, with the charter as the signed-off response and contract with the supplier. The charter therefore integrates and adapts information from the previously-mentioned documents, including information on:

- The mandate
- The background to the program
- The expected benefits
- Any assumptions and constraints
- The program scope
- The risks, issues, and corresponding responses
- The initial delivery assumptions: schedule, resources, budget
- The stakeholders
- Governance, including the role of the sponsor and any oversight committees, as well as the program structure and reporting rules
- Program manager authority, including limits to decision making and conditions for escalation

3.2.3 Delivery of Program Benefits

The program benefits delivery phase is composed of two subphases, as shown in Figure 3.1.

Program Delivery Preparation

The first subphase includes all of the overall planning for the program. The program management deliverable on completion is the approved program management plan.

The program-level deliverables of the program delivery preparation stage are:

- The program infrastructure
- The program roadmap
- The program management plan, including all subsidiary plans

The phase closure condition of the program delivery preparation stage is:

- Approval of the roadmap and the program management plan

Program Infrastructure

The program infrastructure provides all of the specific support required in order to be able to deliver the program. The infrastructure includes:

- Organizational structures, such as a specific program office
- Technical capabilities, including software tools, communications networks, etc.
- Key personnel, including business analysts, technical architects, and other subject matter experts

Program Roadmap

The program roadmap provides a schematic view of how and when the components of the program are planned to deliver the proposed beneficial outcomes.

Program Management Plan

The program management plan is either a document or a linked set of documents describing all of the aspects and rules for the program. It covers areas such as:

- Strategic alignment and prioritization rules
- Benefits management
- Program boundaries
- Scheduling
- Budget and financing
- Personnel planning
- Technical resource management
- Program risk and issue management
- Program communication management
- Program stakeholder engagement
- Program performance management
- Program information management
- Program procurement
- Program administration, including:
  - Structure and rules
  - Program change management
  - Escalation
  - Roles and responsibilities
- Program quality management
- Program transition management

Details of the corresponding templates and forms can be found in *Implementing Program Management* (2013) by Ginger Levin and Allen R. Green.

The approval and baselining of the program management plan are required for closure of the program delivery preparation stage and for entering the execution stage.

Program Delivery Execution

The second subphase, program delivery execution, is responsible for the progressive delivery of the planned initiatives, capabilities, and outcomes. It is repeated until all of the work is compete with respect to the scope and objectives of the program, or until the decision is made to close the program prematurely. A simple representation of this iteration is shown in Figure 3.2.

The program delivery execution subphase controls the initiation and management of the component projects and the integration of their outputs for the development of the program benefits.

The program-level deliverables of the program delivery execution stage are:

- Project-related items
- Integrated outcomes
- Program progress reports
- Program change requests
- Realized benefits
- Operational improvements
- Additional governance documents.

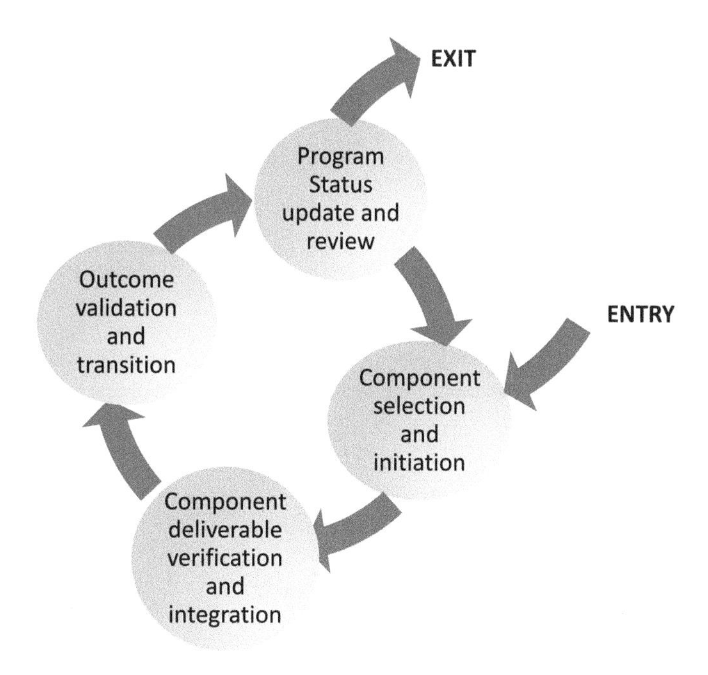

Figure 3.2 The Program Delivery Execution Subphase

The phase condition of the delivery of program benefits phase is:

- Accomplishment of the program management outcomes or a decision to curtail the investment.

The program manager ensures effective control during the delivery of program benefits phase.

Project-Related Items

Typically, the program manager is the sponsor of each component project because the project output contributes to the overall program objectives. Even if the program manager is not the day-to-day sponsor, he or she will be involved in the following:

- Initiating the project:
    - Developing the business justification and project charter
- Conducting project phase reviews
- Resolving escalated issues
- Ensuring validation, acceptance, and transition of project deliverables

Program Progress Reports

The objective of program progress reports is to provide the program sponsor and steering committee with an honest assessment of progress toward the defined program outcomes, with respect to the baselined program plan. For this reason, the main topics address:

- Spend versus plan
- Completion versus plan
- Update on the business case based on the previous two topics
- Risks and issues:
  - Action taken and planned at program level
  - Escalation requests

Program Change Requests

All changes to baselined documents must be supported by a formal, documented change request. The change request document contains two parts: one from the requester and one from the assessment team.

From the requester:

- The request
- The reason for the request; for example:
  - The corresponding value-added, reduction in threat exposure, issue resolution, etc.
- The estimated impact of performing the change; for example:
  - Cost, time, risks, etc.

From the assessment team:

- The technical solution options and their impact
  - The proposed solution
- The recommended decision:
  - Accept, refuse, defer, or request additional information

Additional Governance Documents

The following documents are also normally required:

- The program decision register
- The audit reports
- Any program closure recommendations

3.2.4 Program Closure

The program closure phase is also composed of two subphases or stages. The first stage addresses the outcome of the program, while the other stage addresses the program management process.

Program Transition

This is the phase in which ownership of the result of the program is transferred to the operations group.

The program-level deliverables of the program transition stage are:

- A handover acceptance document, signed by senior operations management
- The archive of program documents

The phase closure condition of the program transition stage is:

- Formal acceptance by the receiving organization

Program Closeout

Once handover is complete, all that remains is to ensure that the organization has captured options for future improvement and that all of the resources are returned, released, or reallocated in line with the program management plan.

The program-level deliverables of the program closeout stage are:

- The final review, with lessons learned
- The release of all resources

3.2.5 Linking the Program to the Benefits

In the same way as for the project management life cycle, the program management life cycle is designed to provide control over the work and the use of resources to accomplish it. It is not directly concerned with the technical aspects of the work to create the deliverables associated with the product scope. For programs, however, there is a layer that fits between the program scope and the product scope, since programs use the products, services, and results to deliver benefits; the extra layer is controlled by the *benefits management* life cycle, described in the next section.

3.3 Modified Benefits Realization Life Cycle

Although the PMI standard does propose a benefits management life cycle, it does not provide a detailed description of the relationship between the two life cycles. This section proposes a solution in which the benefits management tasks are assigned into a benefits realization life cycle aligned with the program management phases.

3.3.1 Defining the Modified Life Cycle

Approach and Scope

The program management approach taken here focuses on the benefits realization management life cycle: Benefits should drive programs, not vice versa, since programs are run in order to deliver benefits. The benefits realization life cycle can be constructed bottom-up, starting from the definition of the benefits realization processes and then grouping them into phases aligned to the program management life cycle.

The goal is not to define the entire life cycle of managing benefits, but only the portion that concerns program management—the realization of the benefits, that is, the creation of the beneficial outcomes and their transition into the operational environment. The longer-term delivery and application of these benefits for the organization is outside the scope of a program. Within the program, it is important to define a clear set of steps for enabling the achievement of the benefits-related objectives. The key steps are therefore described in the next section.

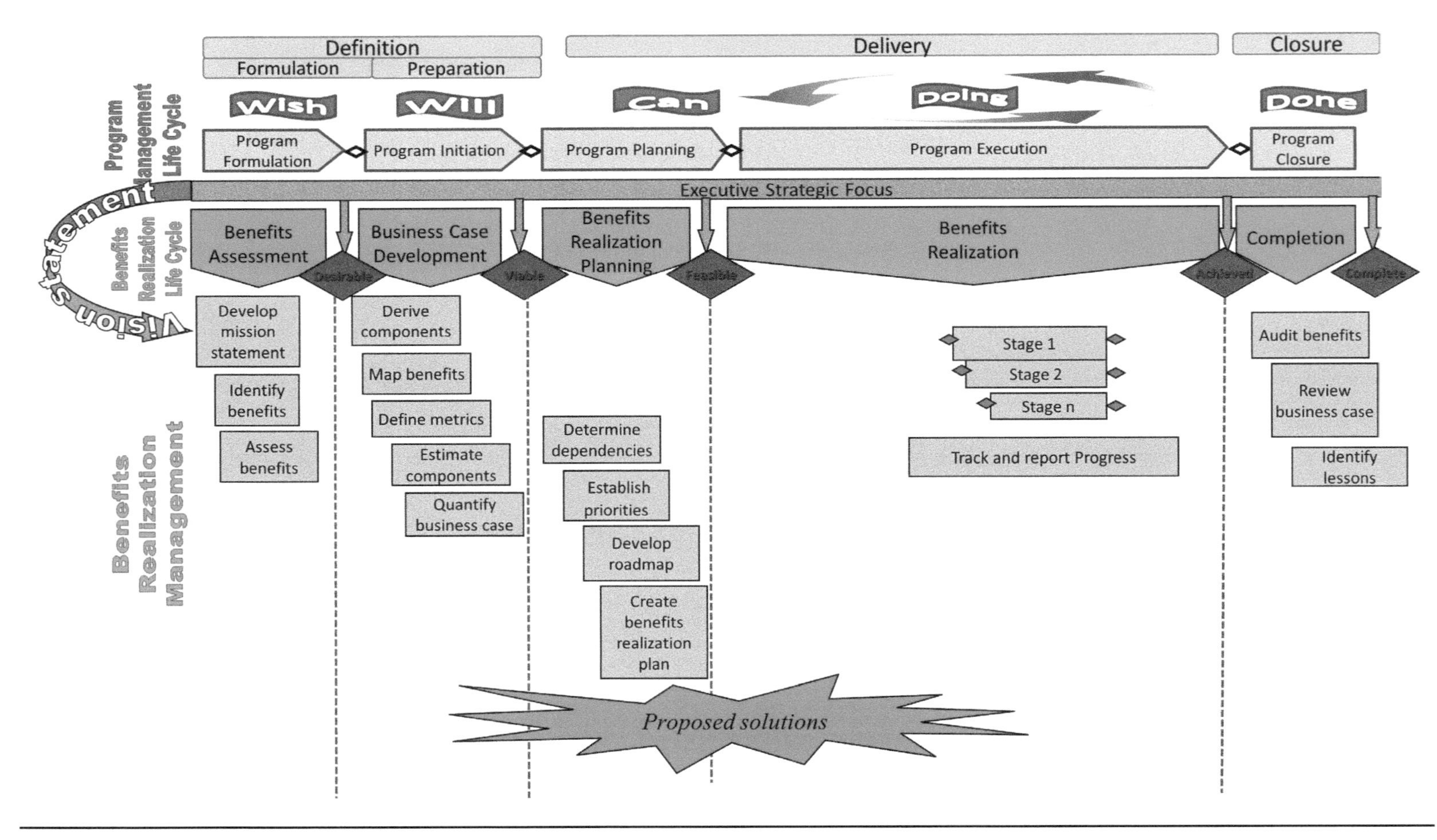

Figure 3.3 Aligned Program Management and Benefits Realization Life Cycles

Benefits Focus Steps

A program needs to take into account the following five levels of focus:

1. Strategic focus
   - Providing *vision*
2. Business focus
   - Ensuring *viability*
3. Benefits focus
   - Achieving *value*
4. Program focus
   - Creating *benefits*
5. Project focus
   - Delivering *capabilities*

A life cycle then needs to be applied to lead the program though these focusing steps, from vision to capabilities. The resulting life cycle is shown in Figure 3.3. The first phase of PMI's benefits management life cycle (benefits identification) has been modified to cover identification and assessment; the second phase (benefits analysis and planning) has been split into two parts: developing the business case and planning how to achieve it. Benefits realization is structured to indicate the potentially staged approach used in many programs; the final phase has been renamed *completion* to indicate its goal more clearly. These phases are explained in more detail in the following section.

3.3.2 Overview of the Benefits Realization Life Cycle Phases

The characteristics of each of the phases in Figure 3.3 are presented, followed by a detailed description of the key activities for each phase.

Benefits Assessment

The benefits assessment phase takes the vision statement and evaluates the general alignment of this vision with organizational long-term goals. It answers the question, "Is this change desirable?"

Business Case Development

The business case compares the costs with the projected benefits, enabling the program steering committee to determine the viability of the proposal.

Benefits Realization Planning

Development of a benefits realization plan provides the information required to decide whether the organization is capable of carrying out the work and achieving the goals.

Benefits Realization

The benefits realization phase is where the creation of the capabilities and the planned outcomes takes place. Its aim is to achieve the vision on which the program is based.

Benefits Realization Completion

Controlled completion requires a retrospective analysis of the work and final approval from the program steering committee.

3.4 Close-Up on the Benefits Realization Life Cycle

This section uses the newly developed life cycle in Figure 3.3 to show how the concepts are developed, planned for, and applied within programs.

3.4.1 Benefits Assessment

The main input to the benefits assessment phase is the vision statement. A vision statement should be written as if it has been created at a future date when the program has been successfully completed, describing what would then be the current situation. The vision statement is written in the present tense, as befits a description of a current situation, and does not explain how the situation was achieved. As mentioned in the previous chapter, for visioning, "tomorrow is the new today."

It is then up to the business change manager to help translate the vision into a set of benefits to be achieved in order to realize the vision. Depending on the size of the program, the role of business change manager can be a dedicated role, or it can be the responsibility of the program manager or the business analyst.

The key output of the benefits assessment phase is a clear definition of each of the benefits required from the program. The first step as shown in Figure 3.4 is to clarify and agree the mission.

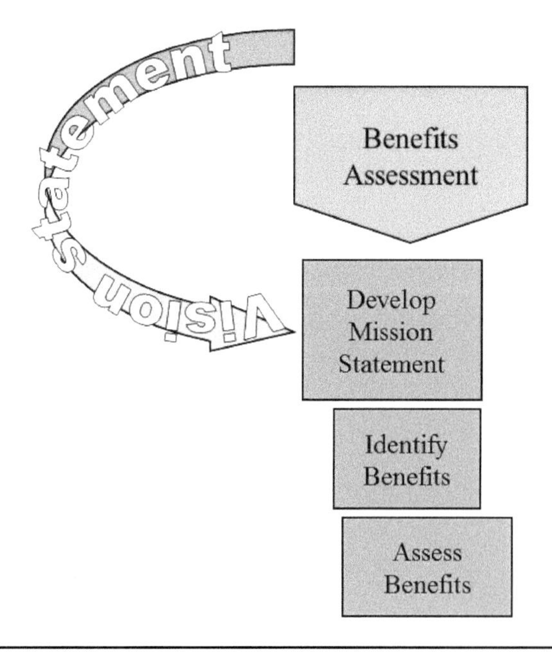

Figure 3.4 Main Steps in Benefits Assessment

Define the Mission

The first step is to translate the vision into a more concrete description of the organizational change required to achieve that vision. Translation of the vision into an action-oriented format is the responsibility of the sponsor or the business change manager and will take the form of a *mission statement.*

The mission statement says what needs to be done (not how to do it) and includes the mechanisms for defining completeness. The mission statement can also provide a high-level definition of the outcomes that need to be created, as well as guidelines to the approach to be taken for realizing the benefits (e.g., quick wins, minimal cost exposure, high-risk first, etc.).

A simple, generic structure for a mission statement is "to <narrative description of goal> by <top-level actions to be taken>."

This structure links actions to benefits in the vision statement.

Identify the Benefits

The goal of the *identify the benefits* step is to determine the relevant benefits that contribute to the overall goal, as well as their interdependencies and any disbenefits that could occur as side effects. The term *disbenefit* is used to describe negative side effects of planned activities; for example, bringing a new model of an existing product to market will lead to a new line of revenue (a benefit) but can reduce the appeal of the older version (a disbenefit).

A graphical representation of the benefits and their interdependencies is provided by the benefits realization map (BRM), described in detail in Chapter 4.

The benefits description is based on an analysis of what will have changed (for the better or for the worse) between the current state and the state described in the vision. Once the benefits have been identified, they need to be assessed.

Assess the Benefits

At this point there is not enough data for the viability (i.e., the business case) to be assessed numerically. However a qualitative analysis in the form of a draft business case can be performed to decide whether the idea is sufficiently promising to pursue. The benefits should be described in purely qualitative terms: the benefits criteria should be defined, along with a definition of the meaning of qualitative assessment terms (such as *high*, *medium*, and *low*) or order-of-magnitude estimates.

This draft business case is used as input to the *Initiate Program* process. If the decision is made to continue, the qualitative analysis is accepted and the subsequent phase launched.

Once the benefits have been identified and assessed in line with the vision and mission, they provide the basic building blocks for the earned benefit technique in parallel with developing the business case as explained in the next section.

3.4.2 Business Case Development

The goal of the business case development phase is to prepare the first part of a business model: a numerical estimate of the benefits. Approval of the business case authorizes the development of a detailed plan for realizing these benefits.

As shown in Figure 3.5, the first step in developing the business case is to determine the set of program components.

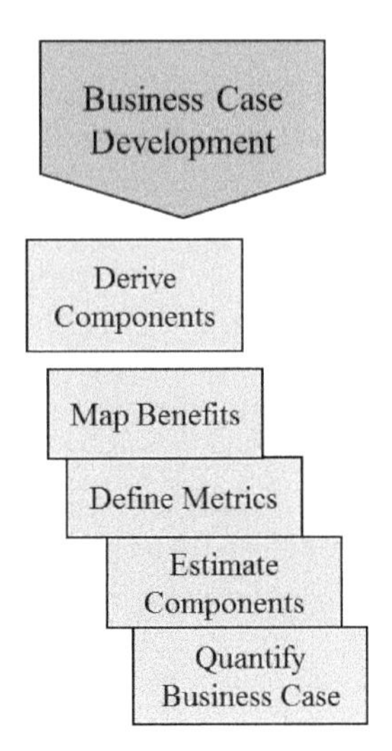

Figure 3.5 Main Steps in Business Case Development

Derive the Components

Once the set of required capabilities has been defined in the mandate, these have to be investigated in order to determine the set of projects, subprograms, and nonproject work (i.e., components) required to deliver these capabilities. Prioritization will take into account the contribution of each component to achieving the strategy. The information on the components and their contributions is shown in the component–benefit matrix derived by mapping the benefits.

Map the Benefits

The goal of the *map the benefits* step is to provide the interdependency diagram between components and benefits. A full description of this benefits realization map is provided in the next chapter.

Define the Benefits

Benefits Metrics Part 1: Units

To allow tracking and consolidation, the qualitative analysis in the previous phase needs to be transformed into a quantitative forecast. This entails defining the units in which benefits will be measured. In the case of the balanced scorecard, for example, there are four sets of units: financial (e.g., dollars), process-related (efficiency and effectiveness measures, such as exceptions per process), customer-related (number of help-desk calls, for example), and learning and growth (ability to change; this could be measured by the speed of adoption of change programs).

Benefits Metrics Part 2: Quantities

The completion criteria and the value of completion with respect to the relevant units also need to be defined.

For each benefit, the following are required:

- Benefit *capability completion criteria*: How to determine whether all of the capabilities required to achieve the benefit are in place
- Benefit *value*: measured in the predefined units

Estimate the Components

Estimating components entails involving technical experts to evaluate the cost and duration of the proposed program components in the form of projects or subprograms as well as any nonproject work. Standard project management estimating tools can be used (e.g., Chapter 7 of the *PMBOK® Guide:* PMI 2013a and *The Practice Standard for Project Estimating:* PMI 2010).

Quantify the Business Case

At this point the quantified benefits can be added to the draft business case to provide a financial feasibility analysis. The full business case is the basis for approval of phase completion and the go/no-go decision for the next phase. It will also be used throughout the lifetime of the program to track progress and alignment with the original strategic intent.

Once the initial draft of the business case has been developed, the viability of each element of the benefits realization map needs to be reviewed based on cost, value, and interdependencies. If relevant, scenario analysis can be performed to investigate whether or not the objective can be achieved in a more effective manner. Chapters 4 through 7 address these topics in theory and with the help of a detailed case study.

3.4.3 Benefits Realization Planning

Although the overall business case has been developed, we do not yet have a clear idea of how to achieve it.

The goal of the benefits realization planning phase is to produce the benefits realization plan defining actions, dependencies, priorities and dates. Once the plan is approved, work can begin on executing this plan.

One key characteristic of programs is the need to understand and manage the dependencies between components. Understanding these dependencies is the first planning step as shown in Figure 3.6.

Determine the Dependencies

Some of the dependencies were defined in the previous step in order to allow us to quantify the business case. However, the way in which the components, their deliverables and the corresponding outputs work together defines constraints on the order in which the components can be implemented. Given these constraints, the overall plan will need to determine the way in which these dependencies influence the achievement of the program's objectives. In order to develop the program roadmap, as shown in Figure 3.7, a number of sequencing and prioritization options need to be defined. These are described in the next section.

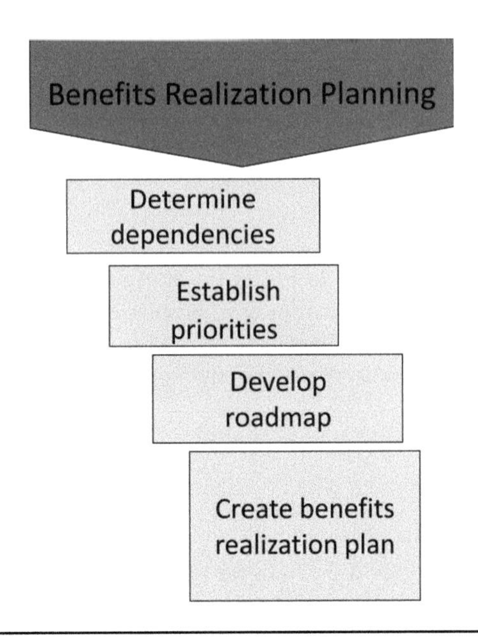

Figure 3.6 Main Steps in Benefits Realization Planning

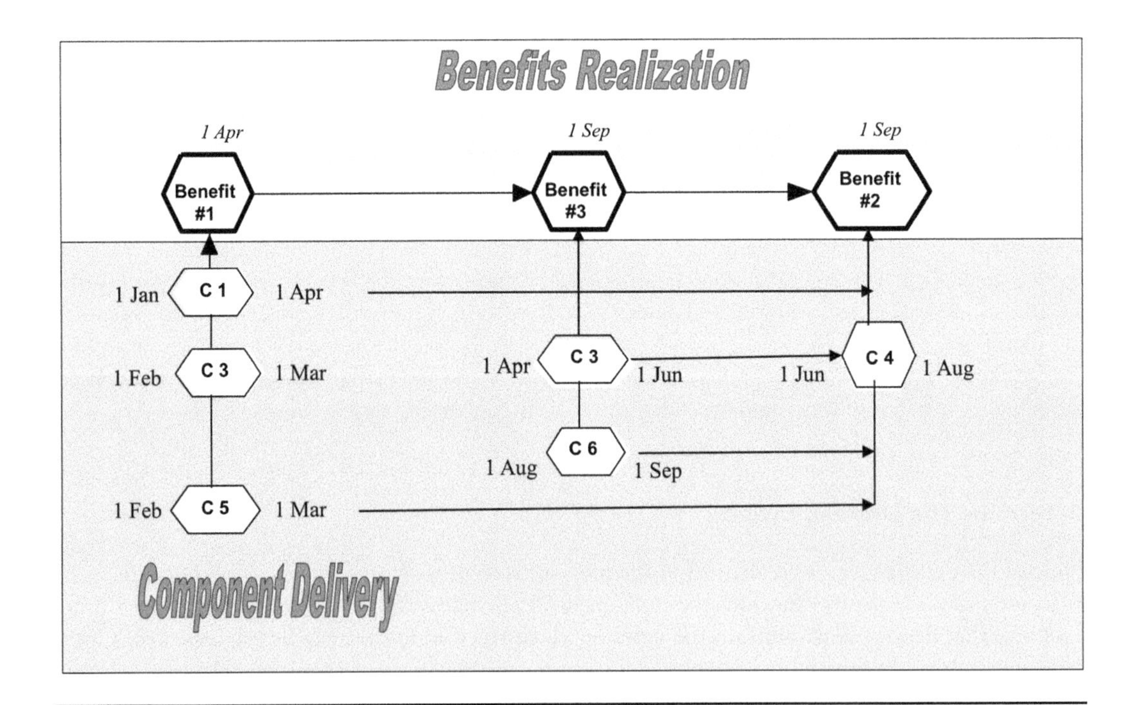

Figure 3.7 Program Roadmap with Typical Components

Prioritize the Components

Prioritization is based on a chosen set of criteria such as dependencies, time, value, and the like. The strategic criteria for prioritization should have been defined as part of the program strategy and described in the program charter. Typical criteria include:

- *Quick wins:* What can we achieve early to prove we are already delivering value?
- *Big hitters:* What will provide an impressive return on investment?
- *Starting with the safe options:* How can we establish credibility before taking risks?
- *Starting with the risky options:* How can we get more time to address any issues that may occur?
- *Resource-linked priorities:* Where will resource availability act as a constraint or an enabler for implementation?

Develop the Roadmap

The roadmap provides a high-level view of the order in which the benefits will be realized along with the schedule of executing the program components and creating the capabilities required for realizing these benefits (as shown in Figure 3.7).

The program roadmap serves as a guide for developing the benefits realization plan described in the next sections.

Create the Benefits Realization Plan

Creation of the benefits realization plan entails carrying out the following steps, which are expanded in subsequent chapters with a case study and worked solution.

- Link the roadmap to the component duration, cost and value estimates
- Establish benefits realization monitoring
  - The monitoring rules define how earned value and earned benefit are applied for the specific program

Benefits realization planning is performed in close collaboration with the program set-up phase. The planning will subdivide the delivery of the program benefits phase of the program management life cycle into successive stages. Each stage corresponds to the delivery of all of the components required for a given benefit. If all of the benefits are only achievable once all of the components are complete, the delivery of program benefits phase will contain a single stage.

The planning algorithm can be described as follows:

A Component sequencing: In priority order, list all components for which all precedence requirements are satisfied.
B Benefit realization milestone determination:
  - For each component from step A, determine whether it completes the work on a benefit.
    - If so, define the corresponding "benefit realization stage complete" milestone.

A2, B2 . . . repeat from A until all components have been addressed.
  - The last benefit milestone corresponds to all benefits realized; stop.

This algorithm will provide an initial roadmap for sequencing the work. However, operational and strategic constraints need to be factored in before a realistic realization plan can be established. Examples of such constraints are:

- Operational constraints
  - Resource limitations
  - Budget limitations
- Strategic constraints
  - Maximum number of projects of a given type at any time
  - Cumulative risk exposure

The roadmap should not be changed arbitrarily during the realization phase. However, as the operational and strategic environments evolve, the detailed plan needs to be reviewed by a formal change control board and as necessary updated.

Now that the life cycles have been aligned, compliance with the program management life cycle requires that the review of the benefits realization plan should form part of the gate review prior to the program benefits realization phase. This review should be carried out by the program steering committee.

Alignment at this point is key to ensuring an ongoing and effective linkage between the management of the program and the management of the business for realizing the benefits.

3.4.4 Benefits Realization

The benefits realization phase is where the preparatory work bears fruit! It can be performed in stages, as shown in Figure 3.8.

The benefits realization phase is composed of one or several sequential or overlapping stages, each of which delivers incremental benefits. Each stage includes initiating and launching the corresponding projects or programs at the start, and integrating the corresponding deliverables at the end of the stage.

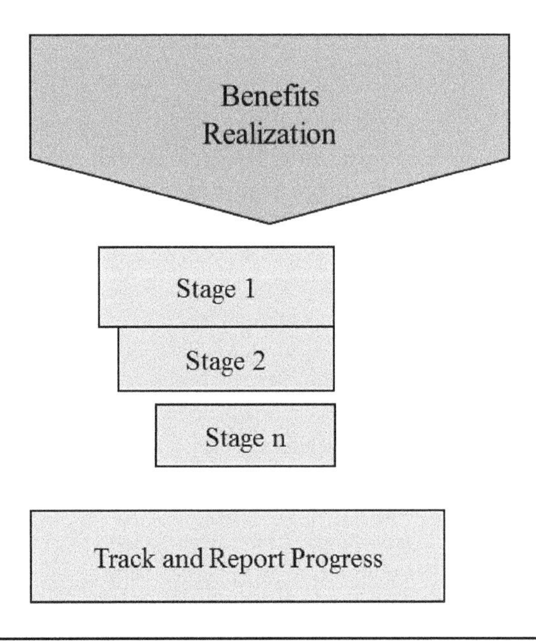

Figure 3.8 Main Steps in Benefits Realization

Although the stages can be tracked individually, tracking of overall progress toward the target outcomes is achieved using the earned benefit approach as explained in Chapter 12.

The program manager ensures that the requisite components are launched as defined in the program management plan and adapts the plan in light of recent developments. These developments can be strategic developments, such as changes in strategy within the organization or in the target environment, or tactical developments, such as components finishing early, resources not being available as planned, and the like.

As the benefits realization stages deliver the planned outcomes and are closed, responsibility for the ongoing management of the capabilities is transitioned to a sustaining organization or is managed within the program until such a time as they can be effectively transitioned.

All of the analysis and planning elements defined in the benefits management phase will contribute to the evaluation of benefits realization progress. That is the domain of the earned benefit method described in Chapter 12.

One additional deliverable from the benefits realization phase is the set of metrics in the form of key performance indicators with which to track whether the results from the program activities are having the planned effect and that the benefits are accruing as predicted.

In subsequent chapters, we step through the life cycle with a worked example in order to understand some of the models and tools that support each of the phases and stages of the life cycle.

Since programs can be large and complex, they need to verify that the final result has reached a stable operating state and no longer needs support from the program team.

For this reason, the benefits realization phase is aligned with the program execution phase and requires approval from the program steering committee before it can be closed. Only the controlled shut-down, as described below, then remains to be performed.

3.4.5 The Completion Phase

Although the work associated with the program delivery is complete, there is still work to do to conclude the program management responsibilities. The first step as shown in Figure 3.9 is to review what was delivered.

Audit the Benefits

Programs will normally terminate before the benefits have all been realized. For this reason, the program should at least be extended until the steady state has been reached long enough for the accrual of benefits to provide an effective indication of longer-term results.

Audit the Business Case

Once the indication of the longer-term results is available and the details of the resources (money, people, etc.) invested in the program are known, these values can be contrasted with the baselined benefits realization map to highlight and help understand any mismatches between the plan and reality.

Identify Lessons (to be) Learned

The reasons for any mismatch between the baselined BRM and reality can serve to improve the modeling for the future. The actuals should be put into the baselined BRM and the BRM should be adapted

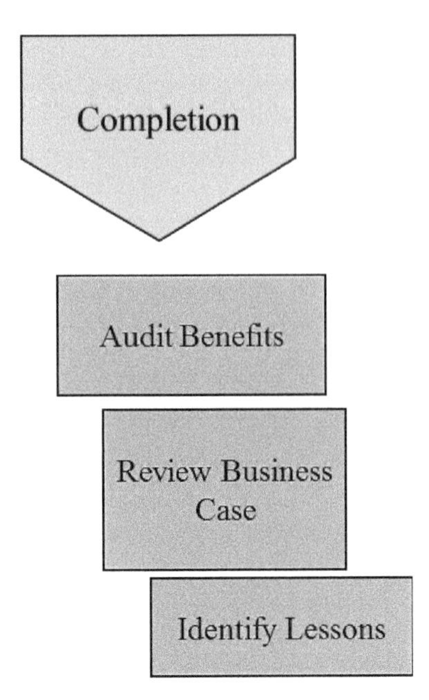

Figure 3.9 Main Steps in the Completion Phase

accordingly to represent the final situation; the description of the changes and the reasons for the changes will provide valuable information for future programs in similar circumstances.

3.4.6 Business Analysis Skills and Tools

All of this analysis, from the benefits assessment right through to the final audit and report, requires a detailed business understanding in order to achieve the correct outcomes necessary to realize the vision. The program manager will therefore require access to the skills of a business analyst.

Since the objective of the business analysis is to generate value-added transformation, and programs aim to deliver benefits, *A Guide to the Business Analysis Body of Knowledge®* (*BABOK® Guide*) (International Institute of Business Analysis™, 2015) is a valuable complement to *The Standard for Program Management*—Third Edition (Project Management Institute 2013c), rather than the *PMBOK® Guide*—Fifth Edition (2013a), as is often suggested. All of the following discussion includes the concept of benefits, and therefore applies the *BABOK® Guide* concepts to a program environment. The starting position therefore is to examine how to align the business analysis concepts into the consolidated life cycle that has been described.

3.5 Integrating Business Analysis and Project Management

One major strength of the PMI standards is their use of the concept of knowledge areas, processes, and planning and control based on clearly defined consistent life cycles.

Version 3 of the *BABOK® Guide* follows the knowledge area approach with the definition of six business analysis knowledge areas (BAKAs). Each BAKA is said to be composed of tasks that fulfill the same role as the processes in the PMI standards. All task descriptions are presented using a well-defined formal structure. In addition, the *BABOK® Guide* defines the competencies required by a business analyst. These are not unique to the BA profession but are required to fulfill the role. Many of them are common to business analysts and project managers.

The BAKAs do not all map easily onto the PMI knowledge areas (PMKAs). In fact, the decomposition into knowledge areas in the *BABOK® Guide* seems almost orthogonal to that adopted by PMI.

For example, there is one BAKA to address planning and monitoring. As such, its scope is more aligned to a process group created by merging PMI's Planning process group with Monitoring and Control. It covers planning for the following:

- Approach to be taken
- Stakeholder engagement
- Governance
- Information management
- Performance improvement

The skills and techniques required for planning the approach cover a number of PMKAs, such as:

- Scheduling
- Budgeting
- Risk analysis and management
- Scope definition

3.5.1 Vive la Différence

The contrasting approach between the *BABOK® Guide* and the *PMBOK® Guide* should not be considered a shortcoming. In fact, it can ensure that *BABOK® Guide* and PMI practitioners complement each other's work by looking at the situation from different points of view.

Despite the highlighted differences, the core concepts of the two areas are closely aligned with each other.

3.5.2 Core Concepts

The *BABOK® Guide* defines six key terms that provide a conceptual framework for business analysis. They are applicable in all of the BAKAs, and are relevant and directly applicable in the program management domain. They are:

- *Change*—in response to a need
- *Need*—which can cause changes; needs also relate to value and to stakeholders in a given context
- *Solution*—to satisfy one or more needs in a given context
- *Stakeholder*—the focus of the need
- *Value*—a measure of the effect of a potential change
- *Context*—environment of the change

3.5.3 Effective Link to Program Management

The clearest link between the *BABOK® Guide* tasks and program management is in the area of requirements management. The synergy provided by program management is apparent in description which refers to "elements working in harmony with one another to support the business requirements" (International Institute of Business Analysis™, p. 148).

To ensure effective interworking between the business analysis environment and the program management domain, there is the need to define a compatible business analysis life cycle. The *BABOK® Guide* does not define any life cycle, so a proposed structure has been developed independently as follows.

3.5.4 The Business Analysis Life Cycle

Building on the program and benefits realization life cycles, a valid and compatible life cycle for business analysis has been developed as shown in Figure 3.10. The diagram also proposes the key roles required for each of the corresponding tasks within the phases of the life cycle. This life cycle alignment provides a solid framework in which business analysis supports program management and vice versa.

The Business Analysis Direction Setting Phase

The business analysis "direction setting" phase supports the identification of benefits as specified in the program formulation phase. The main activities are the definition and development of the business analysis environment and an analysis of the current state and desired future state. The business analyst is directly responsible for leading the work in this phase.

The criterion for closing the business analysis phase is approval of the change strategy document.

Business Analysis Requirements Determination

As pointed out in the *BABOK® Guide,* there are two categories of requirements, both of which are addressed in this phase:

1. The requirements as specified by the business stakeholders as to the desired end-state
2. The requirements of the solution components

In addition, stakeholder engagement during this phase is explicitly defined to support the program in assessing benefits and deriving the components-benefits mappings as the first step in developing the business case.

The key tasks contributing to this phase are those associated with elicitation and requirements management.

This information provides the input required to quantify the business case, as specified in the corresponding phase in the business realization life cycle.

The criterion for closing this phase depends on confirmation of both the elicitation results and the business analysis information, as well as on the approval of the requirements.

Solution Approach

This phase completes the creation of a solution plan.

Traceability between the benefits and the components is key to establishing a working solution that will satisfy the business stakeholders. In this way, business analysis provides a clearly defined approach for refining the business case as the outcome of the solution approach phase. The business case will demonstrate the feasibility of the solution, as required for closing the benefits realization planning phase.

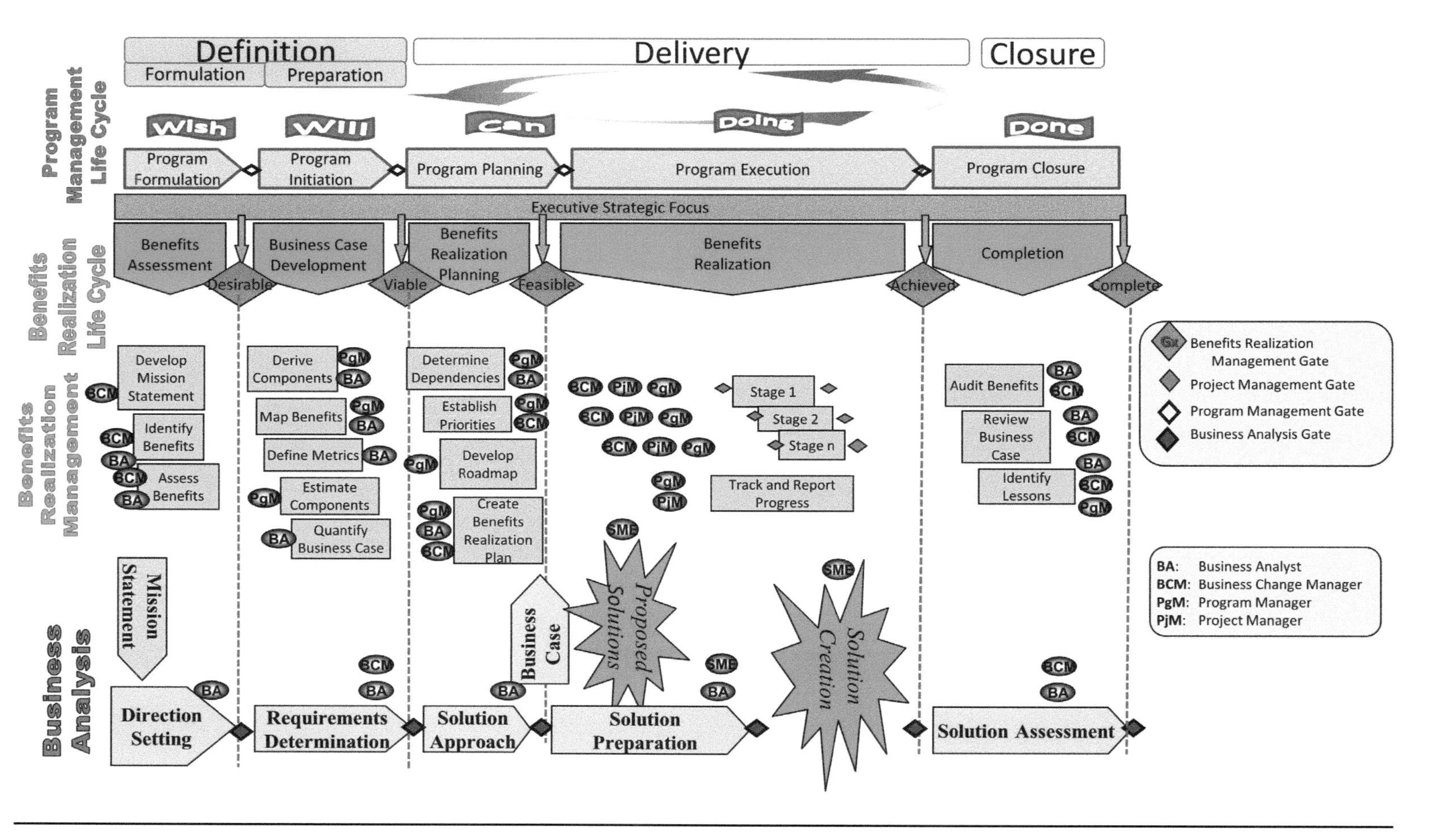

Figure 3.10 Integrated Set of Life Cycles Including the Proposed Business Analysis Life Cycle

Solution Preparation

The business analyst should now work with the subject matter expert (SME) on determining the solution (e.g., an information technology [IT] architect in the case of an IT-based solution), before handing over responsibility to the SME for creating the deliverables that compose the solution.

As shown in Figure 3.10, work on solution preparation is aligned with the benefits realization phase of the benefits realization management life cycle. It is therefore coordinated by the program manager with the involvement of the business change manager. The results of these preparation actions are documented in a solution implementation plan that, once approved, is integrated into the program management plan. At this point, the actual implementation of the plan and the creation of deliverables and capabilities can commence.

Solution Creation

During solution creation, the benefits realization management process to track and report progress is invoked. Program-level tracking and reporting is supported by the business analyst to measure and analyze performance measures.

When all of the components of the solution have been created and integrated into a cohesive solution, the solution creation phase can be closed. All that remains is to close out the work.

Solution Assessment

Once the solution creation is complete, the business analyst and the business change manager collaborate to evaluate the strengths and weaknesses in the chosen solution as implemented, with respect to the initial organizational needs and final vision. This retrospective analysis can be used to identify business analysis performance improvements. The results of this analysis are then integrated into the lessons learned work performed in the completion phase of the benefits realization management life cycle.

3.6 Working with the Business Analyst

In certain circumstances, the program manager can take the responsibility for business analysis as well as for managing the program. However, this will only be possible in small and fairly simple programs, because both roles require a strong focus and can sometimes come into conflict with each other over objectives and priorities. However, because of their focus on benefits, all programs need someone to fill this role.

This collaboration can be summed up in an adaptation of a song from the musical *Oklahoma!*, as follows:

The PM and the BA should be friends
Oh, the PM and the BA should be friends
The PM likes to make a plan
The other serves the business man
But that's no reason why they can't be friends.
Projectizing folks should stick together
Projectizing folks should all be mates
PMs work with BA's data
BAs work to the PM's dates,
But that's no reason why they can't be friends.

Now that we have a definition of the domains and main roles, this approach will be made clearer by applying it to a simple example. The following case study will also be used in subsequent chapters to provide a practical example of the use of the tools and techniques.

3.7 Case Study: QERTS Merger

The example is built around a takeover of one logistics company by another. The goal of the takeover is to merge the companies' processes and services into an integrated whole, while maintaining or expanding the combined customer base. The two companies are completely different:

- Q&E (Quick & Easy), the company being taken over, is dynamic and flexible, if somewhat undisciplined, and aims at a younger set of customers. It has revenues of $22 million per year.
- RTS (Reliable Transport Services), the new owner, is a well-established company with formal and effective processes and working practices that, while reducing the scope for errors, also reduce the scope for adapting to special customer requests. It has revenues of $32 million per year.

The new company will be known as QERTS (Quick, Easy, and Reliable Transport Services).

Executive management is aiming for a 12-month payback period once the corresponding benefits are available. These have been defined by the management team as coming from two areas: increased market share and increased profitability. The corresponding vision statement is that, thanks to the program, "QERTS provides a full range of logistics services in an increasingly profitable manner to an expanding customer base thanks to efficient systems and the involvement of happy employees." The areas on which management wants to focus at this stage are staff, customers, processes, and information.

The decision has been made to run the post-takeover integration as a program. The first steps in the life of this program are outlined below.

3.7.1 Benefits Assessment

Benefits assessment is the first phase of the life cycle shown in Figure 3.3 and Figure 3.10.

Develop Mission Statement

For this example, the mission statement is: "To increase revenue, decrease operational spend and reduce the turnover of staff in QERTS within 12 months of the completion of the program by integrating systems, data and processing capabilities as well as increasing market visibility."

This mission statement then serves as the basis for benefits identification.

Identify Benefits

The mission statement called out the benefits expected by the management team:

- Increased market share
- Increased revenues
- Operational savings
- Staff satisfaction
- Higher profits

Although it is too early to develop a full model for a business realization plan, it can already be seen that some of these "benefits" are interdependent—for example:

- Increased market share may lead to increased revenue but increase total operational costs.
- Increased revenue and operational savings affect profits.
- Staff satisfaction may (indirectly) impact market share by making QERTS a pleasant company with which to do business.
    - Low satisfaction can lead to staff departures, and staff turnover can increase operational costs because of recruitment, training, etc.

This merger was obviously based on a set of top management business objectives; these can be quantified as the first step toward developing the full financial analysis.

3.7.2 Business Case Development

One key input to the business case is available almost from the start of the program: management expectations on the result of the program. These expectations have to be elicited and quantified. For the case study, the management analysis of the result of the merger is specified as follows:

- Market share:
    - An external consulting company has told management that doubling a company's size has been shown to lead to an additional 6 percent of market share because of name recognition, if the correct steps are taken.
    - The sum of the revenues of the two companies is $54 million—an increase of 69 percent with respect to the revenue of Q&E alone.
    - As a result, the extra visibility given the increased size of QERTS should provide at least a 3 percent increase (say, $1.7 million).
- Increased revenues:
    - The additional set of services that QERTS can offer to each customer should allow sales per customer to increase by 10 to 20 percent (say, $8 million).
- Operational savings:
    - Applying the more structured operational processes from RTS to Q&E should save about 10 percent of the revenue earned by Q&E that is spent on operations (say, $2 million).
- Staff satisfaction:
    - Reduced staff turnover overall in QERTS may result in an extra 10 percent of operational savings (say, $200,000).
- Higher profits independent of changes in market share:
    - Based on the previous estimates: $8 million + $2 million + $.2 million (so, about $10 million).

These estimates also demonstrate one of the risks associated with a simple linear analysis of this type, and that is the risk of double-counting benefits (e.g., counting the benefits of reduced staff turnover on its own as well as in the increased profits). The double-counting risk can be eliminated by use of the structured approach explained in the next chapter (using the BRM).

3.8 Summary

A way of managing all of the interdependent project domains in an integrated manner has been developed by aligning the life cycles of program management, benefits realization, and business analysis. The roles of the key experts, including the role of the business analyst, have been explained. The principal case study that will be used in forthcoming chapters has been introduced.

3.9 References

International Institute of Business Analysis™. *A Guide to the Business Analysis Body of Knowledge® (BABOK® Guide), version 3.* Toronto, Ontario, Canada: International Institute of Business Analysis, 2015.

Levin, G., and Green, A. R. *Implementing Program Management.* Boca Raton, FL, USA: CRC Press/Auerbach, 2014.

Project Management Institute. *A Guide to the Project Management Body of Knowledge (PMBOK® Guide)—Fifth Edition.* Newtown Square, PA, USA: Project Management Institute, 2013a.

Project Management Institute. *The Practice Standard for Project Estimating.* Project Management Institute, 2010.

Project Management Institute. *The Standard for Program Management—Third Edition.* Newtown Square, PA, USA: Project Management Institute, 2013c.

Chapter 4

Building an Integrated Business Model

This chapter presents the concepts behind the modeling approach that will provide the consistent link for each stage in the life cycle of a program. As such, it ensures that an "honest" business case can be built, avoiding wishful thinking, double counting, and numerical or political manipulation. A new case study is used as a practical example of the technique.

The principal terms used in this chapter are defined in the Glossary.

The outcome of the modeling is an artifact called a *benefits realization map* (BRM). It looks similar in structure to the benefits map defined in the Office of Government Commerce (OGC 2011) publication *Managing Successful Programmes*. However, it includes additional information and applications, which will be developed in successive chapters.

The creation of the benefits realization map is started in the program initiation phase to support the business case and is then expanded and applied throughout all of the successive phases.

4.1 Business Case Development

This is the second phase of the life cycle developed in the previous chapter, and shown in Figure 4.1. The goal of this phase is to develop the numerical business model.

This phase requires the following six steps:

1. *Derive components:* Decide on the project and nonproject work needed to achieve the organizational change described in the mission statement.
2. *Map benefits:* Provide the interdependency diagram between components and benefits. This diagram is the basis for the BRM that is explained in this chapter.
3. *Define units:* Benefits can only be measured if the criteria and the corresponding metrics are defined (e.g., pure monetary units, net present value, a balanced scorecard, etc.).
4. *Derive metrics:* Define completion criteria and the value of completion with respect to the relevant units.

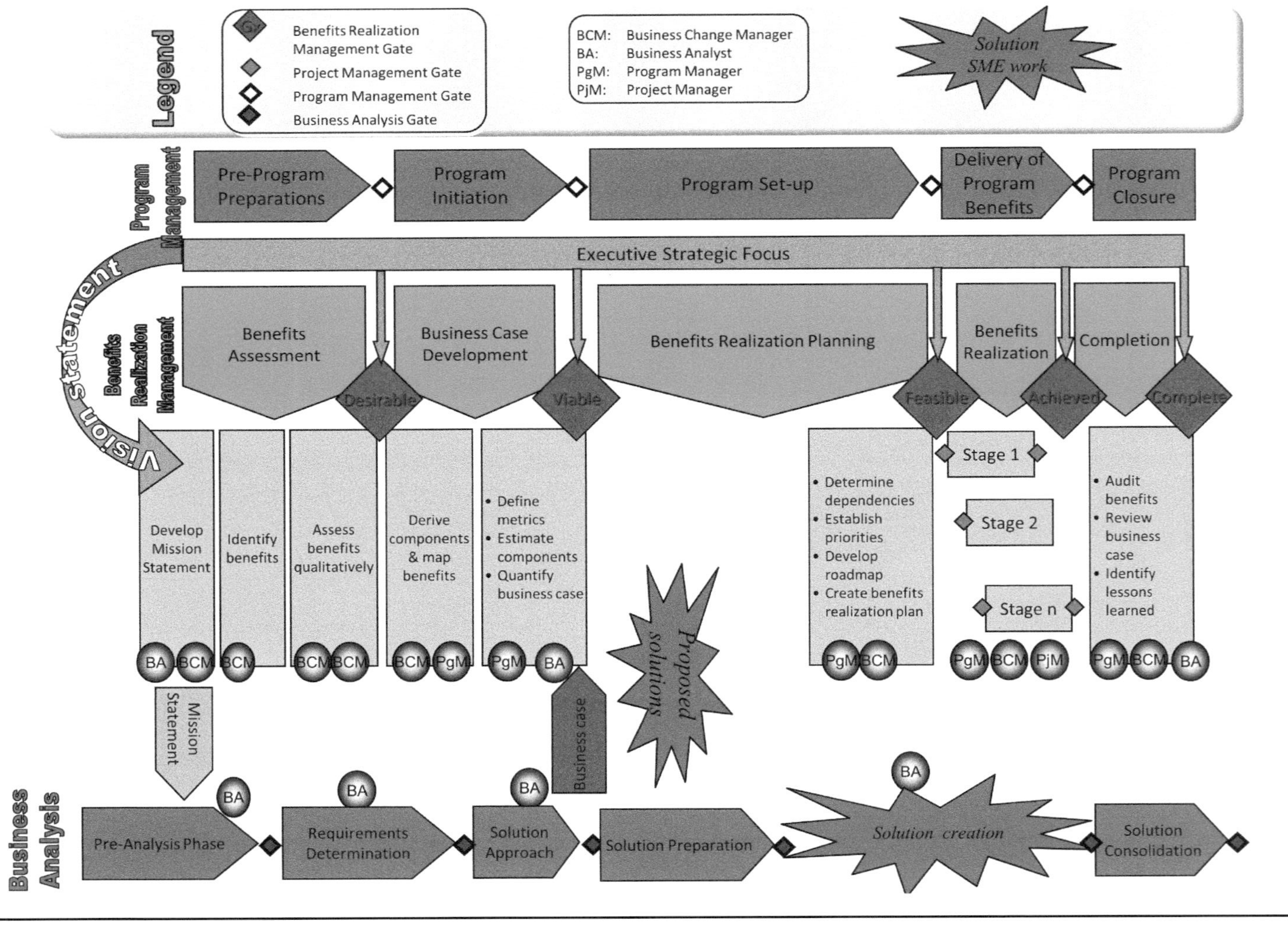

Figure 4.1 The Total Program Management Life Cycle

5. *Estimate component costs:* Work with technical experts to determine the resource requirements—money, effort, etc.—required to carry out each component.
6. *Quantify business case:* At this point, the quantified strategic targets can be added to the draft business case as input to a financial feasibility analysis. Approval for phase completion and the go/no-go decision for the next phase requires a business case showing the viability of the program.

4.2 Contributions and Allocations

The goal of the BRM is to ensure that the forecast benefit is justified with respect to the forecast costs to achieve it. The benefit is described in terms of the contribution of each element of work, while the cost is characterized by the allocation of resources (generally investment) that it requires.

4.3 Description of the Benefits Realization Map

This theoretical description is explained first based on a case study for a business transformation program designed to improve profitability by process redesign. This approach is then applied to the QERTS case study that was introduced in the previous chapter.

A BRM is represented as a set of nodes, corresponding to the *entities*, and their interrelationships, represented by *links,* as explained in the following sections.

4.3.1 Entities

In BRM diagrams, the entities are represented by nodes, and the shape and/or color of the node identifies its category. The theory is explained next, and is then applied to the examples.

Initiatives

The initiatives are the technical elements of work to be carried out in order to achieve the required results of the program. Initiatives will normally be organized as projects but can be subprograms or temporary operational services. The main characteristics of an initiative for the BRM are:

- Cost
- Deliverables or other results ("outputs" in OGC [2011] terms)

The other nodes all depend on outputs from initiatives or from each other as explained in the following sections.

In the case study, the initiatives are as shown in Figure 4.2. Initiatives lead to capabilities.

Technical Capabilities

The term *capability* indicates that it enables the organization to do something (hopefully beneficial) that it could not do before, or at least not as well. Technical capabilities can be considered to be new tools or techniques and are provided by deliverables from the initiatives in order to allow changes to ways of working, often at an operational level (e.g., the ability to consolidate all of our customer information in a single database). Several deliverables may need to be combined to provide a given technical capability.

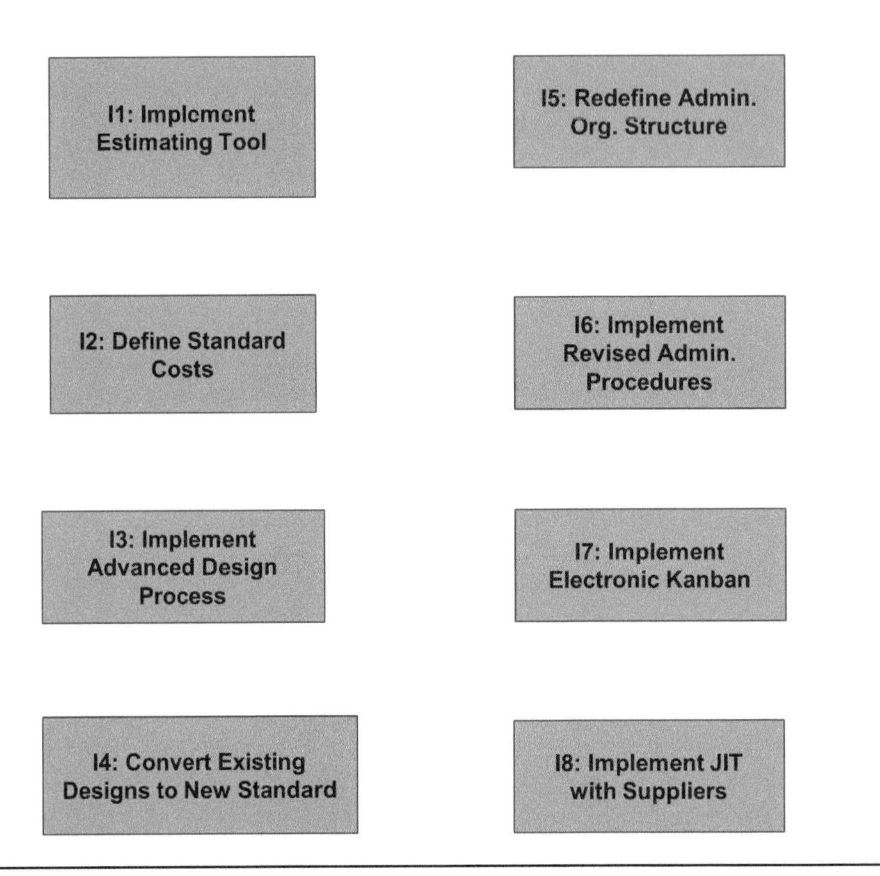

Figure 4.2 Initial Set of Initiatives

Building on the initiatives in Figure 4.2 with the addition of technical capabilities, the map then evolves into Figure 4.3.

A narrative description of the logic behind this diagram, as well as those that show each successive stage of the development, is provided by the full description of the complete BRM for the business transfomation program shown at the end of the chapter in Figure 4.8.

The need for the prior availability of technical capabilities can also be specified as prerequisites for specific initiatives (e.g., "CAD/CAM Tools in Place," as shown in Figure 4.4). Prerequisites can, for example, provide a link to other BRMs or suggest the need for an initial capability audit of the environment.

There is one other category of capabilities: the business capabilities.

Business Capabilities

As opposed to technical capabilities, business capabilities focus on the value-added application of a technical capability (e.g. the ability to address all of our customers). The business potential of the deliverables or technical capabilities (i.e., how they could be applied rather than what they can do) needs to be identified and is described under the heading of business capability; see Figure 4.5.

One important point that this figure shows is that once improved designs are available (BC2), it will be possible to identify the need for a project to optimize the cost of materials (I9), which leads to the business capability of buying them less expensively (BC6). In addition, the figure shows two

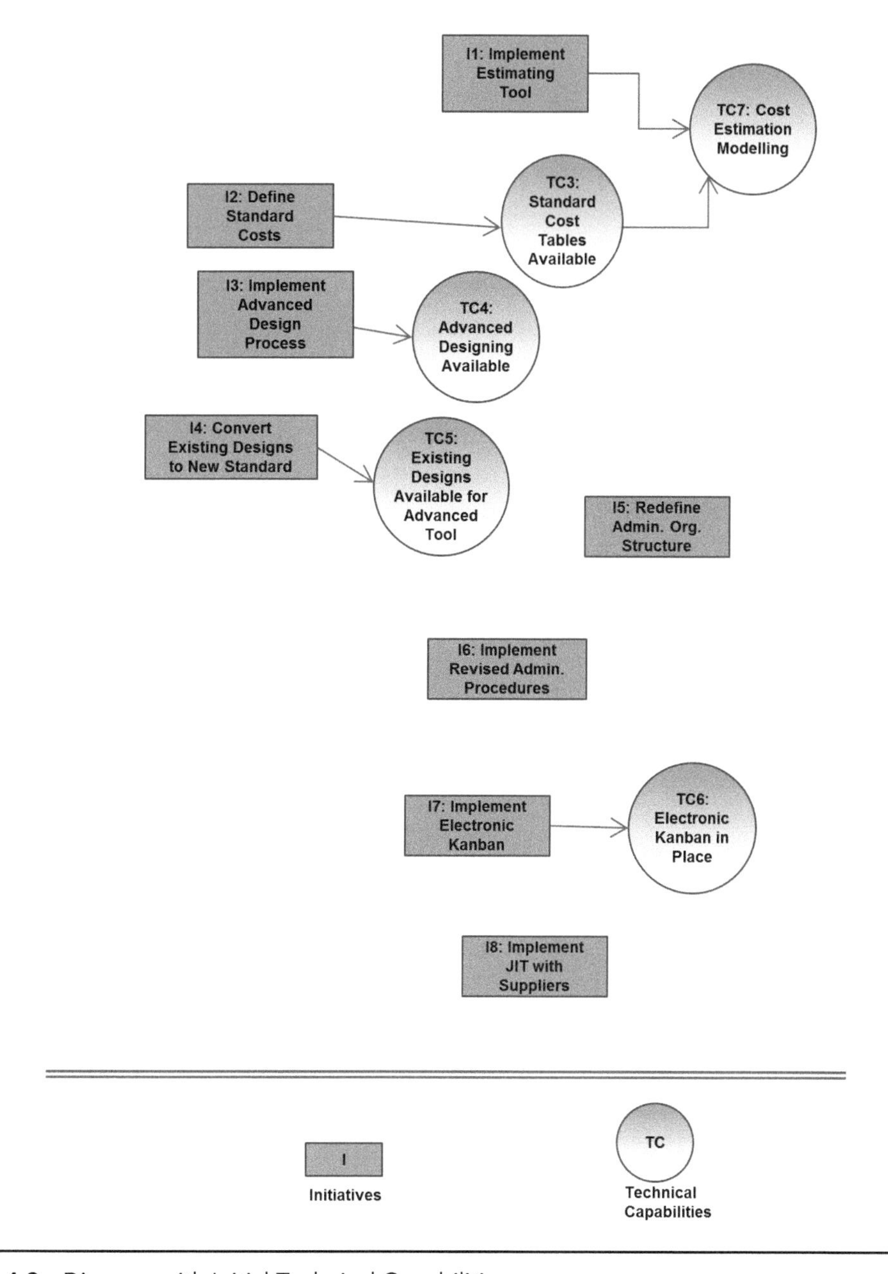

Figure 4.3 Diagram with Initial Technical Capabilities

initiatives that contribute directly to business capabilities: Redefinition of the administrative organizational structure (I5) contributes to optimizing the administration department (BC3); the initiative of implementing just-in-time (JIT) leads directly to the business capability of setting up agreements with suppliers (BC5).

Note that there is always a level of uncertainty in any of these cause-and-effect relationships between capabilities and outcomes, and this can be shown on the map by adding one or more hypothesis statements to the links between capabilities and outcomes, as shown in Figure 4.6. The significance of

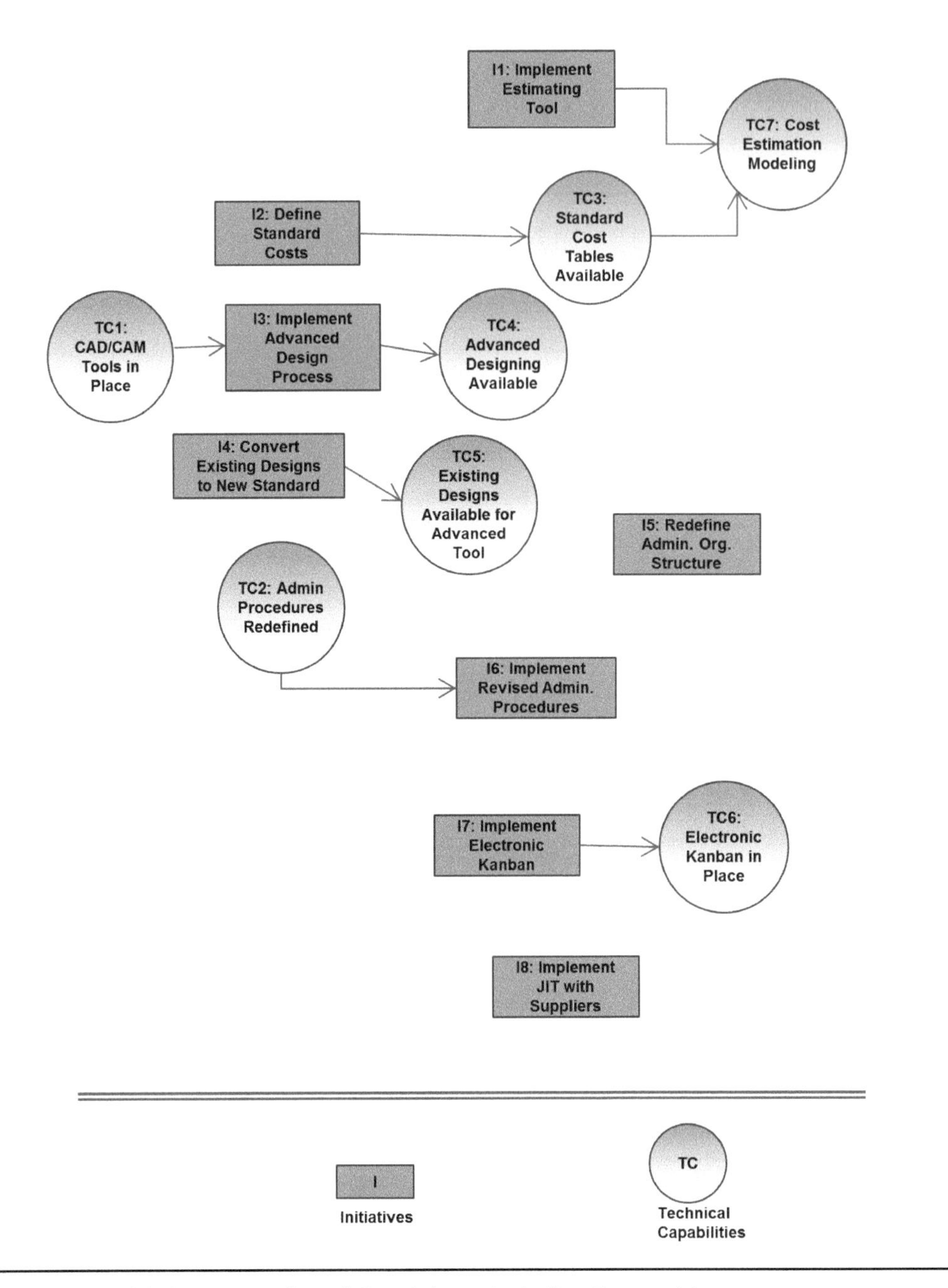

Figure 4.4 Model Showing Technical Capabilities, Including Prerequisites

"100 percent essential" is an assumption stating the corresponding source node that must be achieved for the destination node to become active. For example, TC7 will have no value unless the standard cost tables are available (TC3), even if the estimating tool (I1) is available. Since there is no such assumption on the link from I1, cost estimation modeling (TC7) will be able to deliver some value based on the presence of the estimating tool (I1). Other business-related assumptions are shown on the complete BRM shown in Figure 4.8.

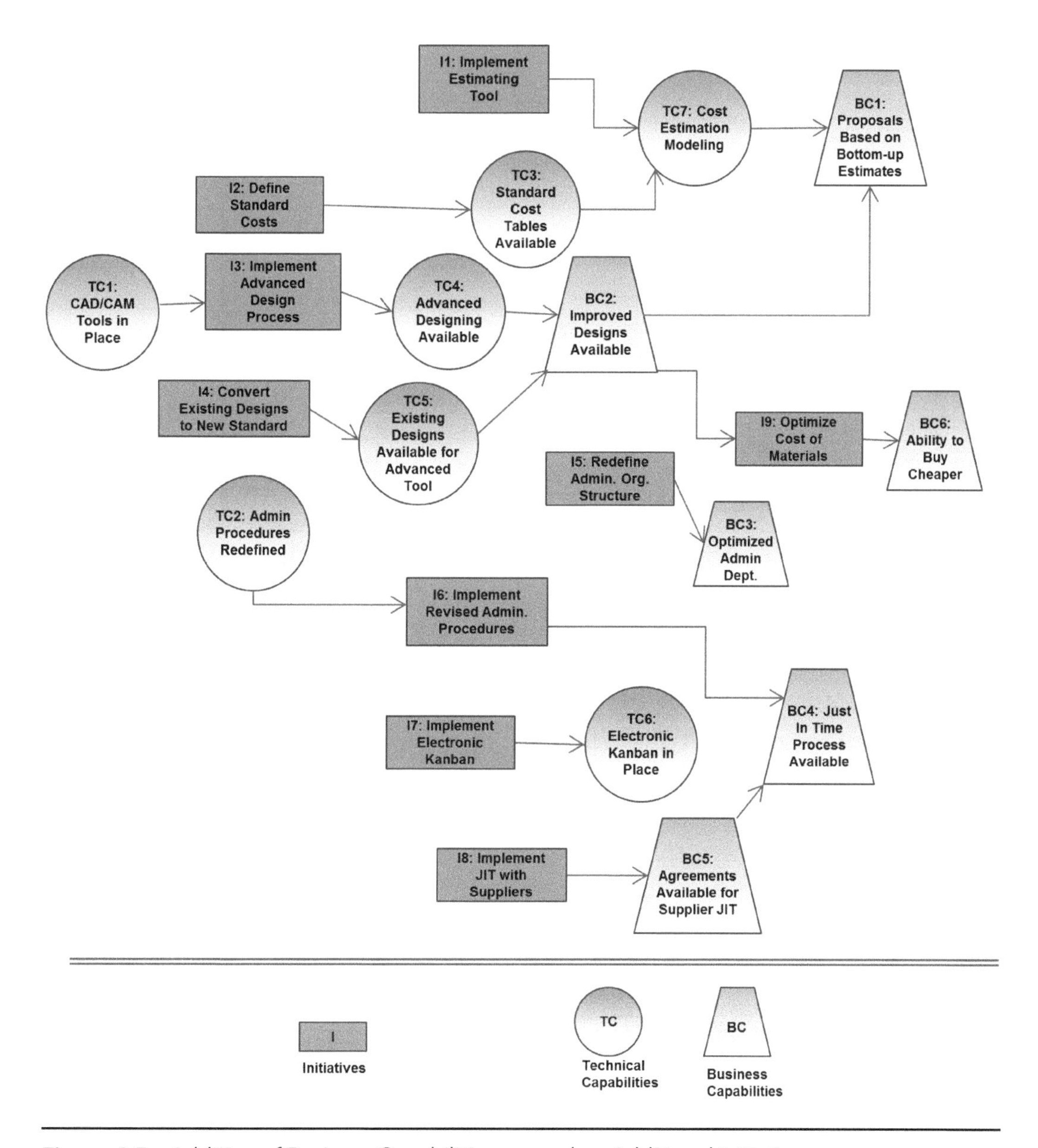

Figure 4.5 Addition of Business Capabilities . . . and an Additional Initiative

Business Outcomes

An outcome is a change in the environment, the business, or the way work is done. Once deliverables or business capabilities are available, they can lead singly or in combination to business outcomes. It is normally the interactions between capabilities and the business environment that lead, automatically, to the outcomes.

The business outcomes have been added to the diagram as shown in Figure 4.7. One important point is that it may be worthwhile to plan the addition of one or more extra initiatives to take advantage of business outcomes as they occur. See *I9: Optimize Cost of Materials*, which can be undertaken once improved designs are available (BC2).

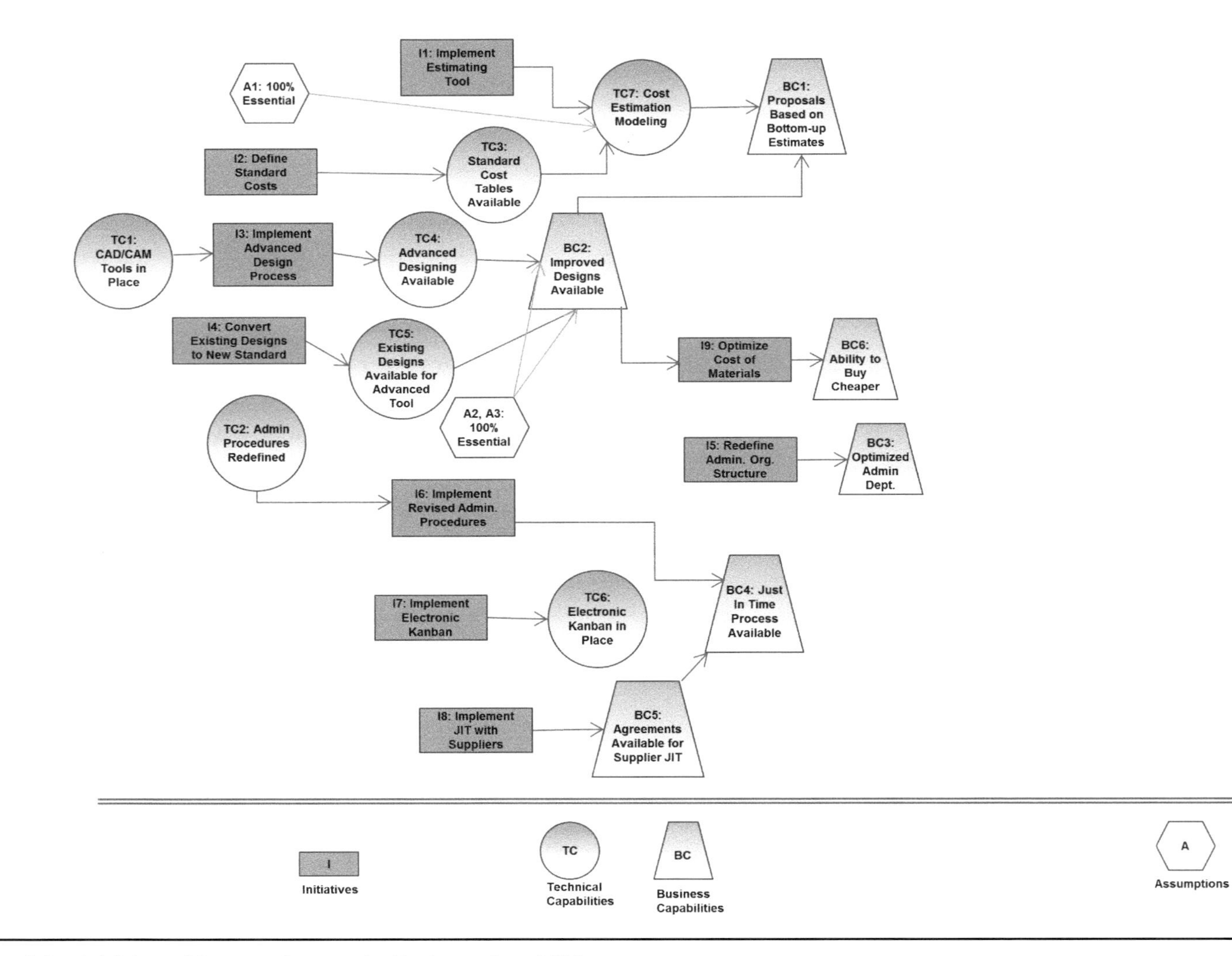

Figure 4.6 Addition of Assumptions to the Business Capabilities

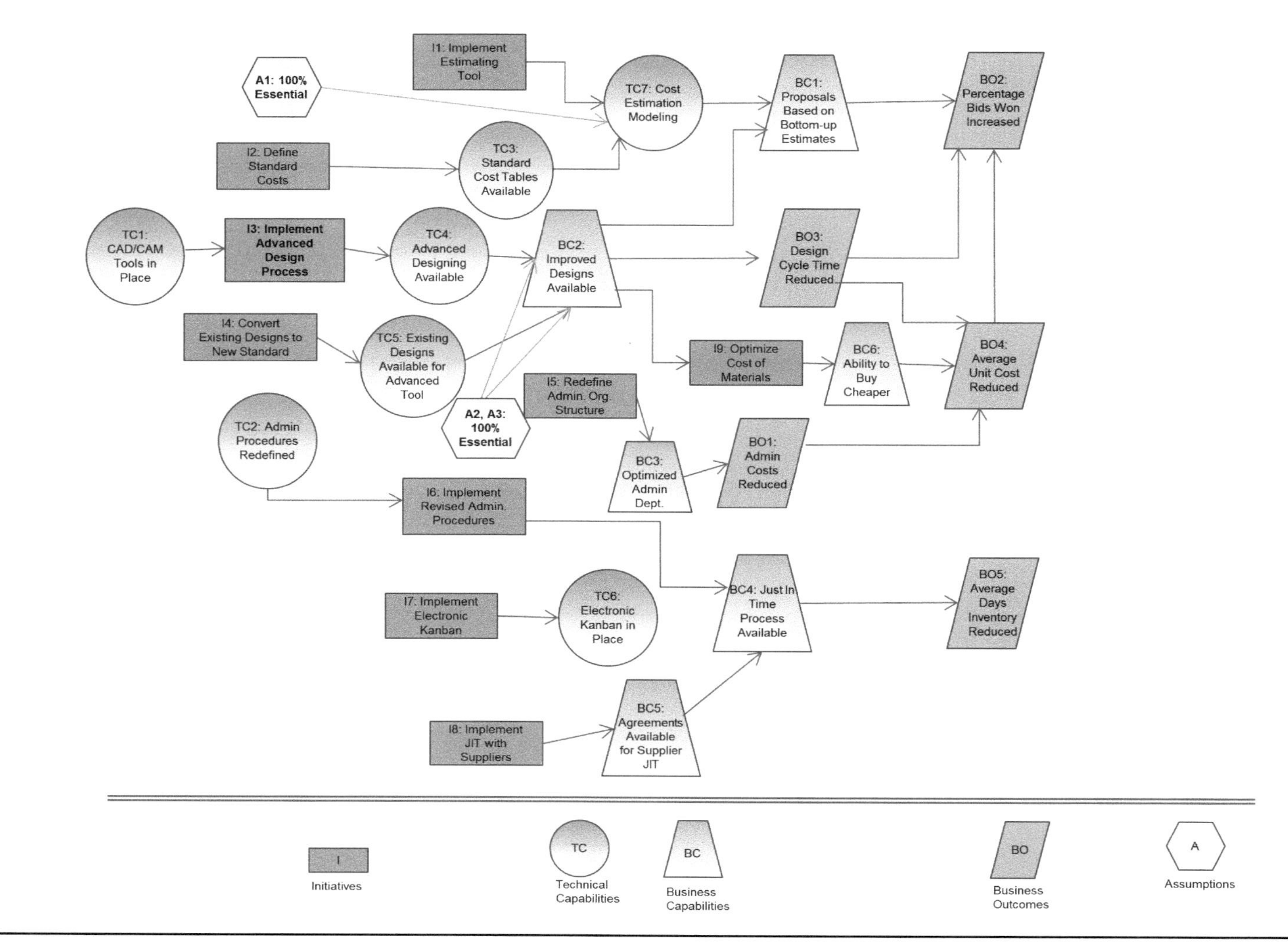

Figure 4.7 The Addition of Business Outcomes

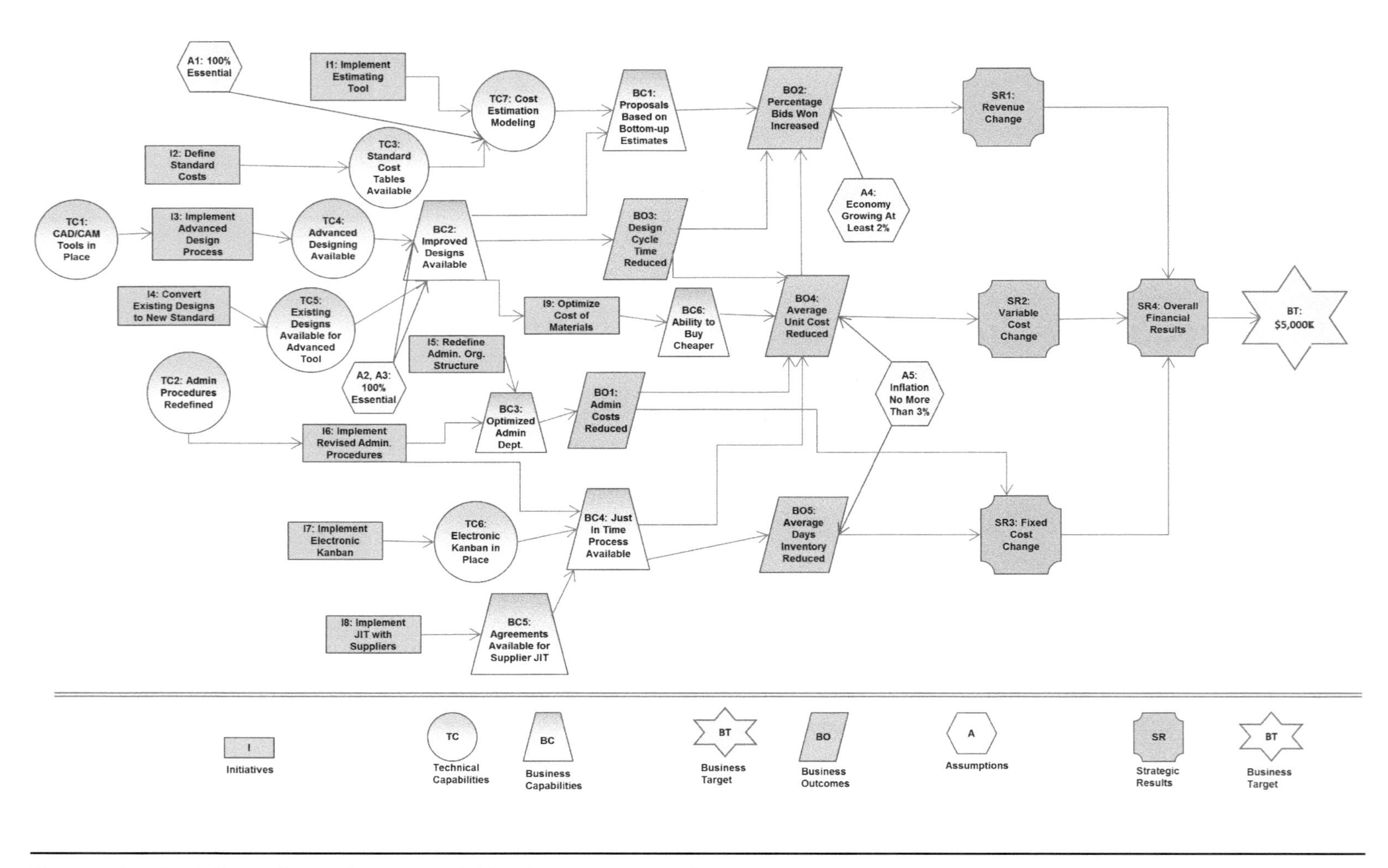

Figure 4.8 The Full Benefits Realization Model for the Business Transformation Program

The development of the map has now reached the point at which it meets the defined business need, so this can now be included.

Strategic Outcomes

The strategic outcomes are those defined in the mission statement. The value of each strategic outcome needs to be quantified in the same units as those used for costing the initiatives. The quantified map can then provide a means to validate the viability of the overall benefits realization approach as proposed.

Business Targets

A business target, or *strategic objective*, is a quantified measure of the value of a strategic outcome.

A full diagram, including the strategic outcomes and the overall business target, is shown in Figure 4.8. This diagram can be used not only to provide a quantified basis for the business case, but also to derive narrative descriptions of the program from different points of view.

4.3.2 From Right to Left

Looking from right to left provides the top-down view of the strategy, and it explains how the final, strategic outcome is built from outcomes, capabilities, and, finally, initiatives, as well as the interactions and synergies between these components. The links are therefore traversed "backward" in the diagram.

This way of reading the diagram provides the value view of the portfolio and would be directed to business managers. Creating the logic of this diagram is the responsibility of a business analyst and must be the first step in the detailed planning.

4.3.3 From Left to Right

The left-to-right view shows the bottom-up implementation view of the situation and quantifies the cost and time required to achieve each of the nodes and the dependencies in the implementation. The links are traversed in the normal direction, from source to destination.

To put it another way, the bottom-up way to read the BRM, starting from initiatives, is:

- What we should do? (initiative, which creates an output)
- What will this output allow you to do? (this identifies the technical or business capability)
- How will the capability change things? (this identifies the business or strategic outcomes)

This way of reading the diagram provides the "investment" (cost and time) view of the portfolio and would be directed to the technical team.

4.3.4 Reviewing the Completed Map

Approving the diagram created by the business analyst is the responsibility of the program manager, who reviews the logic and later quantifies the information.

The example described next illustrates how linking the value (top-down) view and the cost (bottom-up) view provides an *investment* model.

4.4 The Business Transformation Example

The example shown in Figure 4.8 will be used to explain how to "read" the BRM, even in the absence of the preprogram information (e.g., the vision statement, etc.). The power of the BRM representation is a practical proof that one picture is worth the almost one thousand words that follow.

4.4.1 From Right to Left: The Business View

The right-to-left reading of the plan addresses the interests of the business managers, starting from their primary concerns—the impact of the program on their success.

The narrative description of the BRM in Figure 4.8 is as follows:

- Strategic Result SR4: The success metric is *business results*.
  - SR1, SR2, SR3: revenue increase, variable costs, and fixed costs are the three components of the financial results calculations.
- SR1: The amount of revenue increase depends on the Business Outcome (BO2) of winning more bids.
  - The amount needs to factor in Assumption A4, the forecast state of the economy.
- SR2: The variable cost is directly affected by (BO4) the average unit cost.
  - The amount needs to factor in A5, the forecast inflation rate.
- SR3: Fixed costs depend on both (BO1) administrative costs and (BO5) inventory levels.
- BO1: Administrative costs will be reduced depending on the Business Capability (BC3) to optimize the administration department.
- BO2: Percentage bids won depends on:
  - Being able to respond faster because of (BO3) reduced design cycle time.
  - Being price competitive with (BO4) lower unit costs.
  - Being able to develop precise estimates because of (BC1) bottom-up estimating.
- BO3: Design time reduction depends on the availability of (BC2) improved designs.
- BO4: Average unit cost depends on a number of factors:
  - Reduced design cycle time (BO3).
  - Being able to buy for less (BC6).
  - Reduced administrative costs (BO1).
  - Using a JIT process (BC4).
- BO5: Average days of inventory will be reduced by using JIT processes.
- BC1: Bottom-up estimating requires both the Technical Capability (TC7) of modeling cost estimates as well as availability of (BC2) improved designs.
- BC2: To obtain improved designs, you need: (TC4) the capability to use an advanced design process as well as (TC5) having the existing designs available within the advanced design process.
  - Note: If either of these (TC4 or TC5) is unavailable, this (BC2) advanced design capability cannot be provided ("essential link" assumptions A2, A3).
- BC3: Optimizing the administration department requires two initiatives: (I5) redefining the administrative organizational structure and (I6) implementing the revised administrative procedures.
- BC4: Setting up the JIT process by adapting the system that was initially introduced by Toyota for creating a JIT capability, based on a scheduling approach known as Kanban (Ohno 1988). It requires the following:
  - (I6) Implementing the administrate processes to support it.
  - (TC6) Having the electronic Kanban process in place.
  - (BC5) Being able to apply just-in-time with the suppliers.

- BC5: Agreement available for supplier JIT requires an initiative to set the agreements up to use it.
- TC7: Cost estimation modeling requires both (I1) the implementation of the estimating tool and (TC3) having the standard cost tables available (essential: assumption A1).
- TC3: The standard cost table is a capability provided by a specific initiative (I2) to define standard costs.
- TC4: Advanced designing available requires (I3) the corresponding process to be implemented.
- TC5: Existing designs available for the tool requires (I4) the designs to be converted to the new standard.
- TC6: Electronic Kanban availability requires a project (I7) to establish it.
- This leads to the following initiatives:
  - I1: Implementing the estimating tool
  - I2: Defining the standard costs
  - I3: Implementing the advanced design process
    - This process has an external dependency: (TC1) CAD/CAM tools must be available.
  - I4: Converting existing designs to the new standard
    - There is a technical dependency between I3 and I4—the data formats for the modeling process—and that has to be taken into account when developing the schedule and roadmap.
  - I5: Redefining the organizational structure of the administration department
  - I6: Implementing revised administrative procedures
    - Which of course presupposes (TC2) the availability of redefined administrative procedures.
  - I7: Implementing electronic Kanban
  - I8: Implementing JIT with suppliers
  - I9: Redefining the administrative organizational structure

4.4.2 From Left to Right: The Technical View

The left-to-right reading of the plan addresses the interests of the program manager, the technical architect, and the project managers, starting from their primary concerns: what they need to do and what they need to deliver.

The narrative description of the BRM in Figure 4.8 is as follows:

- I1: Implement estimating tool; this step will use (TC3) standard cost tables to allow (TC7) cost estimation modeling.
- I2: Define standard costs; this approach will provide a consistent database for estimating to be used in proposals.
- I3: Implement advanced design process based on existing CAD/CAM tools (TC1) to remove limitations currently affecting our design processes leading to improved designs.
- I4: Convert existing designs to new standard; this conversion will make existing designs available for the advanced tool being implemented in I3 (implement advanced design process).
- I5: Redesign administration's organizational structure; this redesign will streamline the internal processes in order to reduce overhead costs.
- I6: Implement revised administration processes; given a definition of the reworked administration procedures (TC2), these revised processes ensure that those definitions are put into practice in order, with the organizational structure redefinition (I5), to streamline administration and reduce overhead costs.
- I7: Implement electronic Kanban; provide the lean scheduling system required to deliver JIT.
- I8: Implement JIT with suppliers: develop agreements with suppliers for JIT deliveries.
- I9: Optimize cost of materials: apply the improved design capabilities (BC2) to allow the use of cheaper components.

So far, we have taken the existence of the links between entities for granted. However, the links do not simply define a sequencing or dependency relationship. This is explained in the next section.

4.4.3 Linkages Showing Relationships

In the diagrams, linkages are shown as directed lines corresponding to the bottom-up dependencies, such as node A (source node) being a contributor to node B (destination node).

The basic link between entities represents the relationship between these entities. To guarantee the underlying logic of a BRM, constraints are defined below on the way in which the links can be used between entities. In addition, a number of different parameters can be associated with each link, with each parameter measuring a different characteristic of the relationship.

Rules for Links

These rules are provided to help structure the work of building a complete BRM. They indicate good practice rather than provide formal directives.

In addition to the fact that feedback loops are forbidden, only the following relationships are allowed:

- Initiatives
  - Initiatives only connect to technical capabilities or to business capabilities.
  - Several initiatives can contribute to the same technical capability or business capability.
  - One initiative can contribute to one or more technical capabilities or business capabilities.
- Technical capabilities
  - Technical capabilities only connect to initiatives, to technical capabilities, to business capabilities, or to business outcomes.
    - The same rules apply to external (prerequisite) technical capabilities.
  - Several technical capabilities can connect to one business capability.
  - One technical capability can have several outgoing connections.
- Business capabilities
  - Business capabilities only connect to business capabilities or to business outcomes.
  - Business capabilities can have several outgoing connections.
- Business outcomes
  - Business outcomes only connect to other business outcomes or to strategic outcomes.
  - Business outcomes can have several outgoing connections.
- Strategic outcomes (could also be called "strategic results")
  - Strategic outcomes may be an end-point or can connect to other strategic outcomes.

Links not only indicate a logical relationship, but also carry a number of parameters that provide additional information for the source and the destination nodes.

Link Characteristics

The links provide logical relationships between the entities in the BRM. These relationships can be further characterized in terms of the following parameters (the entities at each end of a link are referred to as source and destination, respectively):

Contribution Fraction

- A contribution fraction is the amount of the destination value or capability (measured as a percentage) that is provided by the source.
    - Normally, the sum of all contribution fractions entering any node is 100 percent, since the destination node needs to get all of its value supplied (but see Chapter 6 on disbenefits).

Allocation Fraction

- An allocation fraction is the amount of its cost (measured as a percentage) that the source "charges" the destination for the output of the work carried out by the source.
    - The sum of all allocation fractions originating from any node is 100 percent, since the source node needs to account for all of its cost.

Type of Link

- There are two types of link: contributory or essential.
    - These types become important when considering optimization, tracking, and risk assessment; they are explained and applied in Chapters 6, 9, and 12.
    - Essential links can be indicated on the BRM by a specific format or by the addition of an assumption flag on the link (e.g., assumptions A2 and A3 in Figure 4.8).

Assumption

- Assumptions can be added to any link to explain the rationale for the destination entity.
    - For example: Why can the destination outcome be expected to be a result of the source output?

Likelihood or Confidence Level

- The likelihood or confidence level applies to the potential availability of the destination node.
    - For example: if an assumption has been given, then the likelihood that the assumption is correct or incorrect must be taken into account.

4.5 Estimating the Contribution Fractions

One effective way of estimating the importance of each link is to start from the following team-based, intuitive approach.

4.5.1 Analog Evaluation Approach

The following is carried out for each destination node (N): the participants initially work individually.

For each destination node, the first step is to rank the incoming (source) links from most important to least important. The links are identified by the node from which they originate (the *source node*). Once the links have been listed in this way, on a chart with one row per source, each participant should indicate where (along a line from "very important" to "least important") he or she would situate the corresponding node. Each participant will therefore develop a diagram similar to Figure 4.9 in which each row corresponds to one of the source nodes (in this case, there are five source nodes: C, E, F, G and L).

	Most important									Least important
Node C	☐									
Node E	☐									
Node F					☐					
Node G									☐	
Node L									☐	
	10	9	8	7	6	5	4	3	2	1

Figure 4.9 Ranking Source Nodes

You can then work as a team to compare and adapt the separate diagrams and achieve (as far as possible) consensus on a single diagram or a small number of diagrams. Then you need to quantify the fractions remembering that the contribution fractions entering a node sum to 100 percent.

4.5.2 Quantifying the Fractions

The total of 100 percent should be subdivided between the source nodes on the diagram as follows:

- Graduate the range.
- Evaluate the value of each position.
- Normalize these values so they sum to 100 percent.

If it was not possible to agree on a single quantification diagram in the previous step, the evaluation should be carried out for each diagram, and the diagrams used as a basis for discussion in order to understand the differences and to reach a consensus. If necessary, the results of calculating several versions of the map by using the different sets of values can help understand the impact of the differences in order to come to an agreement. If consensus cannot be achieved, the program manager will take the responsibility to decide the values to be retained, giving a result such as shown in Figure 4.10.

In the example above, the analyst is likely to suggest rounding the numbers to the nearest 10 percent so as to avoid spurious precision. The final set of values of the contribution fractions for destination node N would therefore be:

- CFnc = 30 percent
- CFne = 30 percent
- CFnf = 20 percent
- CFng = 10 percent
- CFnl = 10 percent

The values are now ready to be applied to the map, as will be explained in the next chapter. But first, we need to build the BRM for the QERTS example based on the concepts that have just been developed.

4.6 Applying the Concepts to the QERTS Example

Now that all of the concepts are available, they will be applied to create the benefits realization map for the QERTS case study that was introduced in the previous chapter. The approach that will be used first needs to be explained.

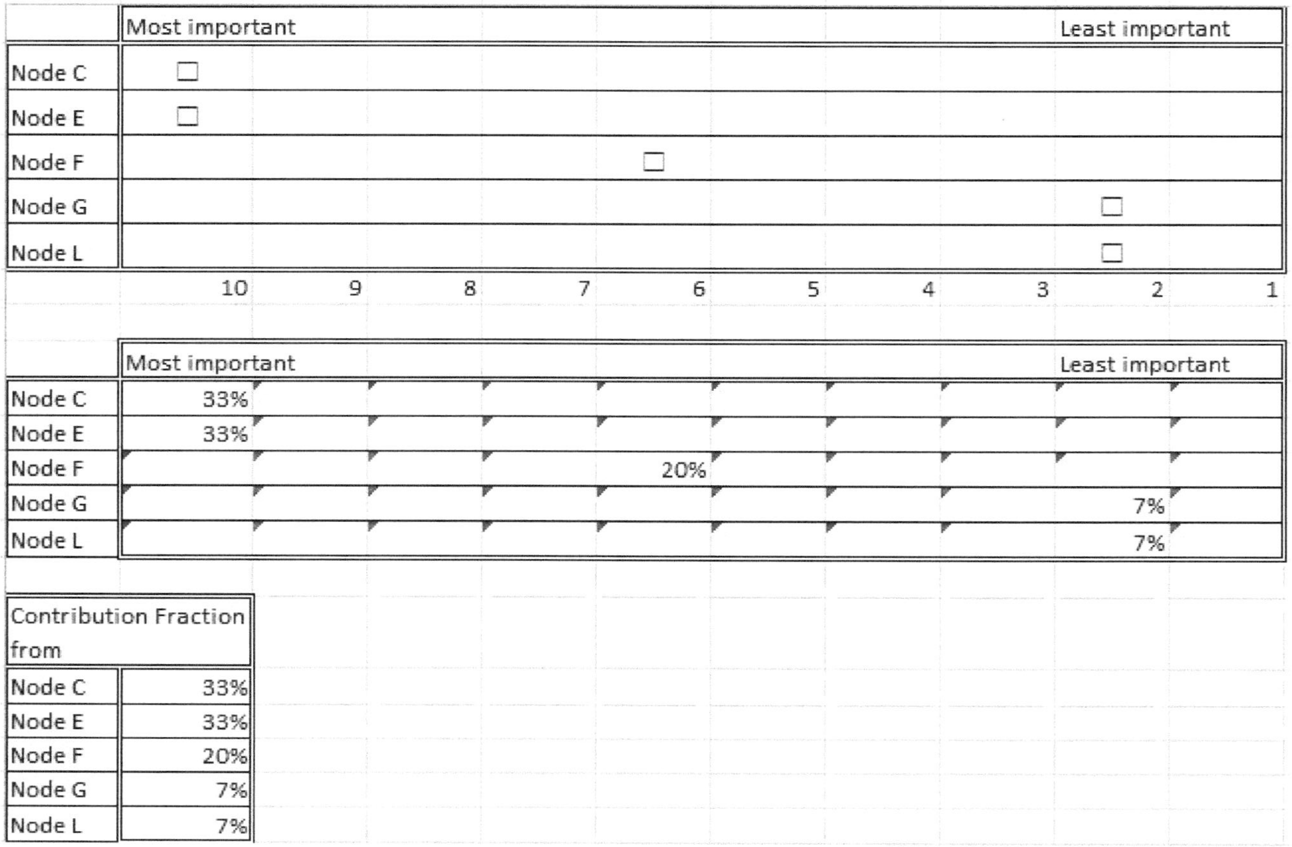

	Most important								Least important	
Node C	☐									
Node E	☐									
Node F					☐					
Node G									☐	
Node L									☐	
	10	9	8	7	6	5	4	3	2	1

	Most important								Least important	
Node C	33%									
Node E	33%									
Node F					20%					
Node G									7%	
Node L									7%	

Contribution Fraction from	
Node C	33%
Node E	33%
Node F	20%
Node G	7%
Node L	7%

Figure 4.10 Quantifying the Contribution Fractions

4.6.1 Business Consultancy Approach for Building the BRM

The explanation of the components of the generic approach presented the elements in a bottom-up, left-to-right manner. However, as shown in the business transformation example, that presentation provides a technology-driven view of the solution. To provide a benefits-oriented view, we need to start from the benefits and work from right to left. The way in which this builds a complete map is developed below. Note that, in order to facilitate reading, the nodes are identified by names and letters. The letters in this example were added later and have been assigned based on the left-to-right view of the diagram.

4.6.2 Start with the Business Targets

The business targets are obtained from the mission statement and should be entered first along with their quantified values (see Figure 4.11).

The next step entails showing the basis for these targets.

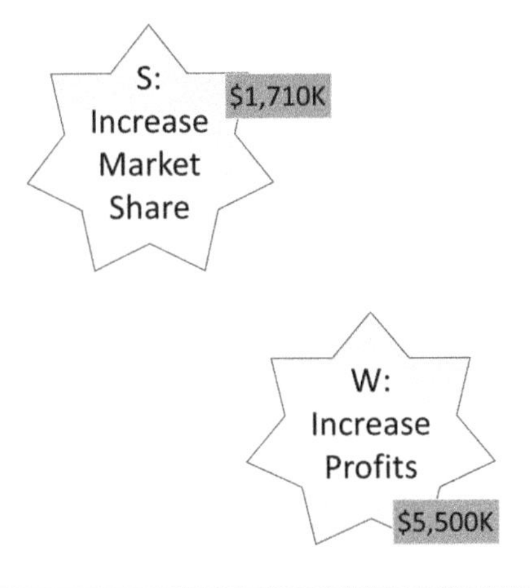

Figure 4.11 First Step in Modeling the QERTS Program: Identifying the Business Targets

4.6.3 QERTS Strategic Outcomes

This step may also be based on the mission statement or may require discussion with strategic management; the addition of the strategic outcomes is shown in Figure 4.12.

The way in which the organizational environment needs to change in order to achieve these strategic outcomes now needs to be mapped.

4.6.4 QERTS Business Outcomes

Some of the business outcomes may be given in the mandate. However, it is now necessary to involve a business consultant, not only to ensure a full identification of all of the required outcomes, but also to identify the relationships between them and with the strategic outcomes (see Figure 4.13):

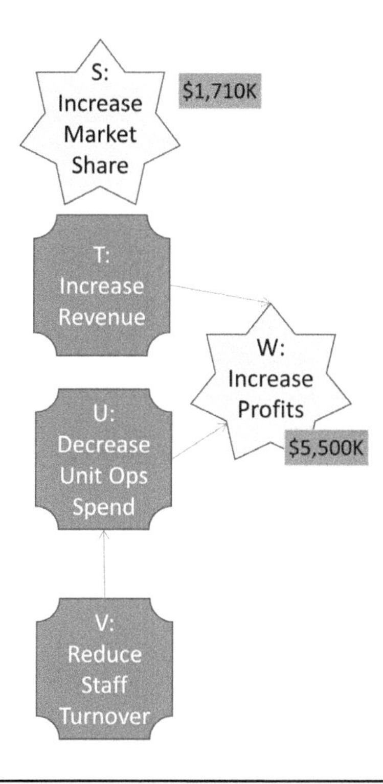

Figure 4.12 Addition of the Strategic Outcomes

- Market share (S) depends on the number of customers (P)—hardly a revelation!
- A reduction in operational spend (U) depends on effective operations (O).
- The number of customers (P) is affected by the business outcome of efficient and friendly service (R). As mentioned in the rules for building the map, business outcomes can depend on other business outcomes as well as on business capabilities, which will be addressed in the next section.
- Efficient and friendly service (R) depends on effective operations (O) and on happy staff (Q).
- Happy staff (Q) reduces staff turnover (V).

Each of these dependencies needs to be validated by the business analyst, and any assumptions must be documented.

The rules for developing the map specify that business outcomes depend on applying business capabilities, so these now need to be determined.

4.6.5 QERTS Business Capabilities

Once again, this work is the responsibility of the business consultant. Determining the business capabilities entails, for each business outcome, answering the question "what needs to be in place in order to create the required outcome?" In addition, any dependencies between these business capabilities have to be determined.

The result for the QERTS map is shown in Figure 4.14. The diagram shows the following:

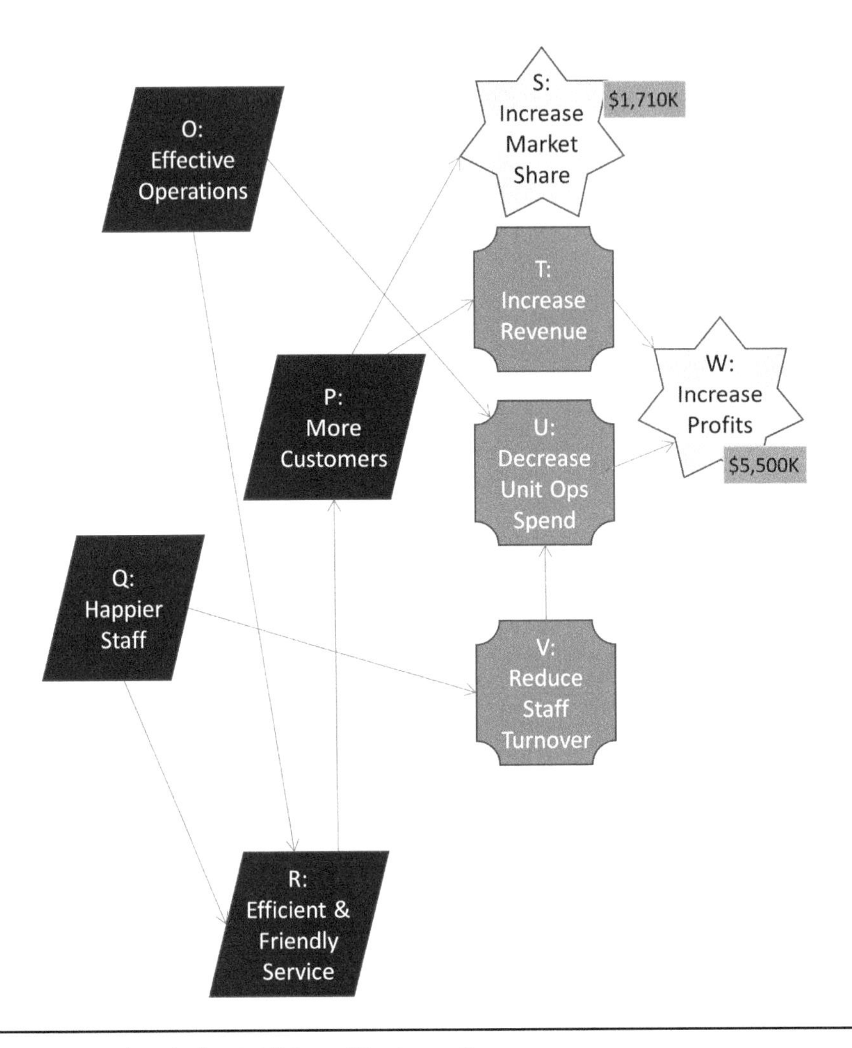

Figure 4.13 QERTS with the Addition of Business Outcomes

- Effective operations (O) requires flexible work practices (J), as well as all stores to be run the same way (L).
- Flexible work practices (J) also contribute to happier staff (Q).
- Happier staff (Q) also depends on management's ability to reward them appropriately (M).
- Happier staff (Q) leads to efficient and friendly service (R).
- Efficient and friendly service (R) also depends on the ability of the staff to apply the processes (N).
- Increased market visibility (K) plus efficient and friendly service (R) lead to more customers (P).

In this example, there are no links between the business capabilities. In certain programs, however, some business capabilities might support other business capabilities, and that possibility must always be considered.

The next step is to identify whether these business capabilities can be delivered directly by executing technical initiatives or whether any specific technical capabilities will first need to be specified.

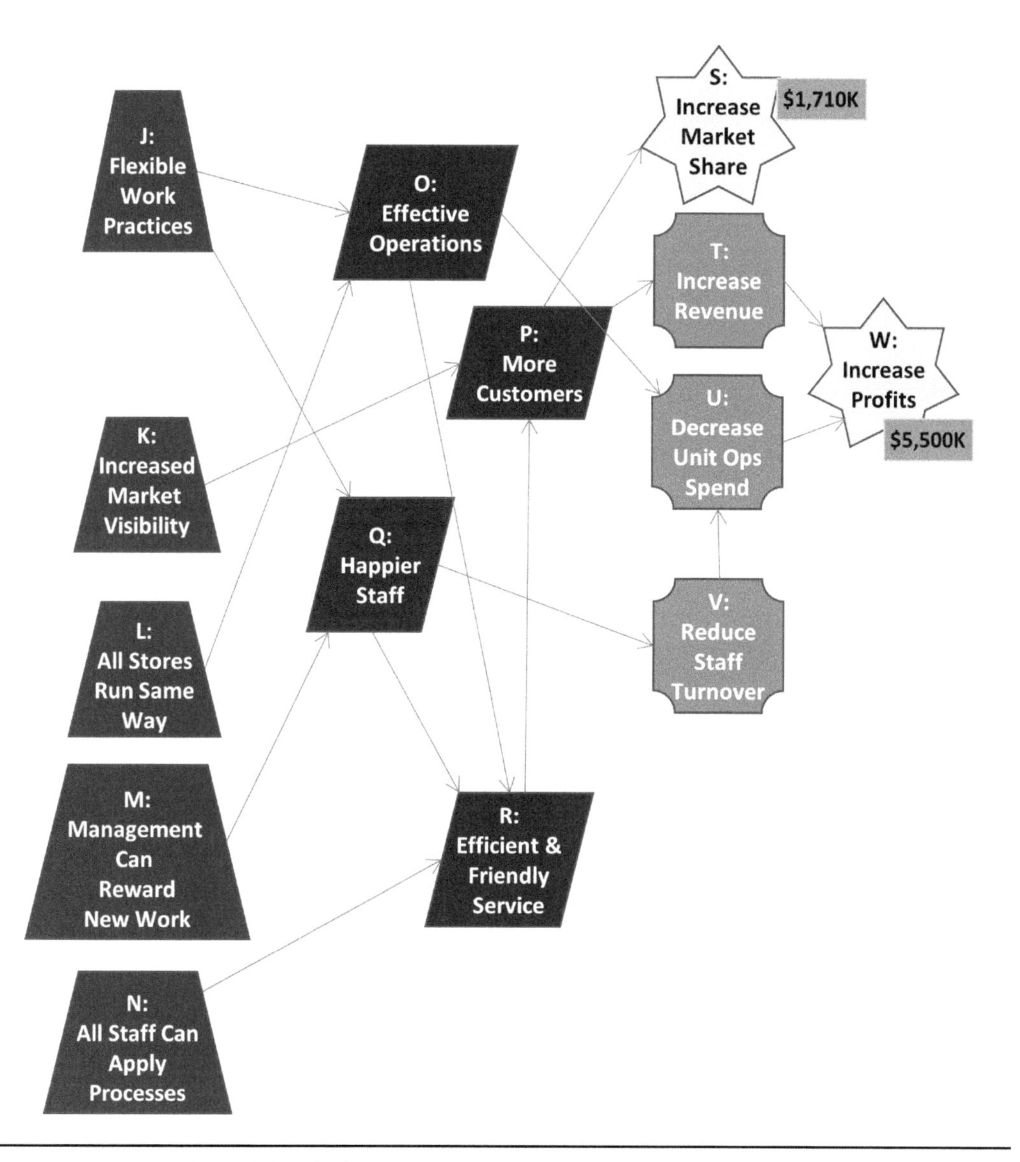

Figure 4.14 QERTS Model Including Business Capabilities

4.6.6 QERTS Technical Capabilities

Deciding on technical capabilities should be a collaborative effort between the business consultant and a technical architect. For the QERTS example, the result is shown in Figure 4.15. Three technical capabilities are identified:

- An integrated customer relationship management system (H) is needed to support the required business outcome of increased market visibility (K).
- This CRM system (H) depends on the availability of another technical capability: the availability of a process model (G).

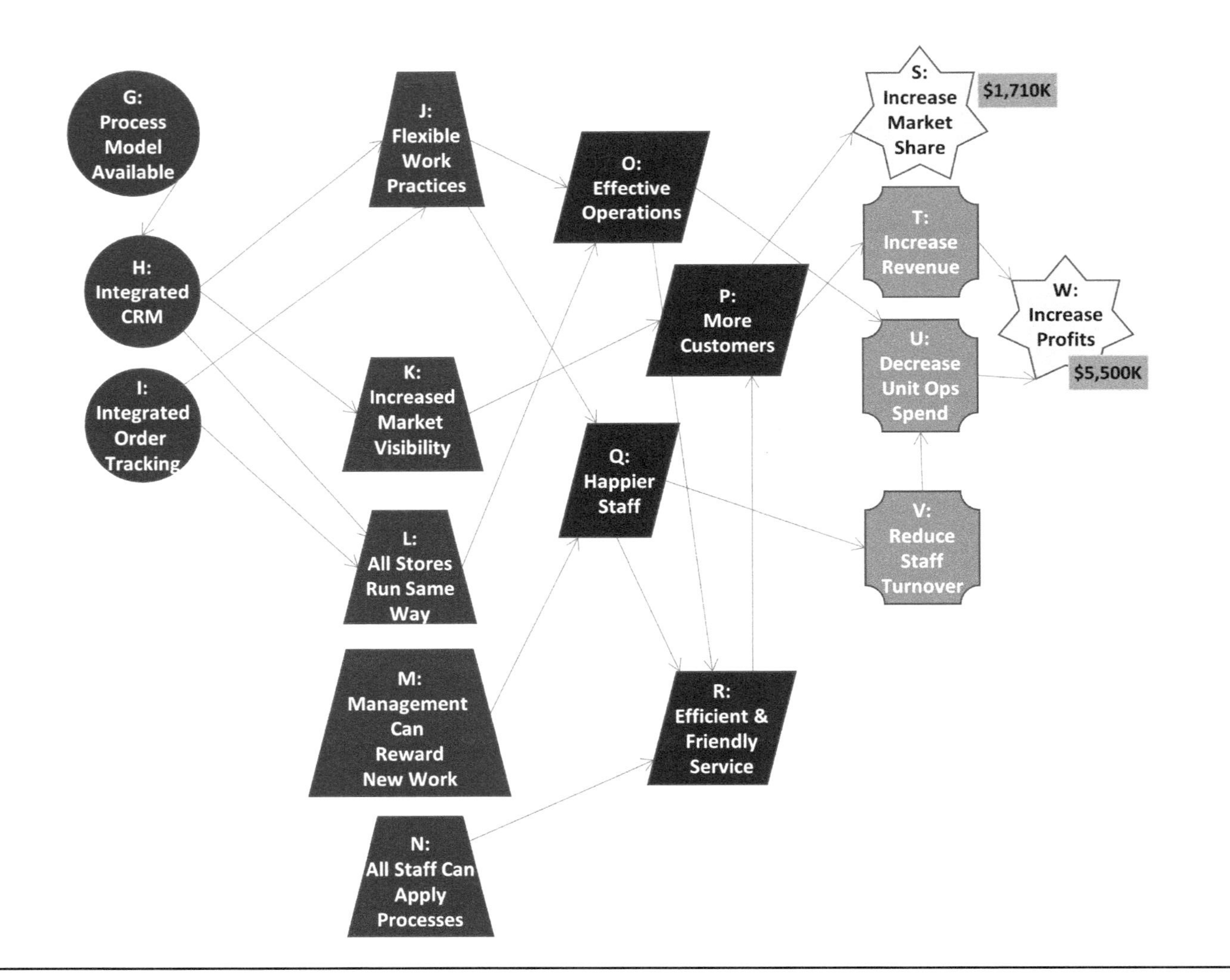

Figure 4.15 Addition of Technical Capabilities to the QERTS Model

- The third technical capability is integrated order tracking (I), which, with the integrated customer relationship management system (H), allows all stores to be run the same way (L). It also contributes to the flexible work practices (J).

The final step is to define the required initiatives.

4.6.7 QERTS Initiatives

This final step also requires close collaboration between the business consultant and the technical architect. The business consultant will specify the requirements on the capabilities of the deliverables; the technical architect will determine the best way of satisfying the specifications. The result is shown in Figure 4.16.

This analysis leads to the following initiatives:

A. Standardize Processes to lead to Process Model Available (G).
B. Integrate Store Records which, with Process Model Available (G), leads to Integrated CRM (H).
C. Adapt Existing Tracking System leads to Integrated Order Tracking (I).
D. Develop Marketing Program which, along with the integrated customer relationship management system (H), leads to Increased Market Visibility (K).
E. Development of Joint Compensation Plan allows management to reward the staff appropriately (M).
F. Training Programs will enable the staff to apply the processes (N).

4.6.8 Completing the Map

Once these initiatives and the requirements have been defined, the business consultant should leave the entire responsibility for costing the work to the technical architect.

Now that the diagram contains all of the links and nodes, the next step for the business consultant is to quantify the contribution percentages, as explained earlier. These two quantification steps lead to the basic quantified BRM shown in Figure 4.17.

The cost ("allocation") of each initiative has been estimated by the technical architect. The total allocation required is the sum of the allocations of each of the nodes A, B, C, D, E and F (i.e., $5.6 million). This compares favorably with the business target of gaining (S + T) $7.21 million. The initial business case therefore looks promising.

4.7 An Honest Business Case

One of the strengths of the BRM development approach is that it insists on the importance of explicitly describing the capabilities based on the required outcomes; this forces the analyst to consider and to describe the hypothesis for the outcome based on the capability. Explicitly describing a business capability and mapping its requirements on the initiatives, for example, will help to validate the logic of the overall BRM before expectations of cost estimates are imposed on business or strategic outcomes.

The fact that the technical architect is tasked to cost the initiatives based solely on the specified capabilities and without interference from the business enables a fair estimation. No prior expectations on an acceptable (but possibly unreasonable) level of cost will have been applied as pressure to the technical architect. The assessment of viability and the optimization of the proposed solution will be based on an initial, unbiased analysis.

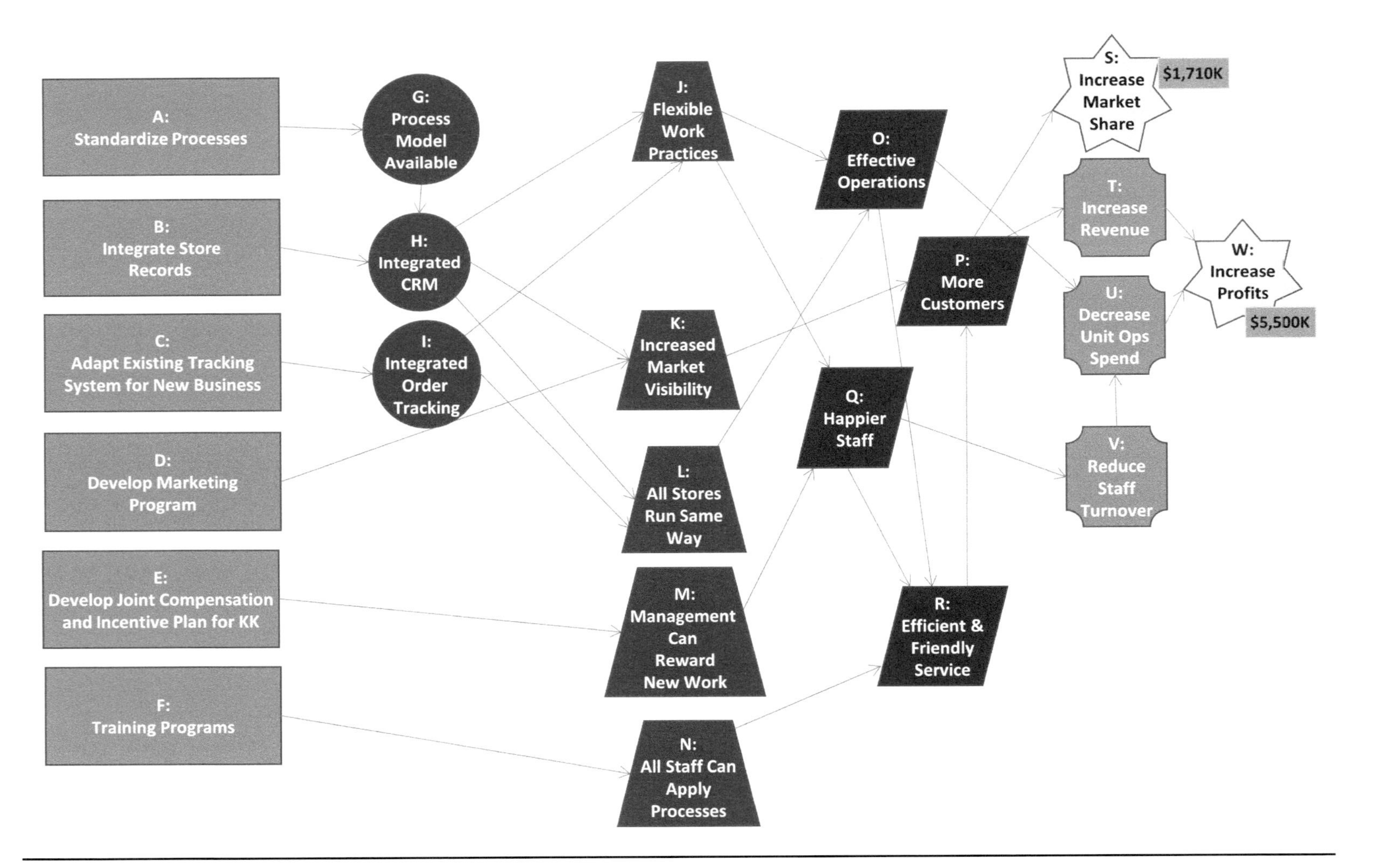

Figure 4.16 QERTS Model, Including the Initiatives

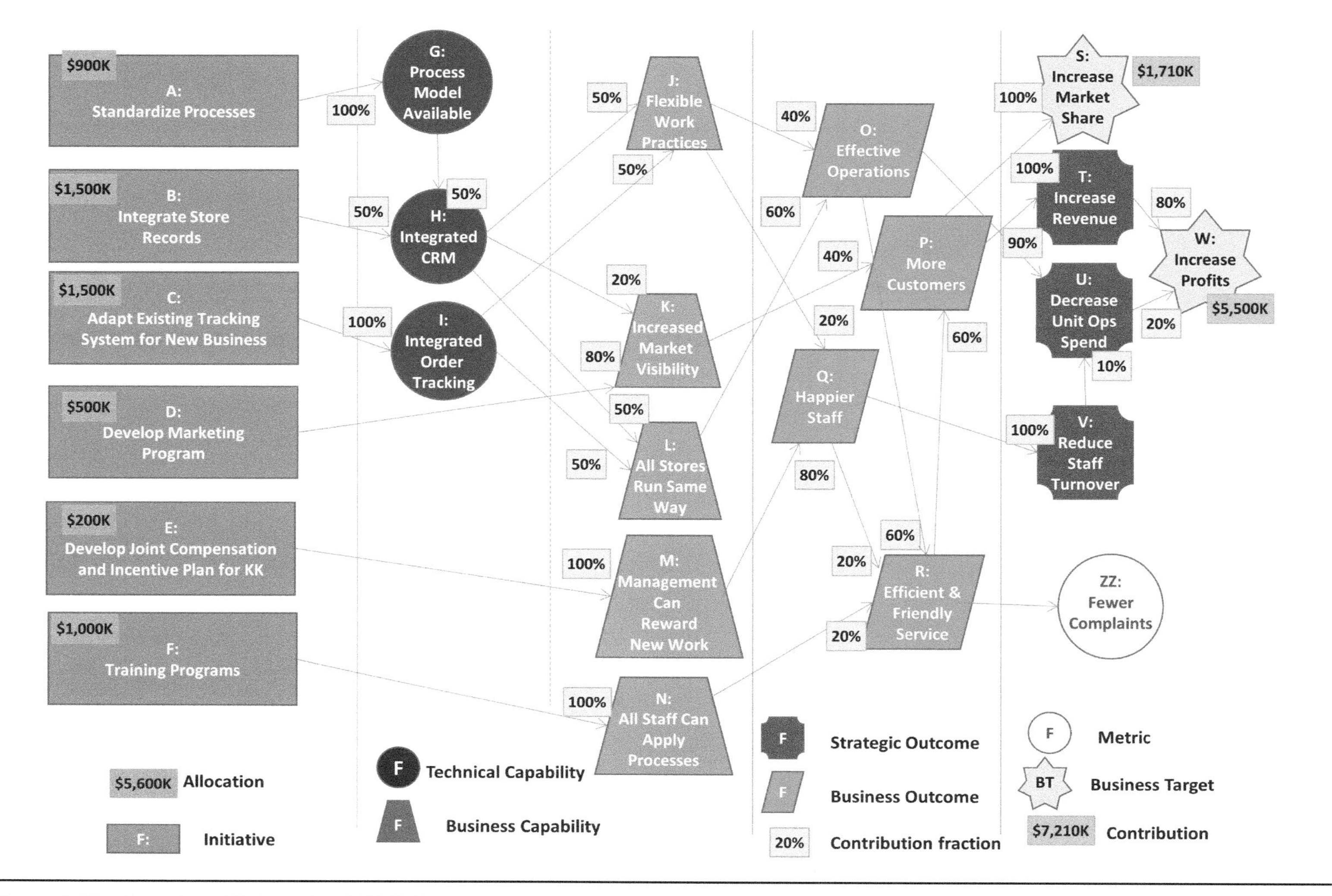

Figure 4.17 Complete BRM for the QERTS Example

The BRM development technique, as described in the example, and the associated algorithms developed in the next chapter form the foundation of the earned benefit program management approach.

4.8 Summary

This chapter has developed the main characteristics of the modeling technique for generating and managing a full end-to-end BRM. The various components of the map have been defined along with their characteristics and numerical parameters. This modeling technique has been applied to the case study in order to develop a BRM to be used as an example in forthcoming chapters.

4.9 The Next Steps

Once the draft business case and the corresponding BRM have been accepted, the remaining quantification of all of the work can be completed.

The next chapter explains how the BRM uses the available information on links and objectives to provide a complete representation of the costs and value of each of the components of the map.

In the interest of simplicity to explain the numerical evaluation algorithm, in the initial examples shown in the next chapters, initiatives will connect directly to business outcomes. Once the algorithm has been developed, however, it is then applied to the full QERTS example, and the strict structural rules are obeyed.

4.10 References

Office of Government Commerce. *Managing Successful Programmes (MSP®)*. Norwich, England: The Stationery Office, 2011.

Ohno, Taiichi. *Toyota Production System—Beyond Large-Scale Production*. New York, NY: Productivity Press, 1988.

Chapter 5

Calculating the Model

This chapter reviews the options for developing a mathematical model that provides a complete and consistent link between the program estimates and all the numerical data required for assessing initial viability and potential improvements, for risk assessments, schedule development, and progress tracking. The analysis of the various approaches is supported by detailed calculations. These calculations are included for completeness, but are not essential reading, as the numerical results are given in the corresponding diagrams. The criterion for selecting the valid approach is explained and defined under the acronym BEER (break-even everywhere requirement). The chapter concludes with the identification of the algorithm that delivers the BEER. The BEER turns out to be the key to calculating, optimizing the model, managing changes, and tracking progress in a consistent and traceable manner. If you would rather understand the effect of the theory on the case study than read the detailed explanations, just go straight to Chapter 7 and come back to this chapter later.

5.1 Picturing the Contributions

As explained in the previous chapter, contributions flow through the benefits realization map (BRM) from right (strategic outcomes) to left (initiatives) to establish the relative contribution of each component of the network, as shown in the simplified diagram in Figure 5.1. For initial simplicity, this is a much-reduced BRM and does not follow the interconnect linkage rules described earlier. It is, however, sufficient to explain how the contributions and the allocations are calculated for each node, based on the corresponding algorithm, the projected value of strategic outcomes, the estimated cost of initiatives, and the contribution fractions.

In normal circumstances (but see Chapter 6 for an exception to this rule), the sum of all contribution fractions for a given *destination* sum to 100 percent since the sources contribute together to the delivery of the node's total value. The product of a contribution fraction by the contribution value is known as the *contribution share* or *share of the contribution*.

The examples show the contribution fractions, contribution shares, and contributions, using the simple naming convention defined in the following section.

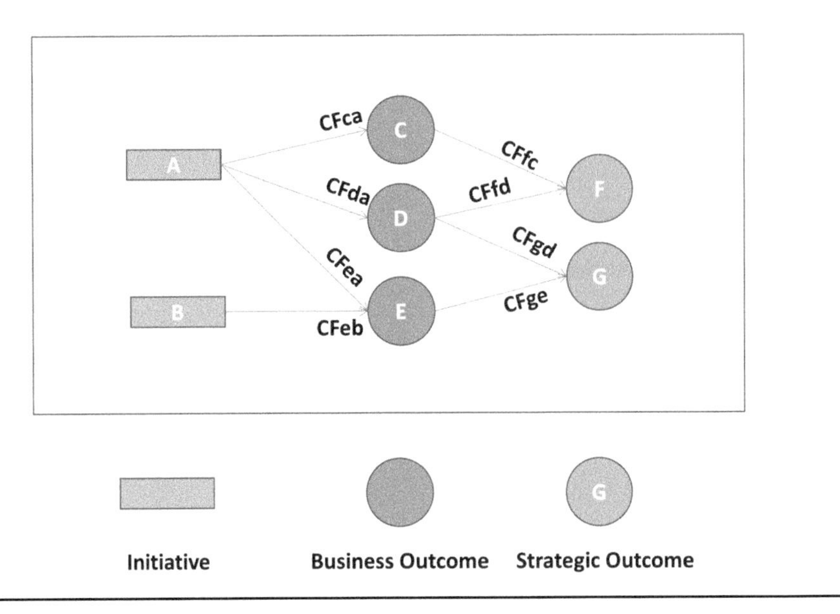

Figure 5.1 Simplified Benefits Realization Map Showing Contribution Identifiers

5.1.1 Identifying the Contributions

The proportions of the contributions (*contribution fractions*) are represented as CF followed by the identifiers (in order) of the destination node and the source node (CFds: the contribution fraction to node D from node S).

Similarly, the contribution share (CS) of a node is shown on the node, and named CSds (the contribution share to node D from node S)

The full contribution of node N is identified by Cn (the total contribution of node N).

The algorithm for calculating all of the contributions in the diagram, right back to the contribution of each initiative, is explained next by use of these conventions.

5.1.2 Calculating the Contributions

A hypothetical example based on Figure 5.1 will be used. The way in which this is quantified across the BRM from right to left is shown progressively below.

Local Contribution Values

The following values used in the example are shown in Figure 5.2:

- Business targets of the strategic outcomes (total contributions of node F and node G)
  - Cf = $40,000
  - Cg = $60,000
- Contribution fractions for node F
  - CFfc = 40 percent
  - CFfd = 60 percent

- Contribution fractions for node G
  - CFgd = 50 percent
  - CFge = 50 percent
- Contribution fractions for node C
  - CFca = 100 percent
- Contribution fractions for node D
  - CFda = 80 percent
  - CFdb = 20 percent
- Contribution fractions for node E
  - CFea = 80 percent
  - CFeb = 20 percent

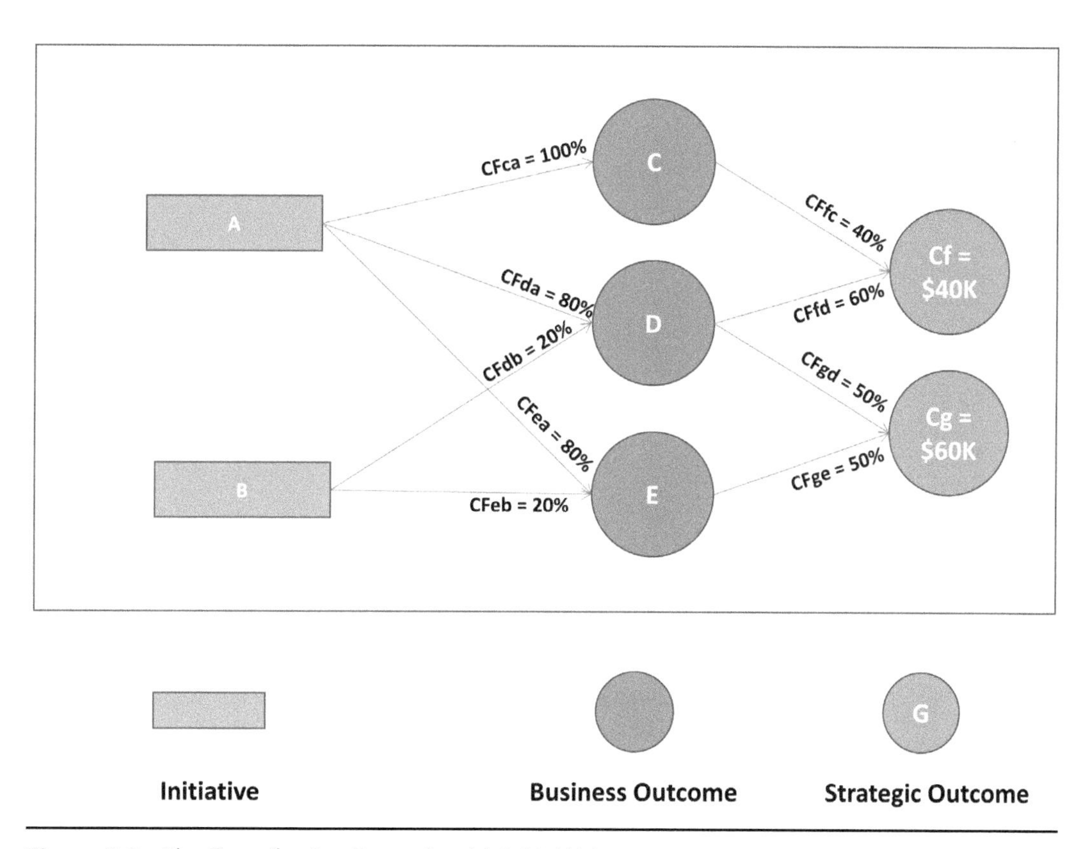

Figure 5.2 The Contribution Example with Initial Values

It is easy to verify in the example that the contribution fractions into any node sum to 100 percent.

To calculate the contribution of Cs of a source node (S) based on the known contributions of the destination nodes (D) (remember we are working from right to left), just sum all of the corresponding contribution shares (i.e., the products of the contribution values of the connected nodes by the contribution fractions specified for the corresponding links). In algebraic terms this is: Cs = Sum of (Cd_i x $CFsd_i$) for all existing destinations d_i for which $CFsd_i$ exists.

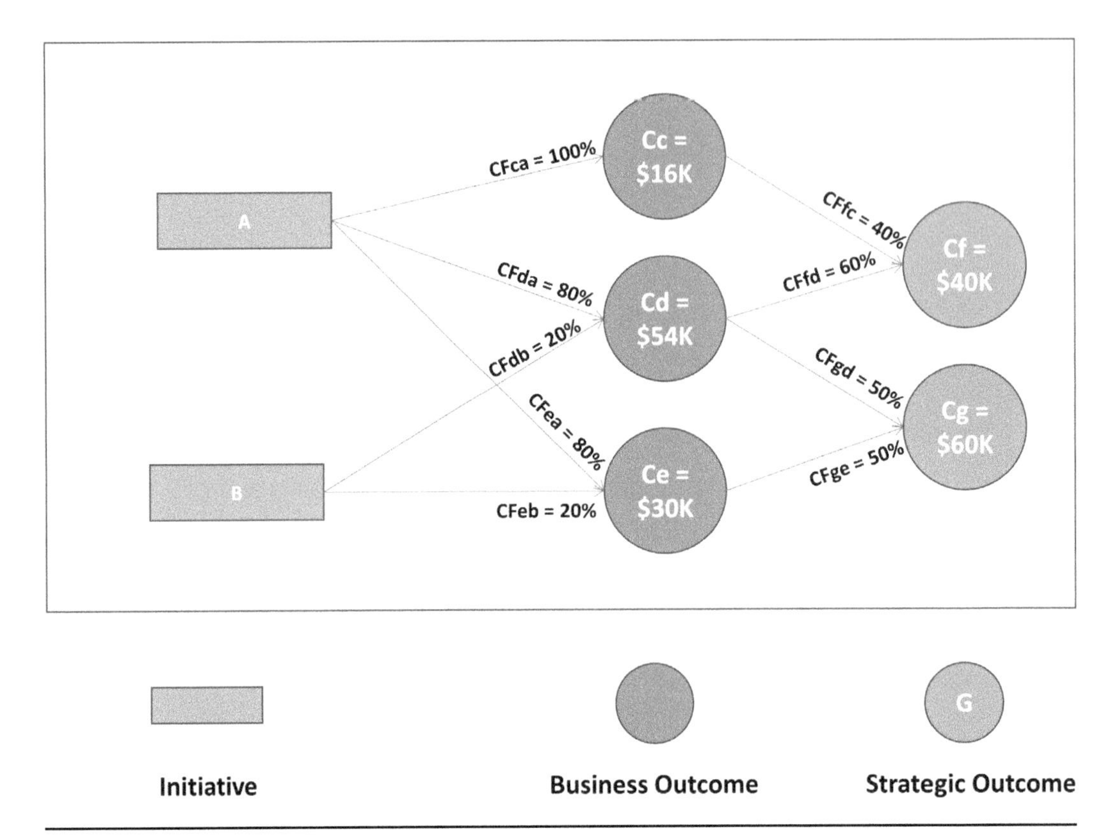

Figure 5.3 The Values of the Business Outcomes Based on the Financial Outcomes

The resulting values can be calculated from right to left starting from the diagram in Figure 5.2 to give the value of the business outcome nodes (Figure 5.3):

- Cc: CFfc * Cf = 40% * $40K = $16K
- Cd: CFfd * Cf + CFgd * Cg = 60% * $40K + 50% * $60K = $54K
- Ce: CFge * Cg = 50% * $60K = $30K

5.1.3 Remote Contributions

The next step entails calculating the corresponding values for the business plans for each of the initiatives. The same algorithm is carried out in a similar manner from the contributions of nodes C, D, and E and the corresponding contribution fractions in Figure 5.3:

- Contribution of node A = Ca: CFea * Ce + CFda * Cd + CFca * Cc = 80% * $30K + 80% * $54K + 100% * $16K = $83.2K
- Contribution of node B = Cb: CFeb * Ce + CFdb * Cf = 20% * $30K + 20% * $54K = $16.8K

The result including the value of each contribution share is shown in Figure 5.4.

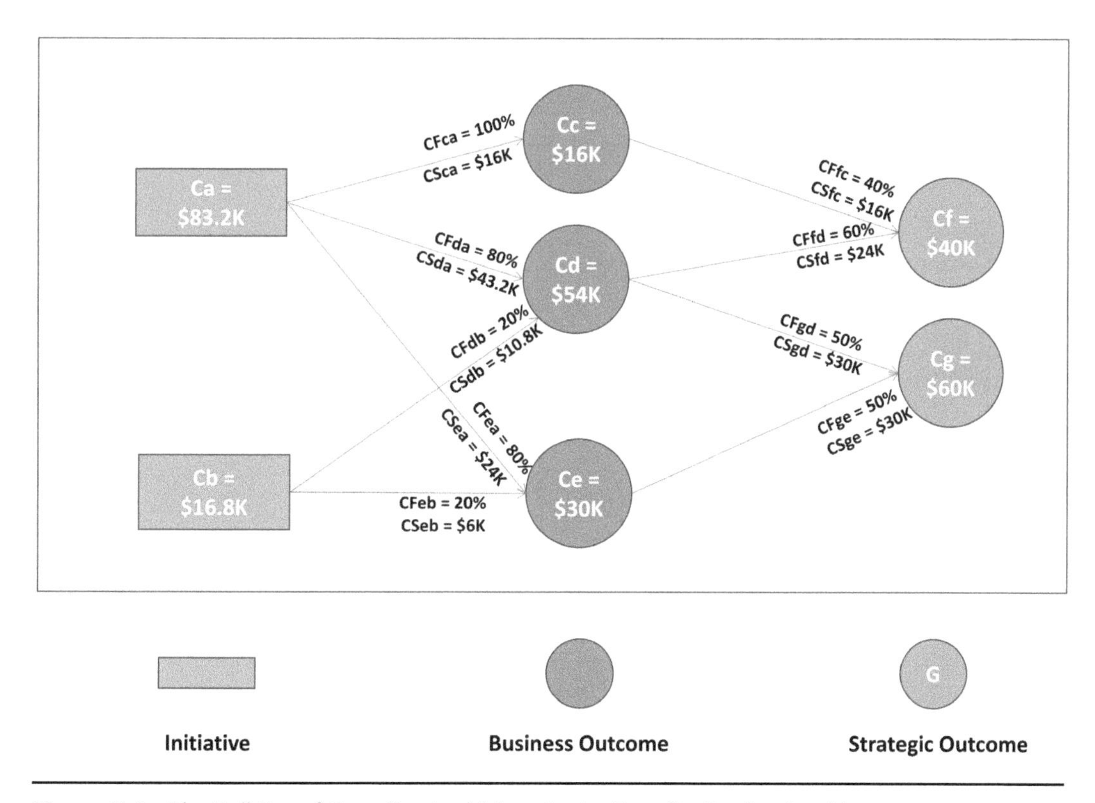

Figure 5.4 The Full Set of Contribution Values in the Benefits Realization Map

In a more sophisticated model, the same calculations would be repeated several times from strategic outcomes to business outcomes, to business capabilities, and so on before reaching the final calculations to give the contribution of each initiative.

Now that we have defined the means of calculating where the value is generated in the model, the next step is to analyze the costs associated with creating this value.

5.2 Understanding Allocations

As defined earlier, the allocation fraction is "the amount of its cost (measured as a percentage) that the source 'charges' the destination for the output of the work carried out by the source."

The concept behind allocations is to determine the correspondence between the investment in the initiatives and the value of each successive entity in the model. To evaluate the value of each entity, allocations are calculated from left (initiatives) to right (strategic outcomes)—in contrast to contributions, which are calculated from right to left. The cost of achieving each entity on the left (the budgeted cost of the implementing the initiative) is used to calculate the cost of the connected nodes to the right (the allocation of the costs for delivering the corresponding contribution).

In a similar way to contributions, the allocation fractions are shown on the relationship arrows as a percentage of the total amount corresponding to the source entity. The product of the allocation by this number is known as the *allocation share* or *share of the allocation*. For allocations, however, it is the set of percentages relative to the source node that sums to 100 percent, rather than the destination node as in the case of contributions.

The allocations and their parameters follow a naming scheme similar to that for contributions:

- The total allocation for node N is written as An.
- The allocation share (AS) from a source S to a destination D is ASsd.
- The allocation fraction (AF) from a source S to a destination D is AFsd.

Naming is one thing; assigning values is another. The challenge to be addressed is how to determine the allocation fraction for each link.

5.3 Selecting the Algorithm for Determining the Allocation Fractions

The question that still needs to be answered is this: How should each allocation be shared between each of the destinations to cover the total amount allocated to the corresponding source? The answer to this question will provide the algorithm for determining the allocation fractions.

There are four possible approaches:

1. By consensus
2. From the contribution fractions
3. From the contribution values
4. From the contribution shares

Each of these approaches is described below, and the reason for retaining the fourth option is then explained.

5.3.1 Agreeing on the Allocation Fractions by Consensus

The same team-based approach as for the contribution fractions could be used. The disadvantages are:

- Extra work for the team
- Lack of direct link between the allocations and the contributions
- Need to reconvene the team in the case of any changes to the model
- Last, but by no means least: risk of "massaging" the numbers to provide spurious justifications for certain favorite initiatives

For these reasons, this approach will not be considered any further.

5.3.2 Using Contribution Fractions to Calculate the Allocation Fractions

The reasoning in this case is that the more a destination node "needs" the current node's output, the more it should pay for it. The way of distributing to total amount of the allocation is to attribute it pro rata of the contribution fractions connected to the node (i.e., the allocation fractions are based on the value of the contribution fraction of the links, normalized to sum to 100 percent). In other words, the algorithm is:

- For each node, sum the contribution fractions over all of its destination nodes, as shown in Figure 5.5 for initiative A.

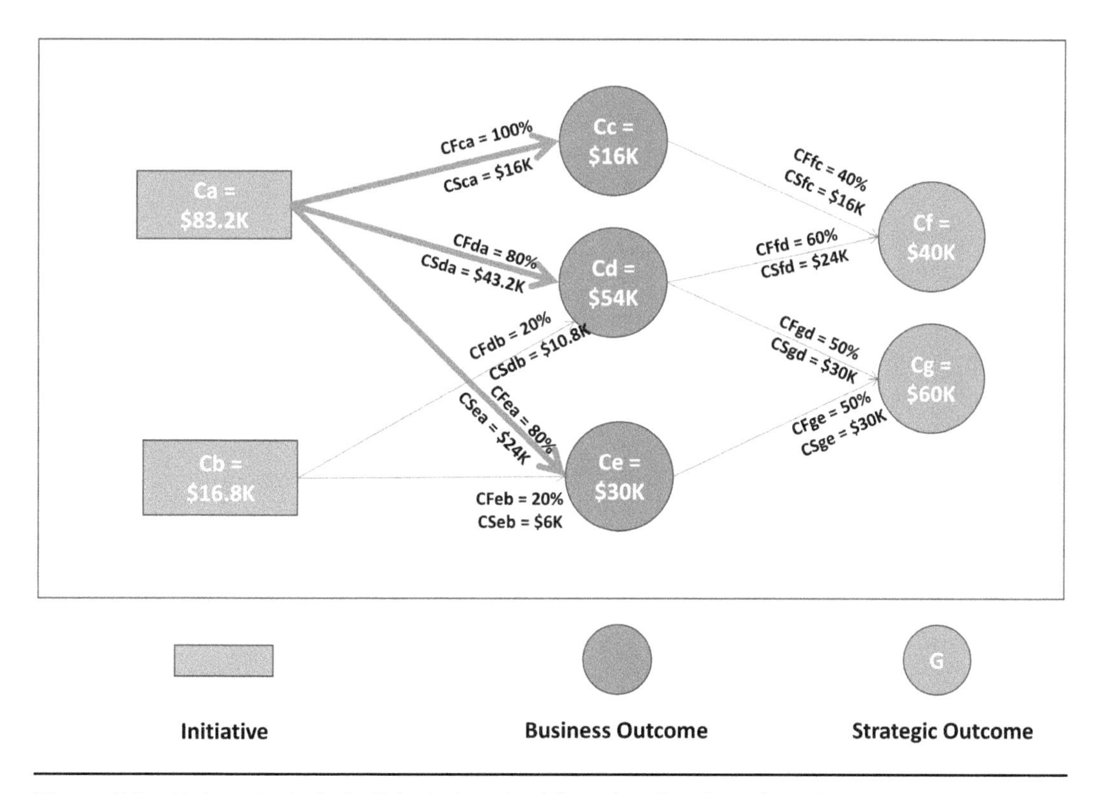

Figure 5.5 Links to Include in Calculating the Allocation Fractions from A

- Calculate the allocation fraction on a link as the percentage of its contribution fraction on that link to the total just calculated.

The result of these calculations is shown in Figure 5.6, taking as naming convention for allocation fractions: *AFsd* where *s* identities the source and *d* the destination. Here are the calculations.

A. Normalization factor for A: (CFca + CFda + CFea) = 100 + 80 + 80 = 260
   - AFac = CFca / 260 = 100% / 260 = 38%
   - AFad = CFda / 260 = 80% / 260 = 31%
   - AFae = CFea / 260 = 80% / 260 = 31%
B. Normalization factor for B: (CFdb + CFeb) = 20 + 20
   - AFbd = AFbe = 50%
C. Only one link from C so AFcf = 100%
D. Normalization factor for D: (CFfd + CFgd) = 110
   - AFdf = CFfd / 110 = 60% / 110 = 55%
   - AFdg = CFgd / 110 = 50% / 110 = 45%
E. Only one link from E so AFeg = 100%

Given the assumption that the program manager has established the cost of the initiatives to be A = $15,000 and B = $50,000, the calculation of the allocations can then be carried out from left to right using the values of the allocation fractions shown in Figure 5.6. The resulting allocation values are shown in black in Figure 5.7.

For comparison, Figure 5.8 shows the diagrams for the Contributions (in white) and the Allocations (in black) on the same diagram.

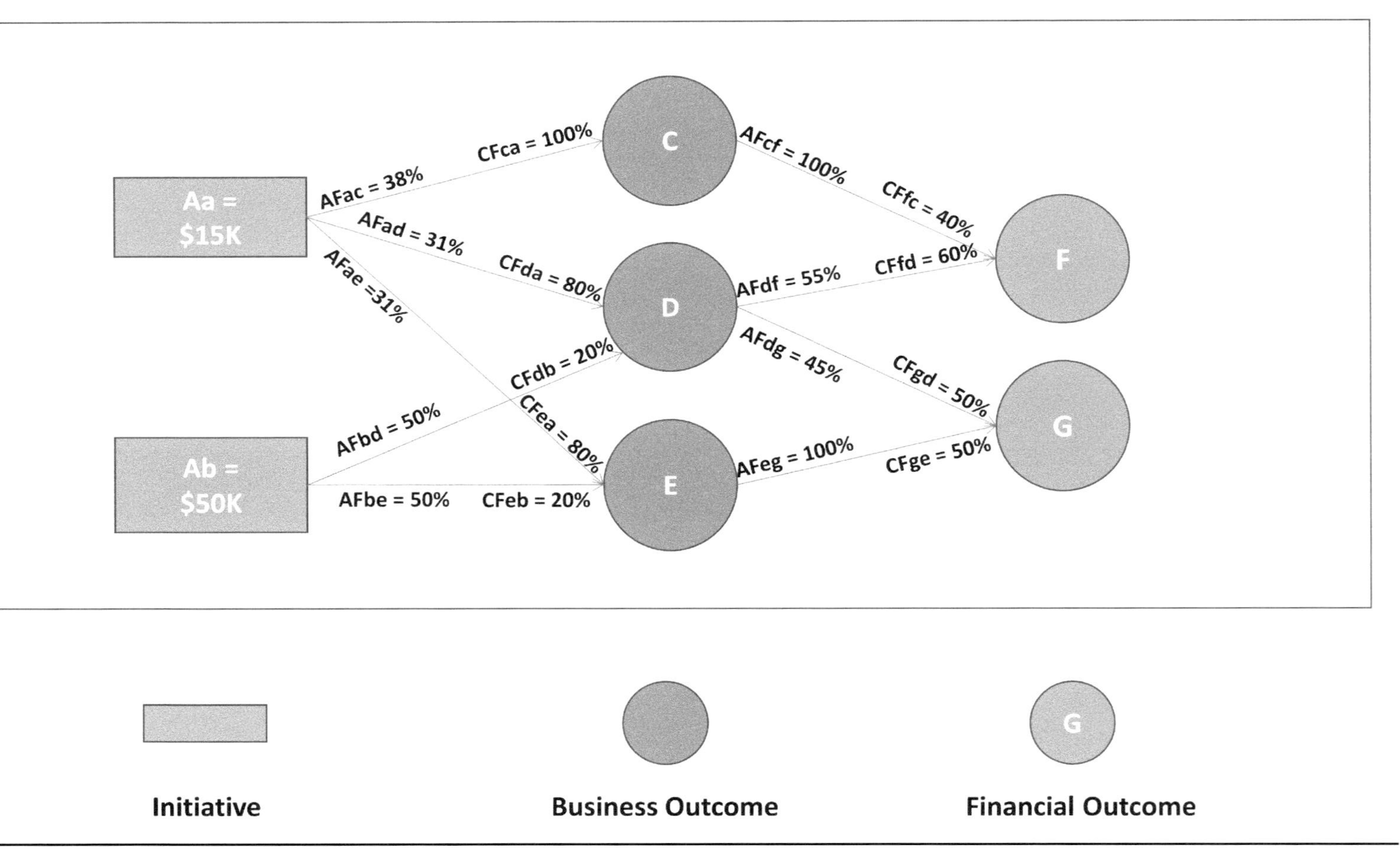

Figure 5.6 Evaluating Allocation Fractions Based on the Contribution Fractions

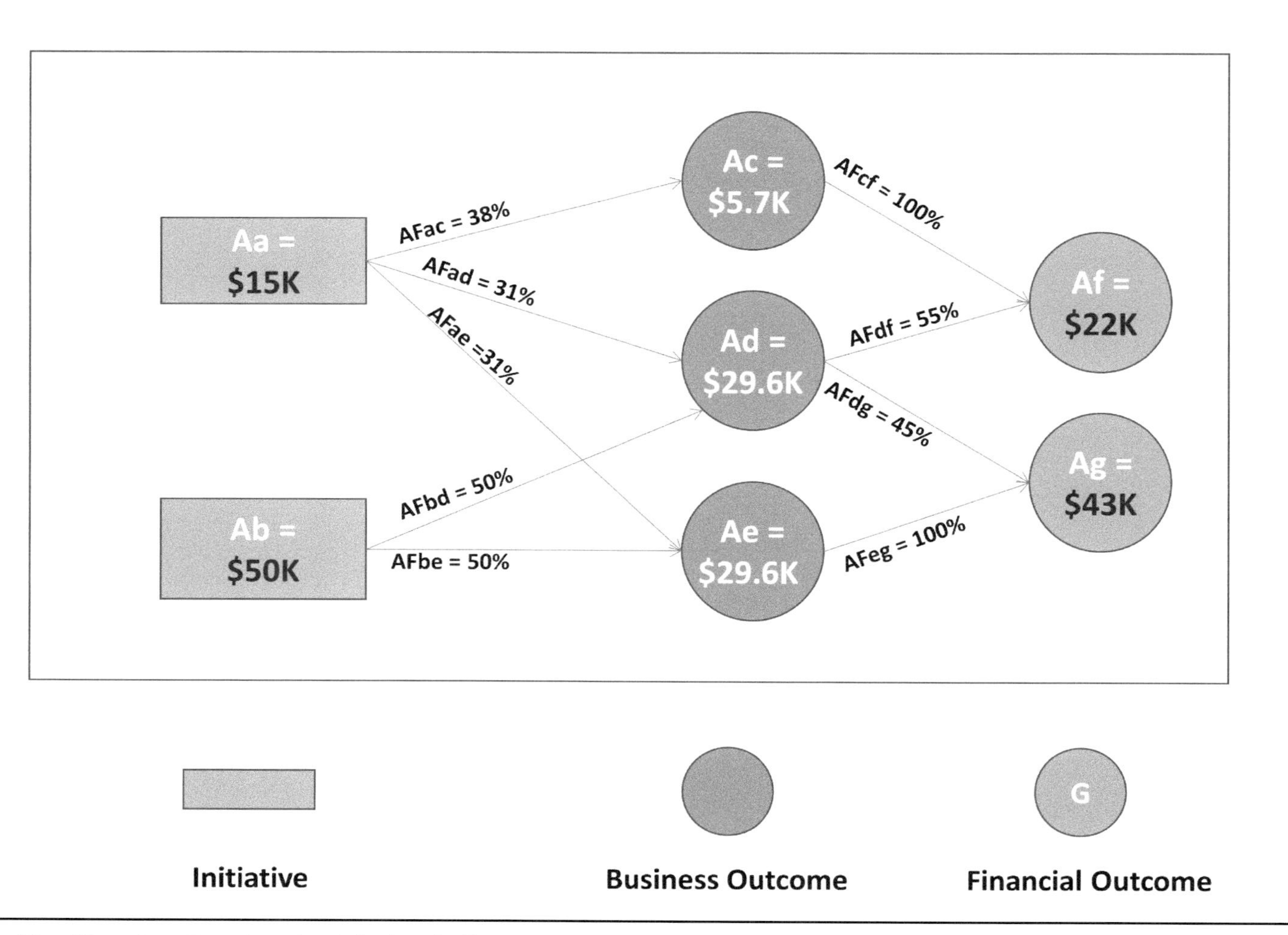

Figure 5.7 The Allocations Based on the Calculated Allocation Fractions

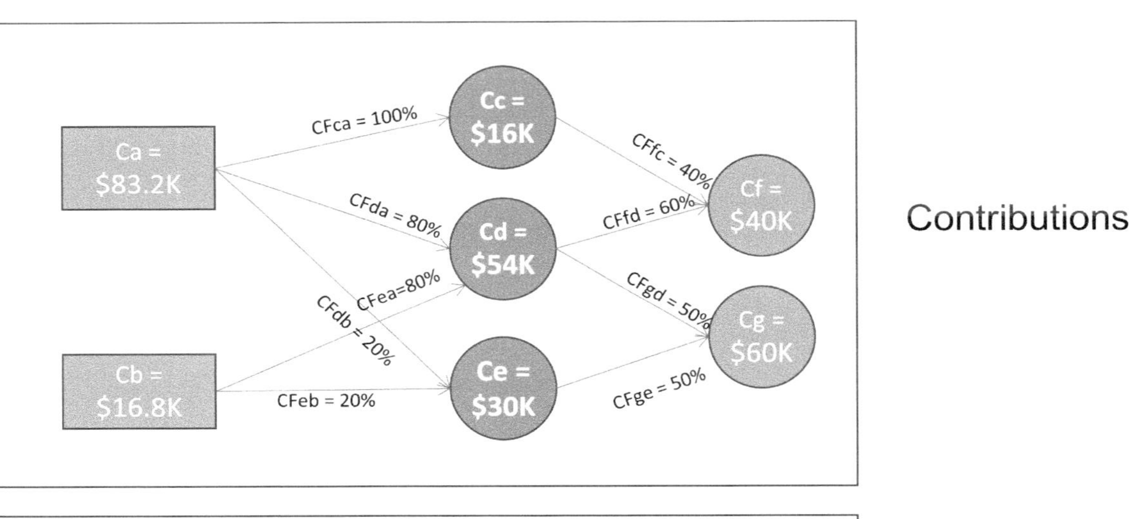

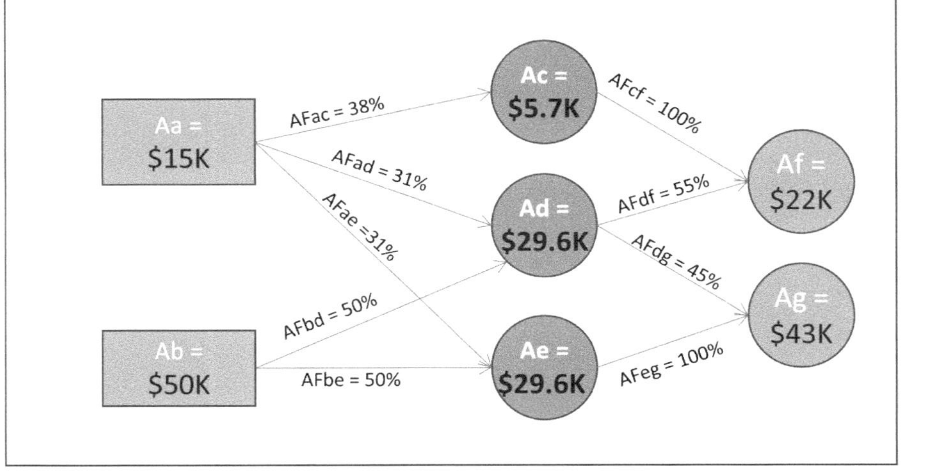

Figure 5.8 The Values of the Contributions and the Allocations in the Example

The strong point of this approach is that it provides a direct correspondence between the business model (i.e., contribution fractions) and the way of assigning allocation fractions. However, it ignores the actual value of the corresponding contributions.

An alternative approach for evaluating allocation fractions is to use the value of the node (i.e., the calculated value of its contribution), as shown next.

5.4 Using the Contribution Values to Calculate the Allocation Fractions

In contrast to the previous approach, the allocation fractions are calculated with respect to the relative size of the contribution value of each connected destination. One way to look at this approach is that the destinations are taxed (their allocation to the costs) at a flat rate proportional to their value (the contribution). So the algorithm for any node is:

- Find the total value to which this node contributes, by summing the contribution values of all of the destination nodes. The set of destination nodes involved for source node A is shown in Figure 5.9.
- Calculate the allocation fraction on a link as the percentage of its contribution value on that link to the total.

Using the values from Figure 5.9, this algorithm gives:

A. Normalization factor for A: (C + D + E) = $100K
   - AFac = C / 100 = 16%
   - AFad = D / 100 = 54%
   - AFae = E / 100 = 30%

B. Normalization factor for B: (D + E) = $84K
   - AFbd = D / 84 = 54 / 84 = 64%
   - AFbe = E / 84 = 30 / 84 = 36%

C. Only one link from C, so AFcf = 100%

D. Normalization factor for D: (F + G) = $100K
   - AFdf = F / 100 = 40%
   - AFdg = G / 100 = 60%

E. Only one link from E, so AFeg = 100%

This approach is compared with the percentage-based approach in Figure 5.10. The contrasting results show that there can be considerable differences between the two approaches—for example, for AFac, where the calculated value from one algorithm is more than twice that of the other.

There is one final approach to be considered that includes the concepts from both of the calculation approaches considered so far—both the contribution fraction and the contribution value. That is to make use of the contribution *shares*.

5.5 Using the Contribution Shares to Calculate the Allocation Fractions

In this case, both the percentage and the value of the contribution are taken into account. This can be seen as taxing the destination (tax = allocation percentage) based on the actual amount of value it

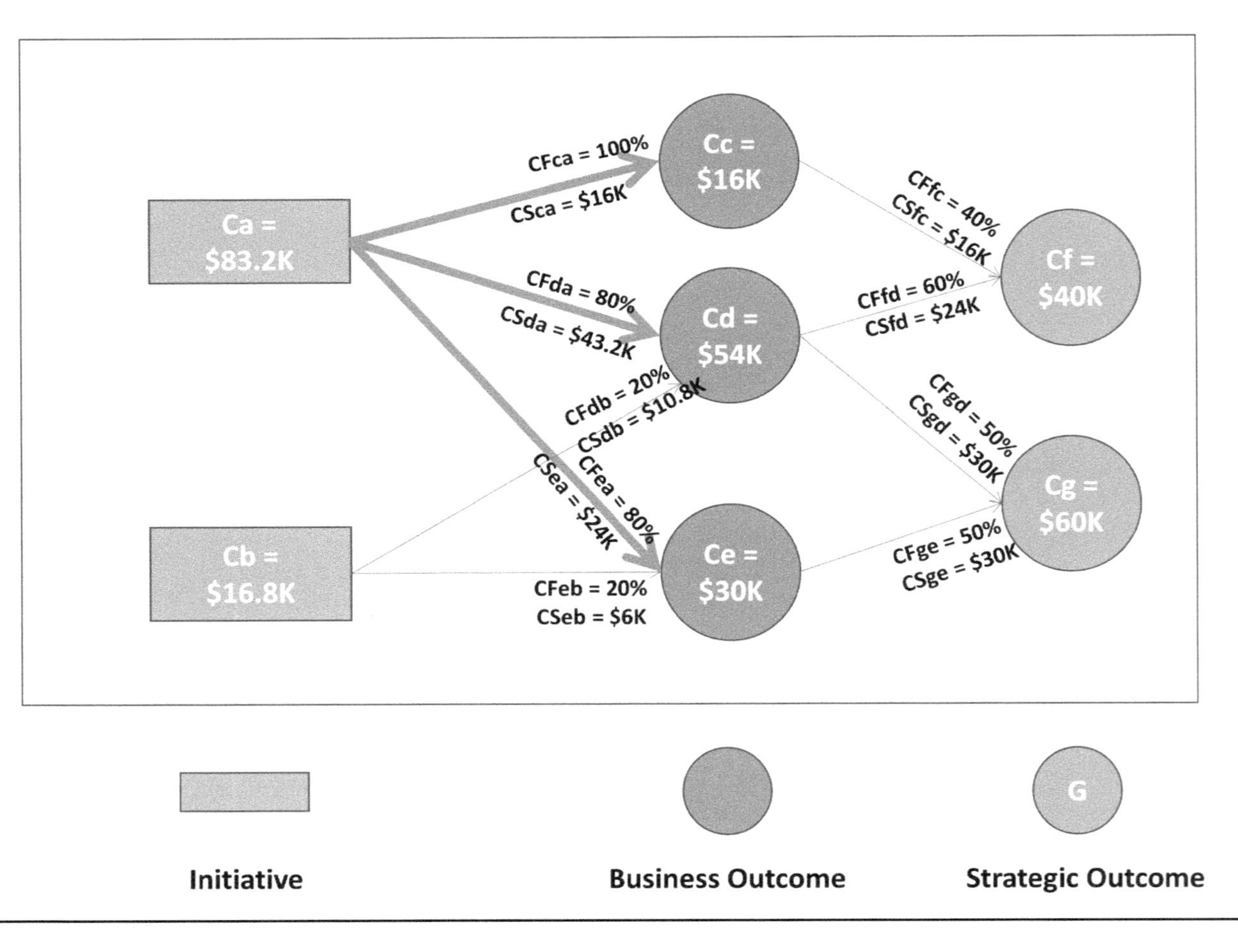

Figure 5.9 Nodes to Include in Calculating the Allocation Fractions from A

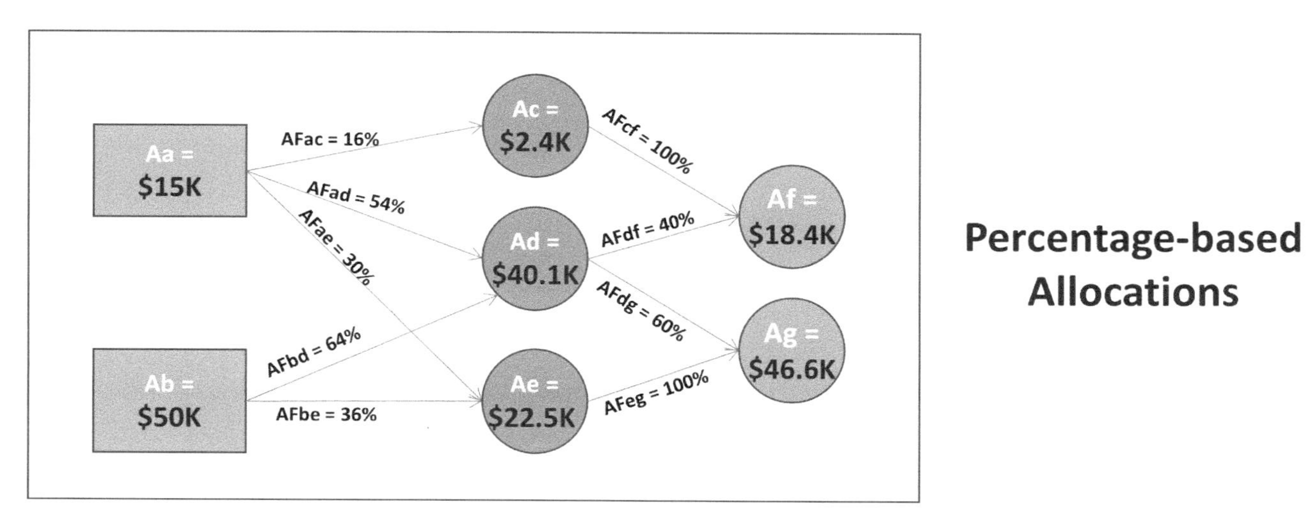

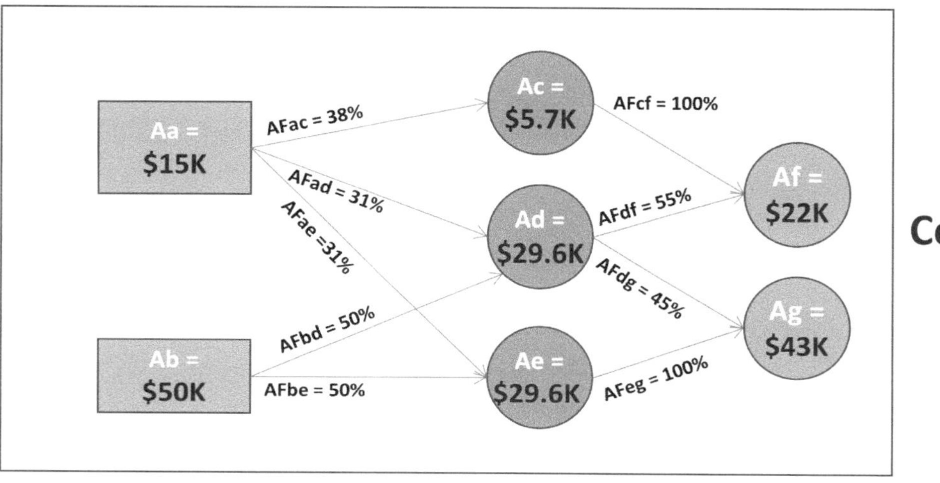

Figure 5.10 Comparison between the Two Methods of Calculating Allocations

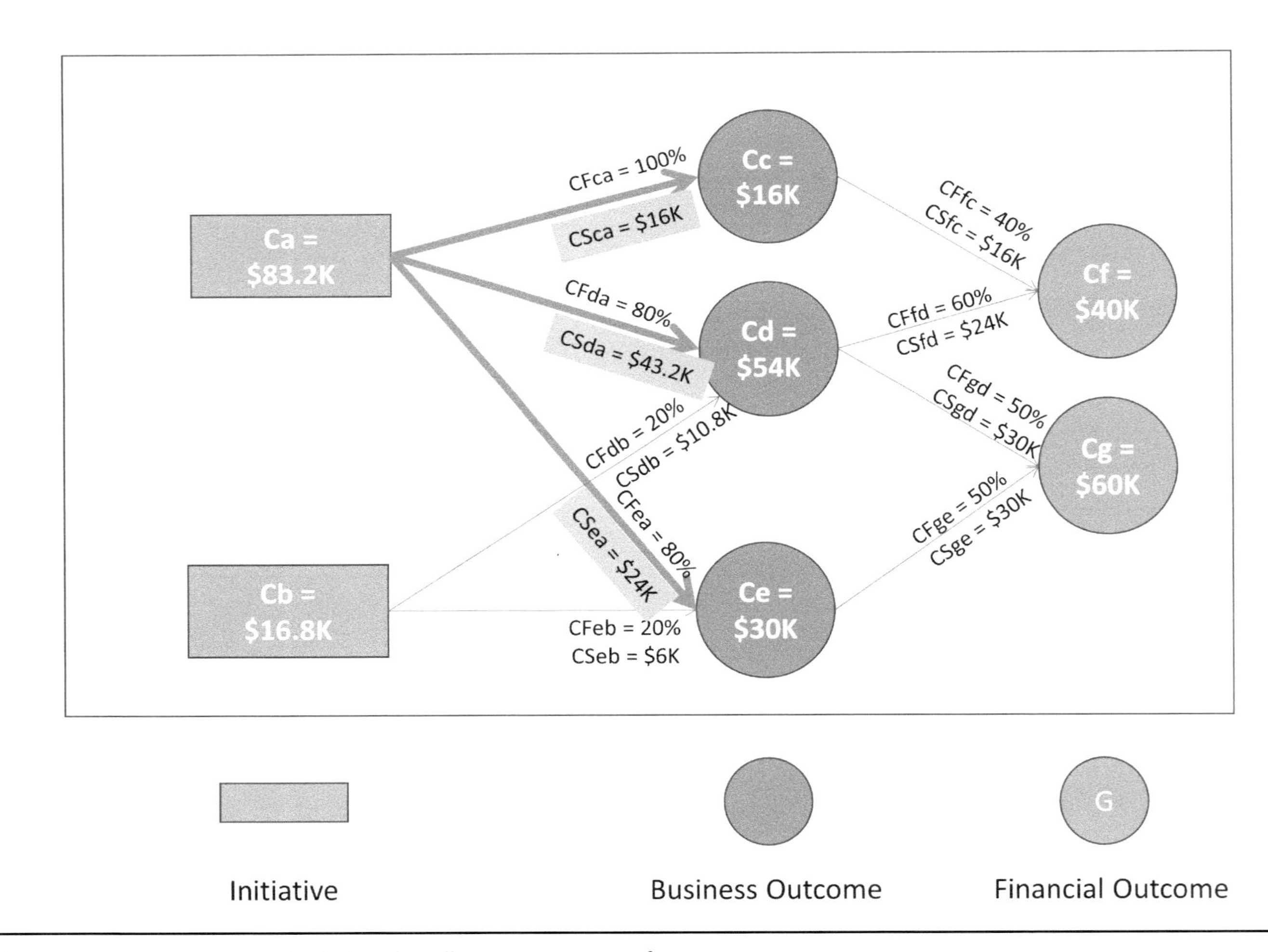

Figure 5.11 Values to Include in Calculating the Allocation Fractions from A

receives from a given node (the contribution share) relative to all of the other nodes to which this destination contributes. As an algorithm, this can be described as:

- Sum all contribution shares from the node (this is equal to the total contribution of this node). The set of destination nodes and their contribution shares involved for source node A are shown in Figure 5.11.
- Calculate the allocation fraction on a link as the percentage of its contribution share on that link to the total.

Starting from Figure 5.11, we see the following:

A. Normalization factor for A: contribution of A = $83.2K
   - AFac = CSca / 83.2 = 16 / 83.2 = 19%
   - AFad = CSda / 83.2 = 43.2 / 83.2 = 52%
   - AFae = CSea / 83.2 = 24 / 83.2 = 29%

B. Normalization factor for B: Contribution of B = $16.8K
   - AFbd = CSdb / 16.8 = 54 / 84 = 64%
   - AFbe = CSeb/ 16.8 = 30 / 84 = 36%

C. Only one link from C, so
   - AFcf = 100%

D. Normalization factor for D: Contribution of D = CSfd + CSgd = $24K + $30K = $54K
   - AFdf = CSfd / 54 = 24 / 54 = 44%
   - AFdg = CSgf / 54 = 30 / 54 = 56%

E. Only one link from E, so:
   - AFeg = 100%

This is shown in Figure 5.12.

5.6 Calculating the Allocations

Taking the fourth option (the one based on contribution shares), the allocations are then calculated based on the allocation fractions shown in Figure 5.12 and give the values shown in Figure 5.13.

The naming convention used for allocations is:

- An: Total allocation for node n
- ASsd: Allocation share for node s (source) relative to node d (destination)
  - Sum over all s_i of ASs_id gives Ad (i.e., sum from left to right)

Given the case where the allocations are A = $50,000 and B = $15,000, the allocation shares are as follows:

A. Allocation for A
   - This is the budget cost of initiative A:
   - Aa = $50K

B. Allocation for B
   - This is just the budgeted cost of initiative B:
   - Ab = $15K

C. Allocation for C
   - Ac = AFac * ASa = 19% * $50K = $9.6K

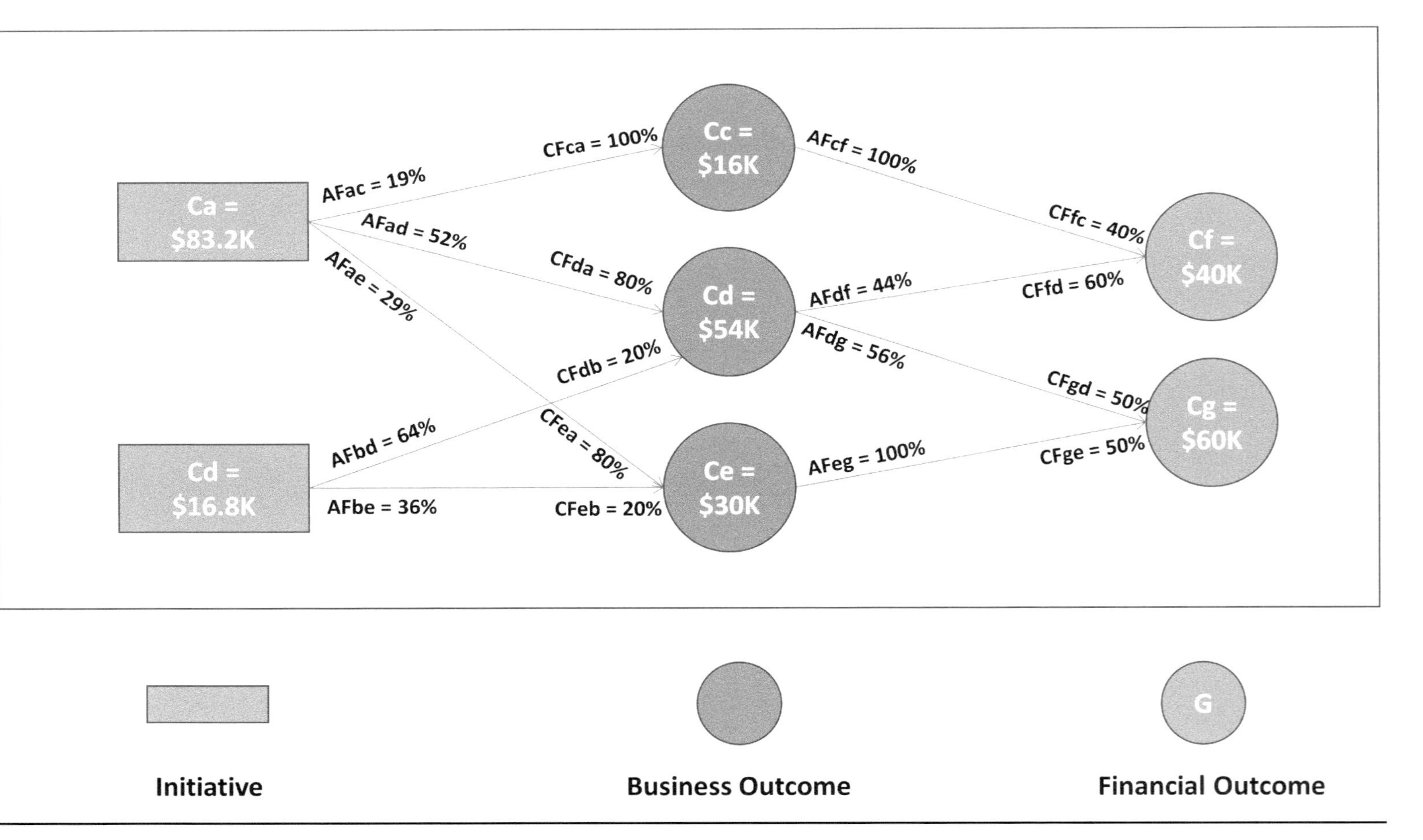

Figure 5.12 The Calculated Allocation Fractions

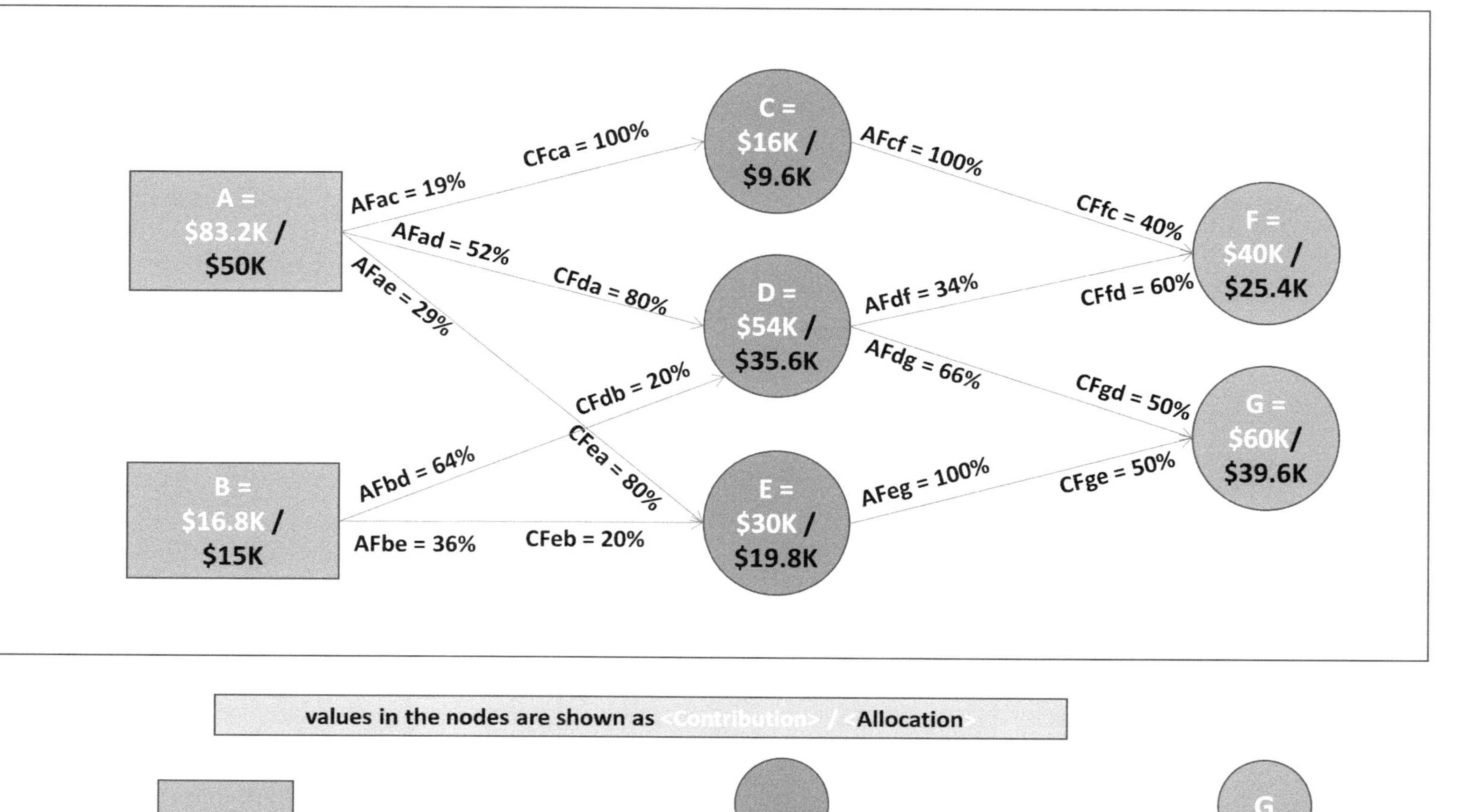

Figure 5.13 The Model Showing Contributions and Allocations

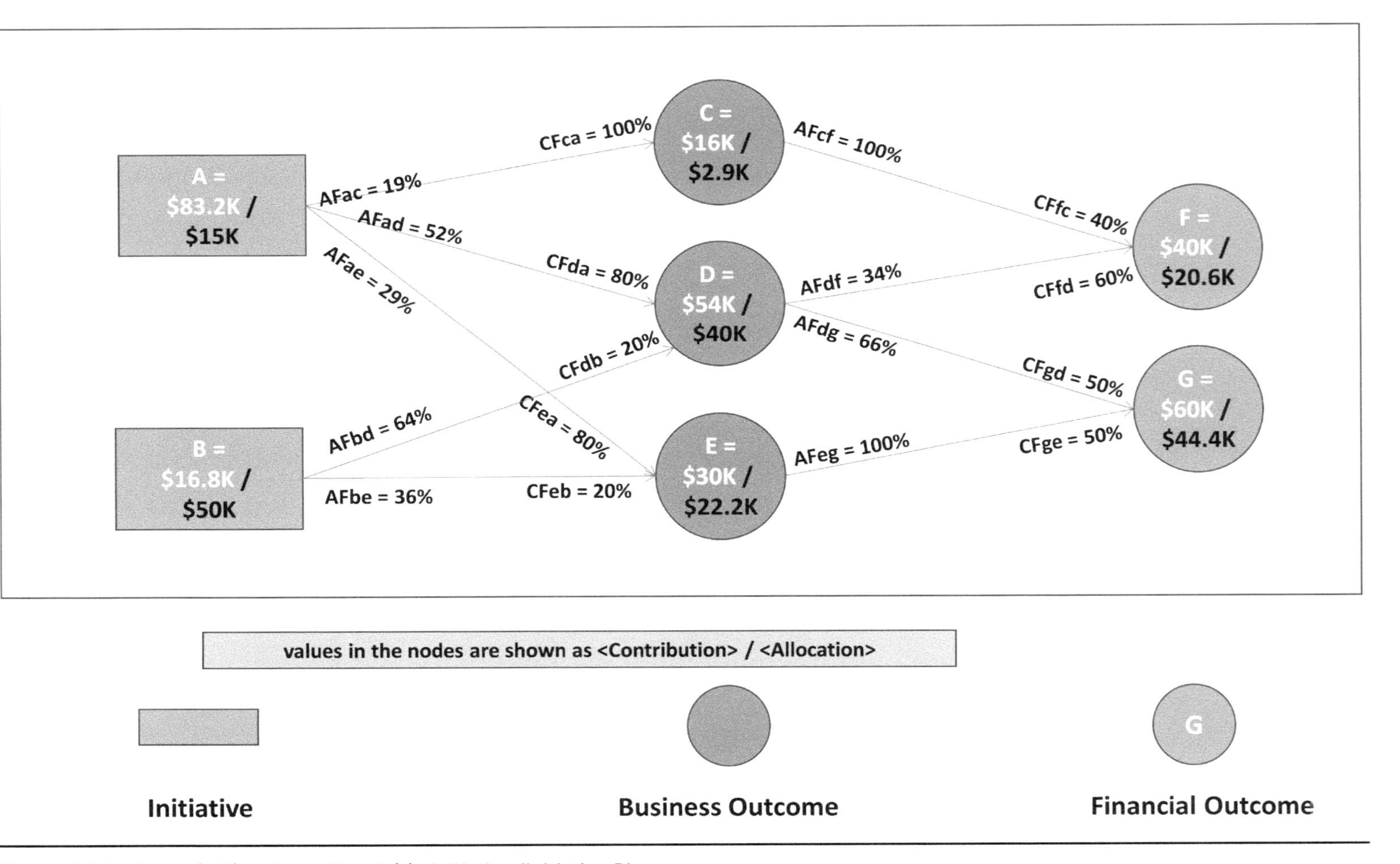

Figure 5.14 Example Showing a Nonviable Initiative (Initiative B)

D. Allocation for D
  - Ad = AFad * ASa + AFbd * ASb = 52% * $50K + 64% * $15K = $26K + $9.6K = $35.6K
E. Allocation for E
  - Ae = AFae * ASa + AFbe * ASb = 29% * $50K + 36% * $15K = $14.4K + $5.4K = $19.8K
F. Allocation for F
  - Af = AFcf * ASc + ASdf * ASd = 100% * $9.6K + 44% * $35.6K = $9.6K + $15.8K = $25.4K
G. Allocation for G
  - Ag = AFdg * ASd + AFeg * ASe = 56% * $35.6K + 100% * $19.8K = $19.8K + $19.8K = $39.6K

The example in Figure 5.13 shows that each node is financially justified: The contribution is greater than the required allocation. However, a second example in Figure 5.14 will serve to show a different situation.

The same basic BRM is used for the second example, but the estimated costs of the two initiatives A and B are exchanged with respect to Figure 5.13. This example therefore has the same overall positive business justification of $65,000 costs for a $100,000 return. In this case, however, the calculation of contributions indicates strongly that initiative B, which costs $50,000 for a contribution of $16,800, appears unjustifiable although the figures for overall program are promising.

This shows the power of using a formal, objective algorithm. Before going any further we therefore need to decide on which of the proposed algorithms to adopt.

5.7 Defining a Decision Criterion

In order to decide which of these approaches should be used as the basis for numerical modeling of the business situation, we need to define an objective criterion on which to base the decision.

To do this, we need to go back to the key objectives for creating and quantifying the BRM: not only to allow the definition of the path linking strategic outcomes to initiatives but also to analyze the financial viability of each step on the way by asking whether each component delivers more value than it costs.

As shown in Figure 5.13 and Figure 5.14, it is not sufficient to base the justification on a simple cost–benefit analysis by comparing the total cost of the initiatives to the total forecast benefit. Even this simple example shows that there can be hidden cross-subsidies between nodes where a profitable one compensates for the smaller losses from one or more other nodes. One goal of the quantified model must be to highlight situations of this type so that they can be analyzed and the solution can be adapted accordingly. A more detailed example of this is given in the case study in Chapter 7.

Rephrasing this business objective using the terms *contribution* and *allocation*, this goal can be restated as follows. The financial aim is to ensure that the contribution of each node is greater by some target amount than its allocation. The break-even case is where the allocations and contributions are equal to each other at all nodes. Logically, therefore, the algorithm should be compatible with this break-even concept as follows: Start with the right-to-left calculations and evaluate all of the contributions; set the allocation for each initiative equal to the contribution value just calculated (i.e., set it in "break-even" state). When you then carry out the left-to-right calculations based on these allocation values, the allocation value calculated for each node by applying the chosen algorithm should be equal to its contribution.

This criterion is described as the break-even everywhere requirement (BEER) of the algorithm. As explained earlier, this can be stated as follows:

If the allocated cost of each initiative is specified as equal to its calculated contribution, then for every node, from the initiatives to the strategic outcomes, the calculated values of the allocation and the contribution will also be identical.

The BEER is a valid constraint, because it ensures that if you start from a perfectly balanced situation in which the initiatives deliver precisely what they cost, all of the steps from the initiatives to the strategic outcomes will also be perfectly balanced. Any other option would be unable to guarantee this result.

It now remains to show that the fourth option—based on contribution shares—exactly delivers the BEER.

5.8 Proof that the Fourth Option Satisfies the Criterion

Let us consider an initiative **I** with calculated contribution value **C**. It is connected to destination nodes *Di* with contribution values **Ci**. Now, calculate the allocations **Ai** for the nodes *Di*.

As specified above for the option based on contribution shares, and starting from initiative **I** with contribution value **C**, the allocation fraction to a given node *Di* with contribution value **Ci** is either 0 if it does not contribute, or **Ci/C** if it does (the sum of these **Ci** is equal to **C**). It is now obvious that if we are in the break-even case, where the allocation of this node is equal to its contribution **C**, then the allocation to node *Di* will be **Ai** = (**Ci/C** x **C**) = **Ci**. The fact the **Ai** = **Ci** proves that, for every node for which the cost (allocation **Ai**) is identical to its value (contribution **Ci**), the same holds for all of its direct successors ($\mathbf{A_{i+1}} = \mathbf{C_{i+1}}$). By recursion, starting from the initiatives and working from left to right, the algorithm based on contribution shares ensures that this is the case for every node. This algorithm has therefore been shown to deliver the BEER.

This completes the initial explanation of the model. Now that we have defined the algorithms, they will be applied in the following chapters to case study examples, with potentially unexpectedly revealing results.

This algorithm has been implemented as a generalized set of Excel formulae. The Excel tables and formulae are described in Chapter 15. They have been used for the worked case study examples.

5.9 Summary

This chapter has described how the components of the benefits realization map work together. It has described contributions, allocations, and the corresponding fractions of these to allocate to each node in the model.

The various options were analyzed for calculating the allocations that should be charged to each node, based on the calculated contributions. The break-even everywhere requirement as the criterion for success—the BEER—was defined and applied to select the preferred option.

Chapter 6

Disbenefits and Essential Links

This chapter adds to the modeling technique described in the previous chapter by taking two major considerations into account. First, not all outcomes are entirely beneficial; these negative effects are known as *disbenefits*. Second, some benefits cannot occur at all if one or more of the contributing capabilities or outcomes is missing; this source–destination constraint is known as an *essential* relationship. The way in which these factors are included in the benefits realization map is explained, along with the corresponding algorithms. The addition of these considerations completes the set of tools required for a numerical analysis of the case study. If you would rather understand the effect of these ideas on the case study before reading the theory in detail, go straight to Chapter 7 and come back here later.

6.1 Allowing for Disbenefits

The examples up to this point have only shown cases in which every outcome is beneficial. Unfortunately, in the real world, clouds not only have silver linings, but can also rain on your parade.

The following example in Figure 6.1 shows that the mapping technique works just as well when one or more of the financial outcomes is negative. In this example, F is a disbenefit with a contribution that is therefore negative, estimated at –$20,000, whereas G is a benefit worth $80,000. The initiatives have been costed at $30,000 for A and $10,000 for B. The contribution fractions are the same as in the previous examples (e.g., Figure 5.2).

The calculations based on the allocation share algorithm give the following values (as shown in Figure 6.2). The calculated values are presented without the detailed working in this case, as they are calculated by the Excel routine that executes the algorithm described in the previous chapter.

6.1.1 The Contributions in the Disbenefits Example

Working from right to left, the contributions are:

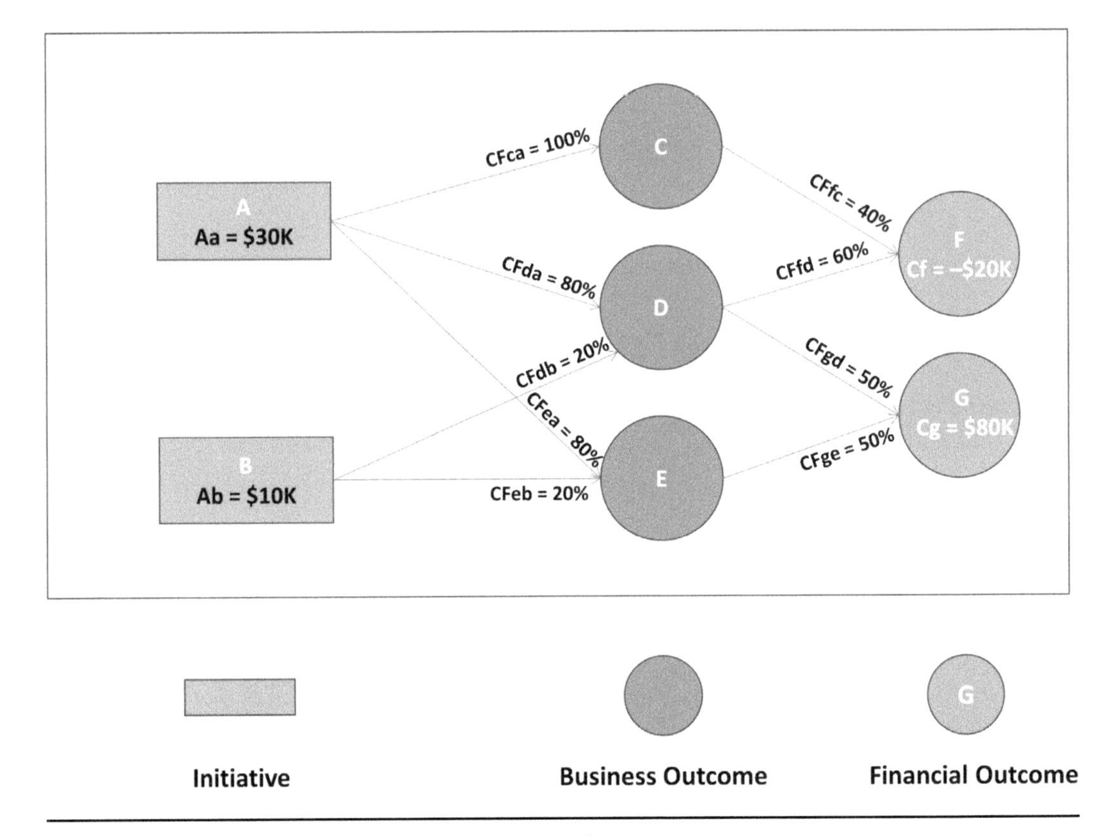

Figure 6.1 A Simple Disbenefits Example (Example 6-1)

- The financial outcomes
  - Contribution of G by hypothesis is a benefit of G = $80,000
  - Contribution of F by hypothesis is a disbenefit of F = –$20,000
- The business outcomes
  - Contribution of E by calculation E = $40,000
  - Contribution of D by calculation D = $28,000
  - Contribution of C by calculation shows a disbenefit of C = –$8,000
- The initiatives
  - Contribution of B by calculation B = $13,600
  - Contribution of A by calculation E = $46,400

6.1.2 The Allocation Fractions in the Disbenefits Example

The calculations in this case do, as shown below, raise an interesting question because some of the fractions turn out to be negative. We need to understand what that means. Using the taxation metaphor that we adopted to explain the concept of allocations, a negative allocation fraction can be considered as a loss leading to a tax credit—that is to say, rather than the source charging to the destination node for its (beneficial) effect, the destination node requires a subsidy to compensate for the damage from this specific source; this amount will then need to be shared between all of the other nodes served by the source node. If the source node causes disbenefits to all of its successors, it is fairly obvious that it would be better to see if it can be eliminated entirely from the plan since, by definition, it provides no benefit.

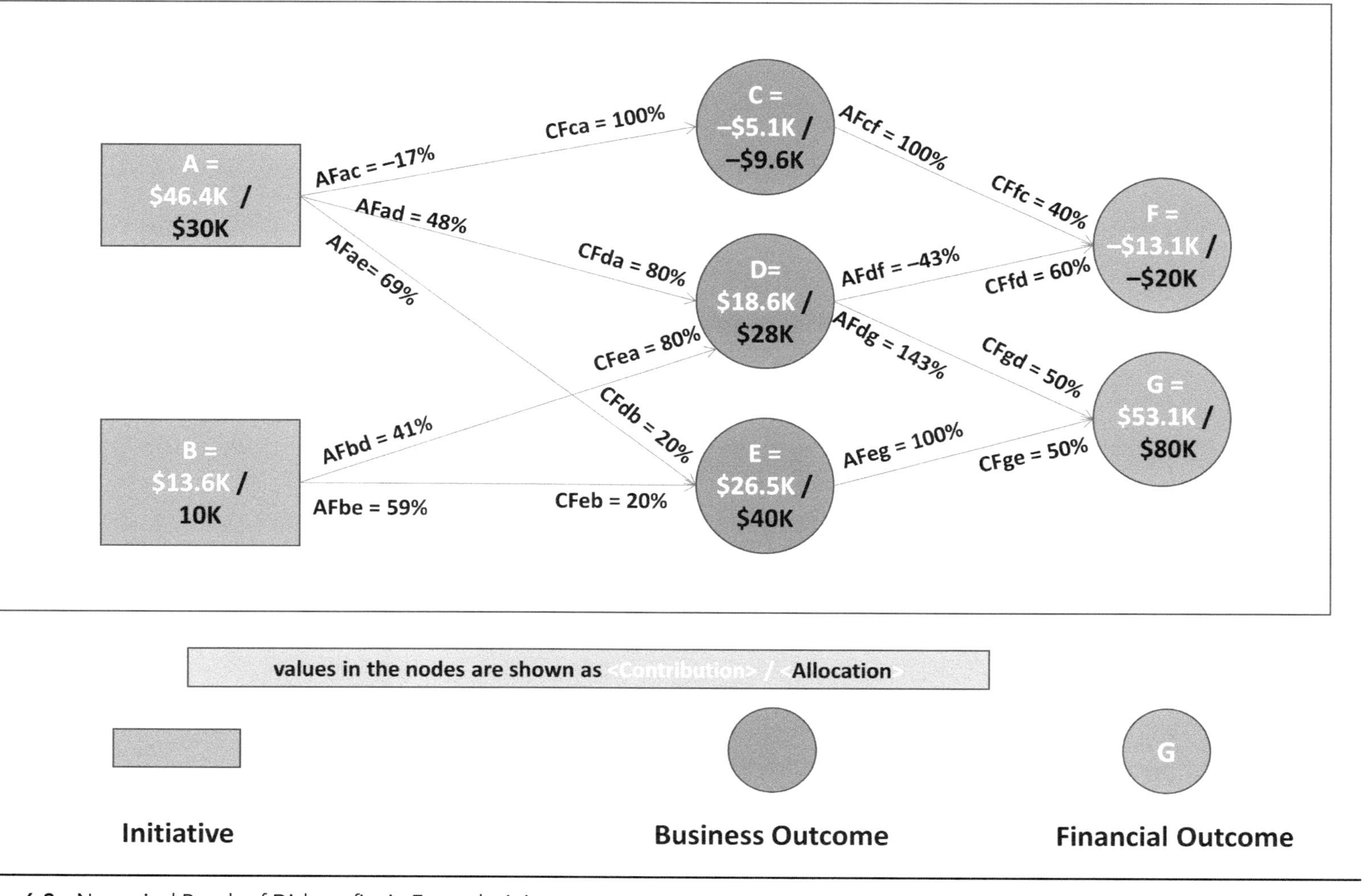

Figure 6.2 Numerical Result of Disbenefits in Example 6-1

Once again, leaving out the fine details of the calculation, the values of the allocation fractions are as follows:

- From node A
    - AFac = –17% ← *Note: Negative value*
    - AFad = 48%
    - AFae = 69%
- For node B
    - AFbd = 41%
    - AFbe = 59%
- For node C
    - AFcf = 100%
- For Node D
    - AFdf = –43% ← *Note: Negative value*
    - AFdg = 143%
- Only one link from E, so
    - AFeg = 100%

6.1.3 The Allocations in the Disbenefits Example

The allocation for an initiative is equal to its budgeted cost.

- Allocation for initiative A
    - Aa = $30,000
- Allocation for initiative B
    - Ab = $10,000

Applying the allocation fractions, starting from nodes A and B, leads to the following values:

- Allocation for C
    - Ac = –$5,100 ← *Note: Negative value*
- Allocation for D
    - Ac = $18,600
- Allocation for E
    - Ae = $26,500
- Allocation for F
    - Af = –$13,100 ← *Note: Negative value*
- Allocation for G
    - Ag = $53,100

6.1.4 Discussion on the Disbenefits Example

As can be seen from the results of the calculations shown in Figure 6.2, the disbenefit identified as node F has had an effect on nodes C and D. The effect on node C is that its contribution to the financial result is also negative (i.e., it is a disbenefit in its own right). For node G, the cost of F's disbenefit reduces its cost/benefit ratio due to the increased allocation fraction required by D with respect to G (143%) in order to balance to books for D's "subsidy" to F.

Another point that needs to be investigated is the meaning of a benefit/cost ratio in the case of a disbenefit. In our example, for node F, the cost (i.e., the allocation subsidy) is –$13,100, whereas the benefit (negative contribution) is –$20,000. A simple cost/benefit calculation would therefore show a healthy value of –20 / –13.1 = 153%. Should we be happy? Probably not—but why not? Given the definition of contributions and allocations, one way of looking at negative values such as these is that a negative allocation is, in fact, a contribution (i.e., it takes away money from the corresponding node), and a negative contribution is an allocation (since, as a subsidy, it brings money into the node). This additional concept means that in our benefit/cost analysis, negative costs should be considered as benefits and negative benefits as costs. Use of this new convention makes our benefit/cost calculation for node F to be 13,100 / 20,000 = 66 percent. This result agrees with the common-sense view that node F is at best a necessary evil since it brings in only $66 for every $100 spent.

6.1.5 Percentage Disbenefits

In the example just described, the disbenefit was quantified as a (negative) financial value. The alternative approach would be to indicate the effect of a disbenefit as a negative contribution fraction. The worked example, based on the QERTS case study in the next chapter, will discuss this approach in more detail.

6.1.6 Are We There Yet?

Now that this analysis is complete, one might think that the business cases for the entities can be developed in detail. However, one additional concept is required. The final example in Chapter 5 (reproduced below as Figure 6.3) indicates that initiative B costs more than it contributes. However, in certain cases, it may turn out to be necessary to keep it. This concept of *essential contributions* is addressed next based on a new example.

6.2 Essential Contributions

6.2.1 First Option

The benefits realization map (BRM) shown in Figure 6.3 clearly indicates that initiative B costs $50,000, which is much more than it delivers in value as a strategic benefit to the business environment ($16,800). Normally, therefore, the decision should be made that initiative B should not be approved. This decision should, however, be analyzed further.

Let us assume that the overall endeavor could possibly be restructured to leave initiative B out entirely, even if it means modifying the scope of A to some extent (let us call the new initiative A') in order to deliver the combined value of A and B. The easiest way of doing this without redrawing the BRM is by setting the allocation of A' to ($15,000 + $50,000) = $65,000, and of B to zero, and then modifying the contribution fractions as follows: Set all (both) of the contributions fractions from B to zero and adjust the other contribution fractions accordingly by renormalizing the remaining contributions fractions to the affected nodes back to a total of 100 percent. In the current case, there is only one source remaining for each of the affected nodes, which makes that calculation straightforward (as shown in Figure 6.4).

The BRM can then be recalculated (as shown in Figure 6.5) using the contribution share approach (the break-even everywhere requirement, or BEER, algorithm) described in the previous chapter to determine the allocation fractions.

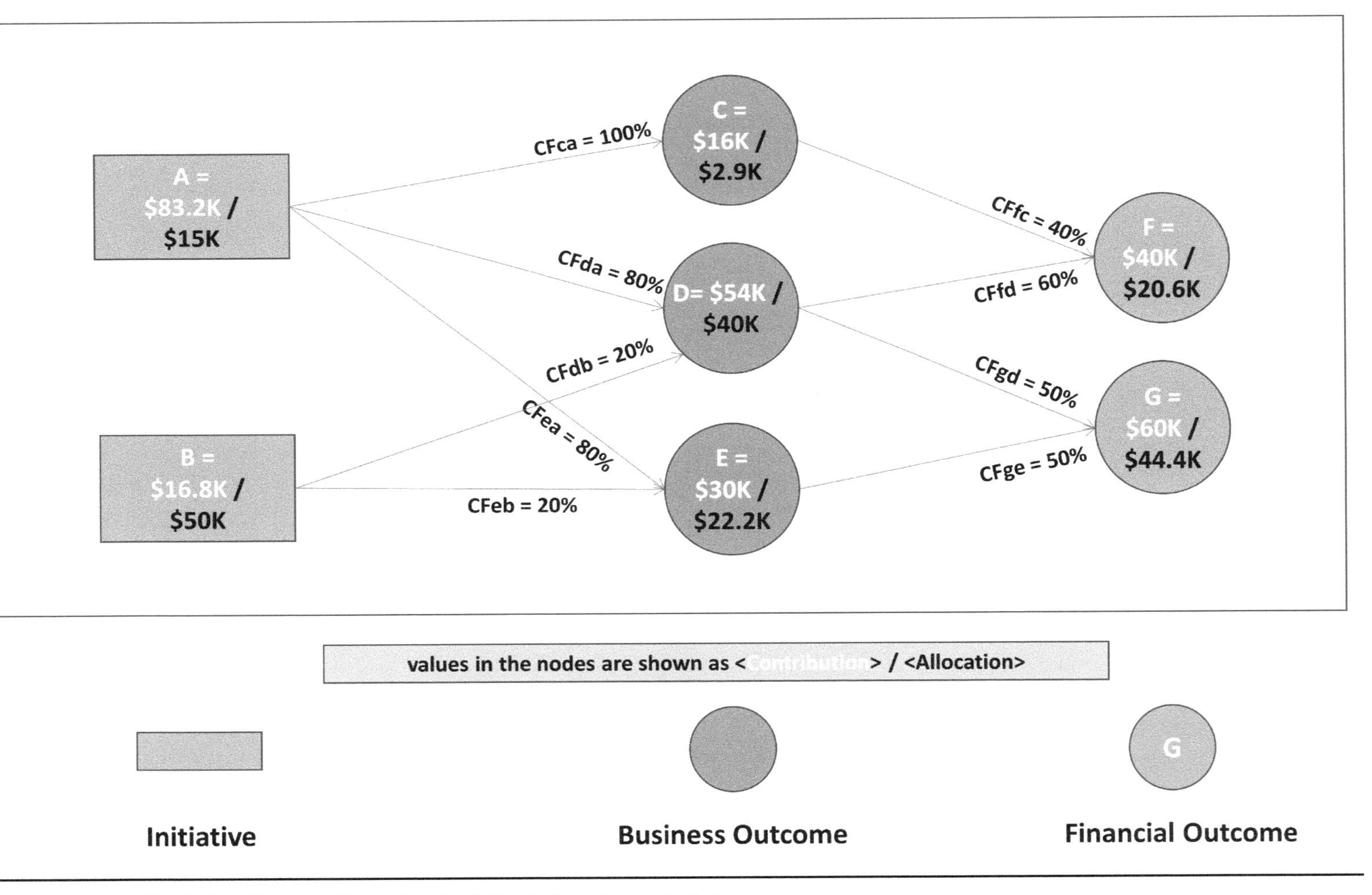

Figure 6.3 Simplified BRM Showing "Loss-Making" Node B in Example 6-1

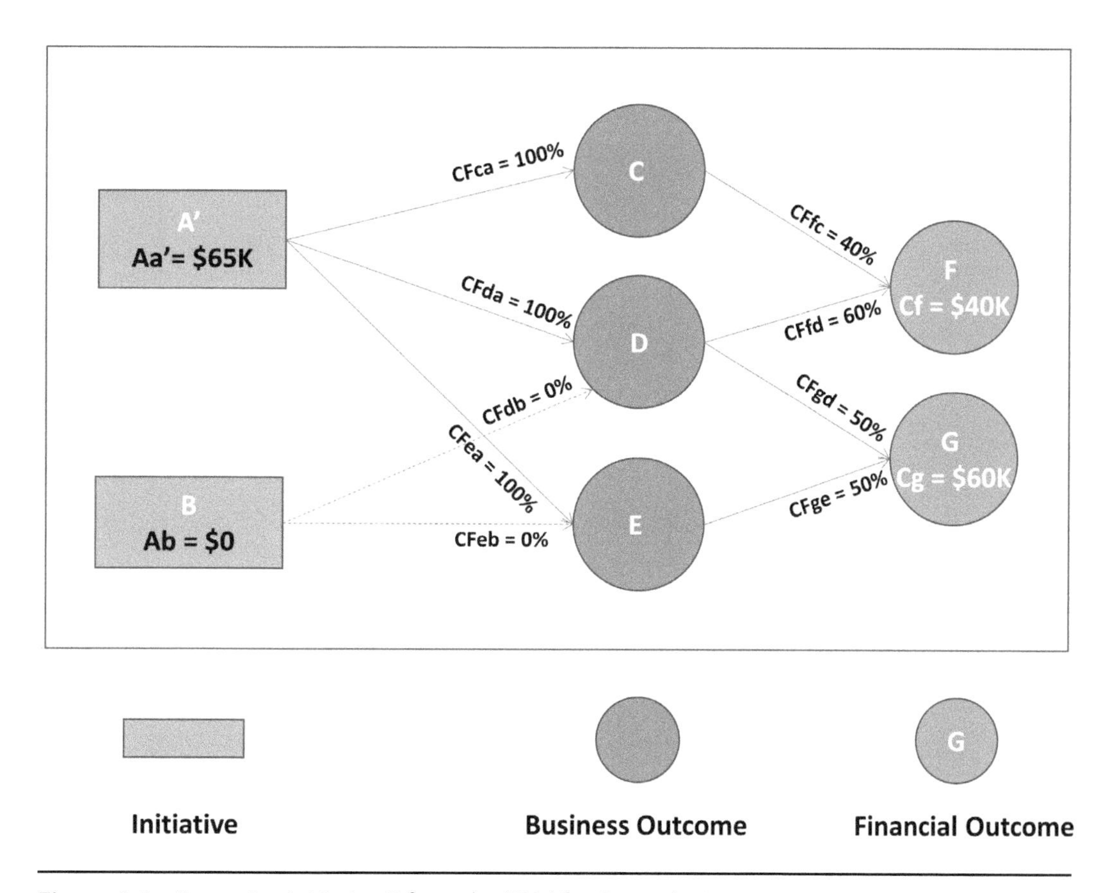

Figure 6.4 Removing Initiative B from the BRM for Example 6-1

This remodeling makes at least three important assumptions that must be verified:

- That node D can actually function in the absence of the capabilities of B; although B may not contribute much value to the strategic outcome, it might provide a capability without which E cannot operate.
- That the final financial outcomes are unchanged by the change of structure resulting from the removal of initiative B.
- That the redefined initiative A' can be implemented for the specified price.

The reason for insisting on these assumptions is that similar ones are often made without being validated when important, real-life programs are restructured, leading to major overruns or even to abandoning the program after large sums have already been spent. In addition, in some real-life cases, it can happen that the new initiative A' is simply created by amalgamating the original initiatives A and B so as deliberately to hide the initial assessment of the nonviability of B; this can make A' too complex to manage, as well as undermining the integrity of the BRM and being an unethical approach to program justification.

In the case where B is really needed despite the fact that the BRM indicates that its contribution is not financially viable, rather than attempting to adjust the scope of initiative A to make up for the removal of initiative B, an additional characteristic of internode links should be taken into account: We can define two categories of links—contributory or essential.

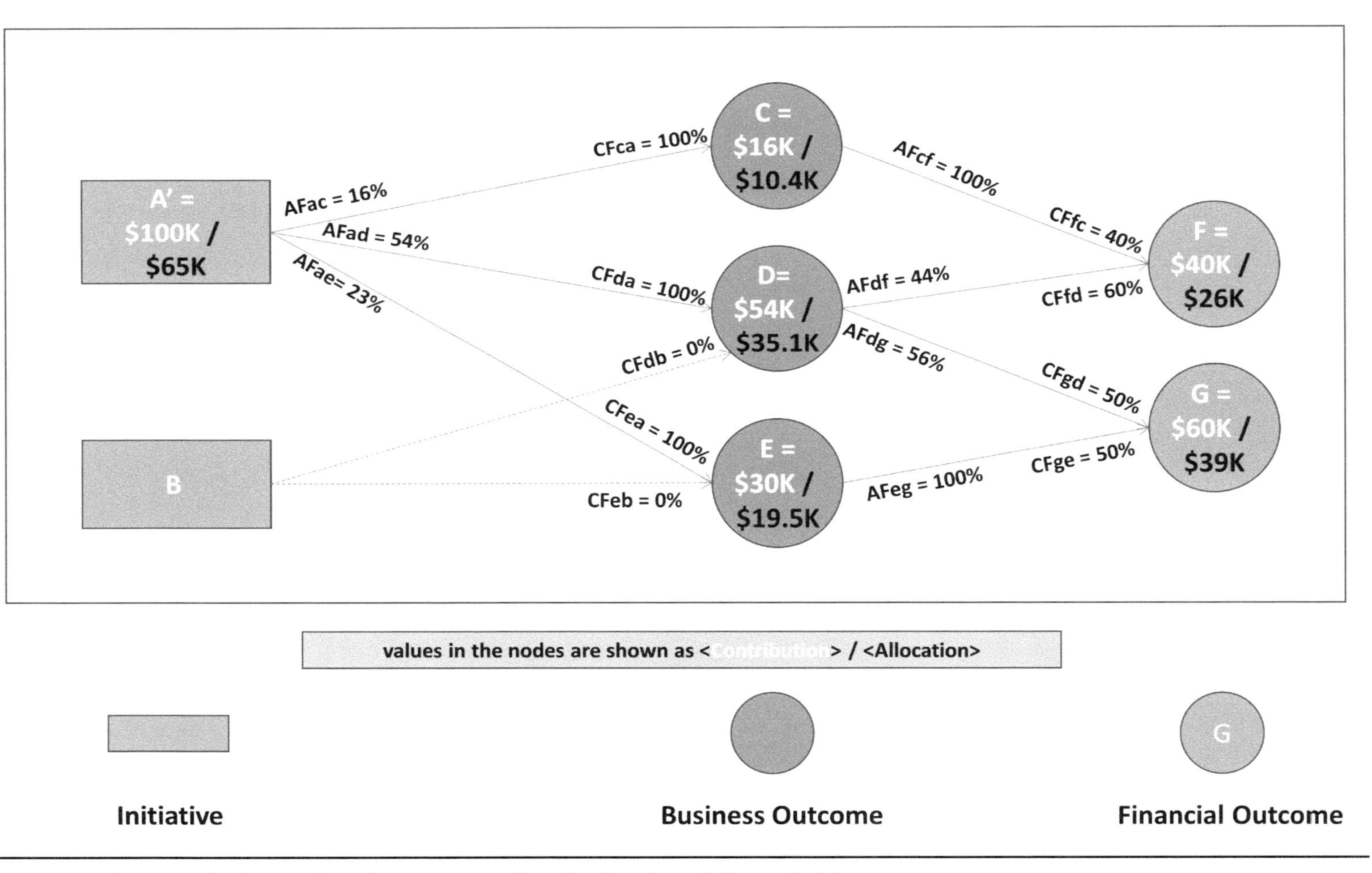

Figure 6.5 The Reworked BRM after Eliminating the Contributions from B Example 6-1

Each contributory link adds its value (its contribution share) to the destination node, whatever the state of the other links. As explained in Chapter 4, a link from a node that is an absolute requirement for the destination to exist is called an *essential* link. From the point of view of the destination, the corresponding source is therefore called an essential node.

Seen from another point of view, the contribution of a node depends on the "AND" of the *essential* shares from the source nodes, whereas, for *contributory* nodes, the value is based on the "OR" of the *contributory* shares.

6.2.2 Essential Node and Essential Contribution

The term and concept of the *essential node* are proposed to address the situation in which the destination node cannot become operational unless the source is available—that is, the source is a prerequisite, even if its actual contribution share, once it is available, is small. Such a source node is known as an *essential node* and the corresponding link as an *essential link*.

To indicate this on the BRM, you can use a specific color or format for the link (as in Figure 6.6 in section 6.4.1) or the *assumptions* construct, attached to the same handle as the incoming link from the essential node, with the tag *100% essential*.

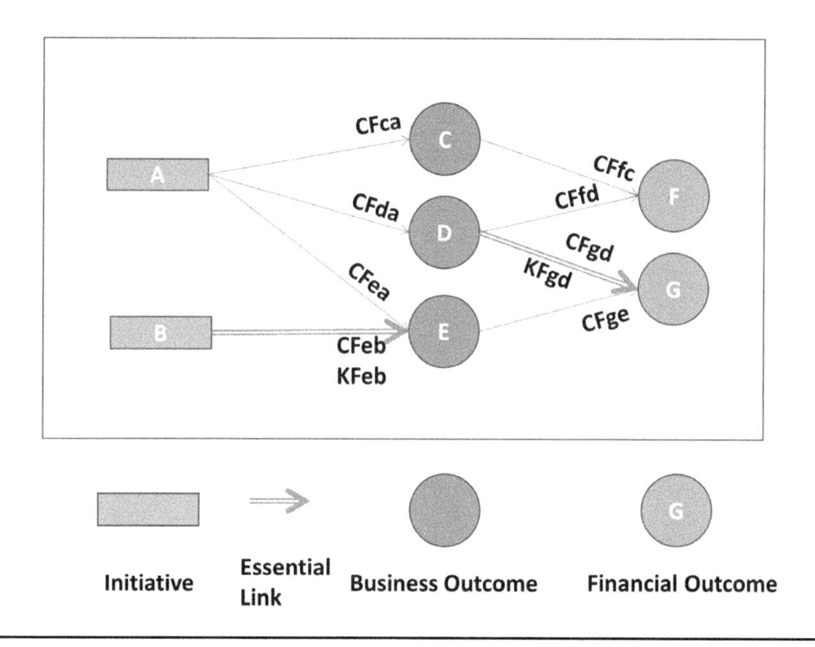

Figure 6.6 Simplified BRM Showing Contributions: B Is Essential to E, and D Is Essential to G

Although, as just mentioned, the essential node's contribution share may be small, the amount of the loss if it is not available is equal to the contribution of the destination node. For this reason, this value is termed its *essential contribution share* for that link. The sum of all of the essential contribution shares for a given node is its *essential contribution*.

Essential links influence four major program management activities (explained in the following section):

1. Prioritizing initiatives
2. Developing the financial justification for an initiative
3. Performing a risk analysis of the network
4. Optimizing the BRM

6.3 Applying the Concept of Essential Links

6.3.1 Prioritizing Initiatives

Where a node has several inputs, it can normally start to function (up to the level of the corresponding contribution values) once any of the source nodes is available, unless any of the uncompleted source nodes is *essential* to it. In the case of essential links, to minimize work in process (i.e., to avoid contributory capabilities being provided before they can be used), all of the essential source nodes must be scheduled to complete at least by the time that one or more of the other sources becomes operational. In general, to minimize the waiting time before the output of a node can be used by its destination node, the following set of constraints should be applied in the order given:

- For essential source nodes, the constraint is that the other source nodes should not be complete before all of the essential nodes Si are complete. This constraint is specified to reduce potential work-in-process, thereby providing a "lean" approach to scheduling.
- The network schedule diagram for contributory nodes should specify that the destination node D will not be complete until all of the essential source nodes Si are complete. Technically, in the case of multiple essential links into a given node, this can be ensured by defining an "essential milestone" to replace these links as predecessor to the given node. The essential milestone is defined as depending on a "finish-to-finish" relationship from each of the preceding essential nodes Si. Resolving this type of constraint allows efficient parallel working across the BRM from left to right while allowing the full achievement of benefits realization milestones to be determined in a Kanban-like lean approach.

This set of constraints should be applied when calculating the network schedule diagram for the program and for the predicted cumulative contribution curve, as will be described in Chapter 8 for scheduling and Chapter 12 for earned benefit.

Once these considerations have been taken into account, prioritization should also factor in the value of each node as provided by its business justification in order to minimize the payback period.

6.3.2 Developing the Financial Justification for an Initiative

To develop the business justification in case of an essential node, an *essential network* should be evaluated. In the standard business justification, the initiative's contribution is compared to its cost (allocation) to establish a return-on-investment value.

For essential nodes, however, the *essential threat* should also be considered to provide a loss-avoidance ratio (LAR = [essential contribution] / [allocation]) based on the loss of value if this node is not available. The process for carrying out the analysis is discussed in more detail in section 6.4 (Creating an Essential Network).

In the case where there are several essential links in the BRM, several essential networks, one for each essential node, may need to be developed.

Essential links also give rise to a specific set of risks. The general approach to risk management is given in Chapter 9, but details of the specific category of risk due to essential links are given in the following section.

6.3.3 Performing a Risk Analysis of the Network

There are two categories of risks to be considered: those that could occur while the capabilities are being developed (the implementation phase) and those affecting the operational phase. For essential nodes, the major focus needs to be on the threat they could pose during the implementation of the program. However, they remain as a specific category of threats in the operational phase.

The Implementation Phase

Any threat that could impact an essential node affects its destination node directly. If the threat would delay or prevent the capability of the source node, the destination node will inherit the effect directly; the impacts of the effects need to be linked to this common source.

The Operational Phase

Any threat that could cause the interruption of removal of the capability of an essential node will immediately remove the corresponding downstream capabilities from the operational environment.

6.3.4 Optimizing the Benefits Realization Map

Although the initial analysis of the BRM before the specification of any essential nodes may indicate that some initiatives are not financially justifiable, no action should be taken to adjust the apparent overall cost–benefit result by eliminating these initiatives before an analysis of essential links has been completed. The reason for delaying the decision is that the initiatives to be removed may be, or may lead to, essential nodes. In the case of essential nodes, their removal could cause greater loss of value than the savings made by eliminating them. A detailed explanation of this point is provided in the following section.

6.3.5 Discussion on Contributions and Essential Nodes

The key to understanding the conceptual basis of essential nodes is that a capability may make a partial contribution to the value of another capability. For example, each hand only contributes 50 percent of applause (this is the *contributory contribution*) but can be 100 percent necessary in order for the resultant capability to occur (this is the *essential contribution*: the sound of one hand clapping is certainly significantly less than 50 percent of the total capability of two hands). Essential contributions therefore need to be taken into account at each phase of the program and addressed accordingly. More details on this issue are given in the following section, based on Example 6-1 in Figure 6.7. The way in which the calculations should take essential links into account will then be given based on a slightly more complicated example (Example 6-2 in Figure 6.14).

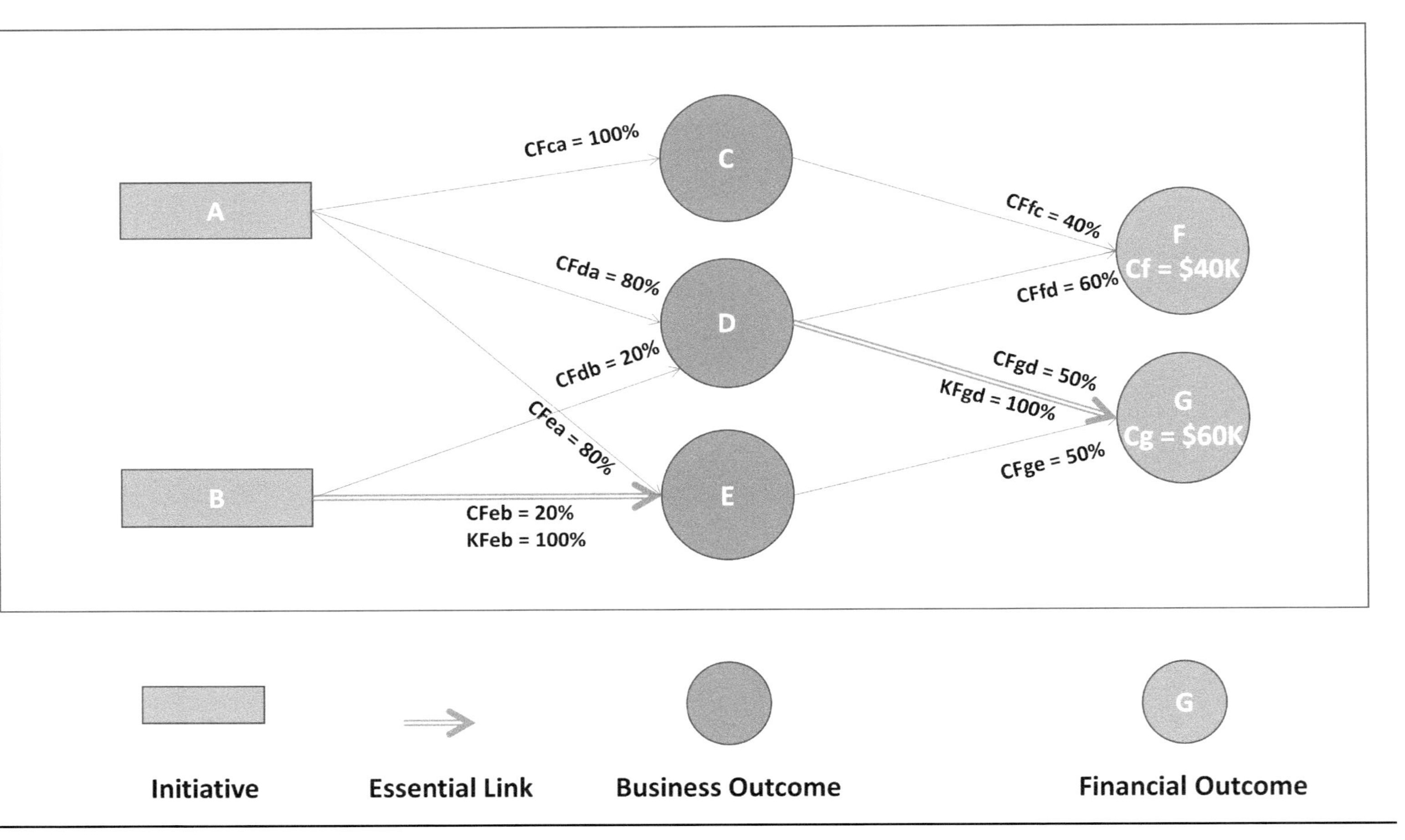

Figure 6.7 The First Contribution Example, with Initial Values for Example 6-1

6.4 Creating an Essential Network

6.4.1 Picturing the Essential Contributions

As explained in Chapter 5, contributions are calculated from right (strategic objectives) to the left (initiatives) through the network. However, logically, the value of the contributions flows through the network from left (initiatives) to right (financial outcomes). A simplified diagram is shown in Figure 6.6.

The values of the *contributory* contribution fractions are represented as CF followed by the identifiers (in order) of the destination node and the source node (CFds); for the essential contributions, the first letter has been replaced by K and the link drawn in bold. Because of the "all or nothing" characteristic of essential links, all essential link fractions (KF) are set to 100 percent, as shown for KFeb in Figure 6.7.

6.4.2 Calculating the Contributions

Two examples based on Figure 6.6 will be used. The way in which the calculations are built across the BRM from right to left is shown progressively in the following sections.

Local Contribution Values

The values used in the example are shown in Figure 6.7.

The resulting contribution values can be calculated from right to left starting from the diagram in Figure 6.7 to give the value of the business outcome nodes (Figure 6.8). Note that there can be two values: the *contributory value* based on the contribution fraction and the *essential value* based on the essential contributions (in dark type in Figure 6.8) for example, the values given for D are as follows:

- Contributory value Cd = Cf * CFfd + Cg * CFgd = $40,000 * 60% + $60,000 * 50% = $54,000
- Essential value Kd = Cf * CFfd + Cg * KFgd = $40,000 * 60% + $60,000 * 100% = $84,000

These calculations confirm what you could see by inspection: Removal of node D would not only remove the 60 percent contribution to node F but also make node F disappear with a loss of $60,000. There are other "upstream" effects; if outcome G disappears, then node E has no reason to exist because its only influence is on node G. The links from A to E and from B to E would also disappear, thereby reducing the value of those initiatives. A more detailed analysis on a slightly more complex example is given later in this chapter, but the current example can already provide additional insights.

Remote Contributions

One important point about essential contributions is that they should only be considered with respect to their direct effect on the node that is the destination of the essential link and not then diffused back through the source of the essential link. The essential effect is limited to the essential link because the essential contribution measures the potential *loss* corresponding to the opportunity cost, and not an inherent value that is due to its contribution. For this reason, in Figure 6.8, Ka = Ca, although Kd is greater than Cd. There is an exception to this rule in the case of linked essential contributions (see the following section).

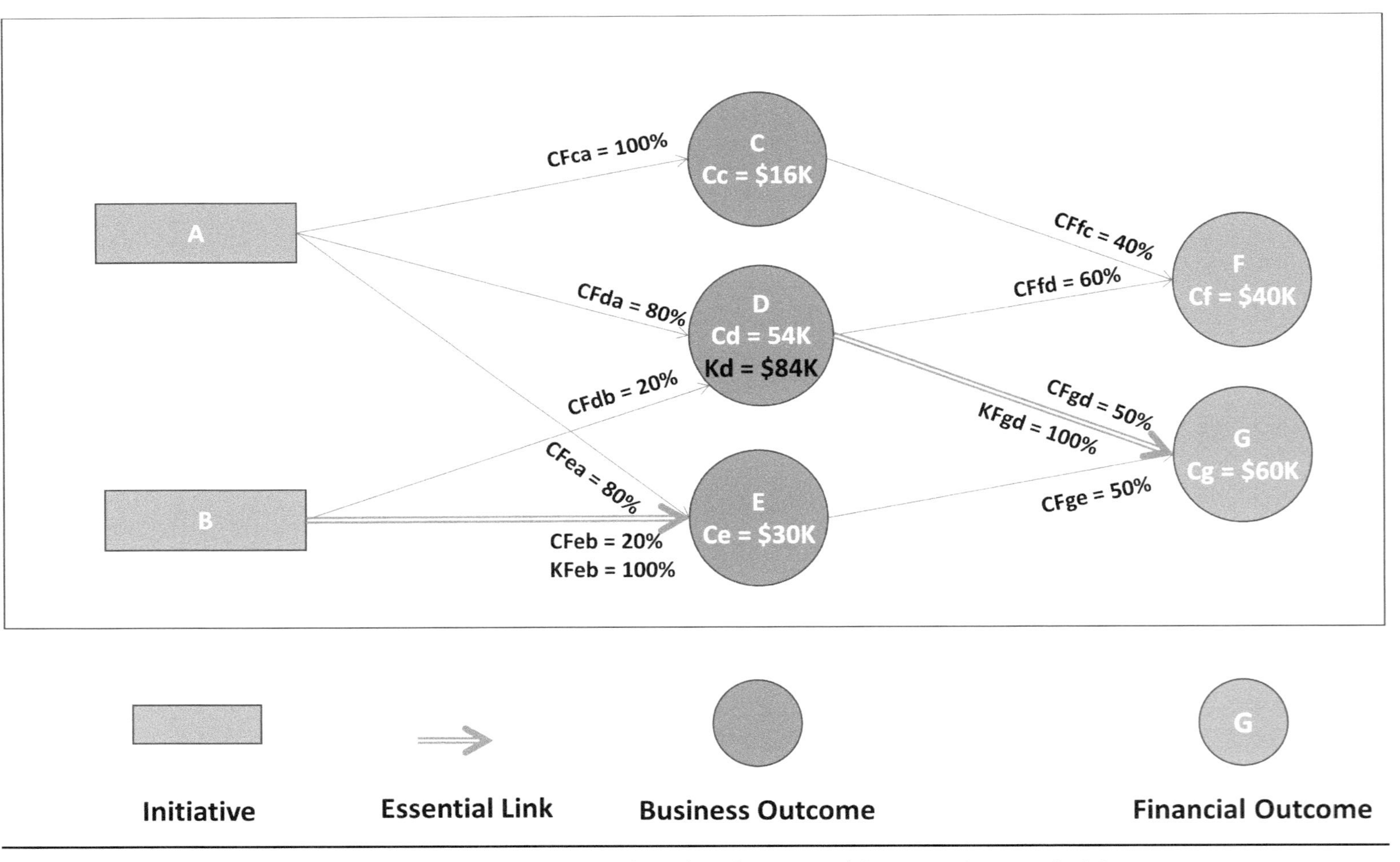

Figure 6.8 The Contribution Values of the Business Outcomes based on the Financial Outcomes in Example 6-1

The value of the essential contribution of B, taking into account its essential link to E and its contributory link to D, is:

- Kb = Cd * CFdb + Ce * KFeb = $54,000 * 20% + $30,000 * 100% = $40,800

These results are shown in Figure 6.9.

It is also possible for essential nodes to be directly interconnected by essential links. That case is explained in the next section.

Linked Essential Contributions

In the case where there is a link between essential contributions, more analysis is needed. This is based on the modification of Example 6-1 shown in Figure 6.10, in which B-D and D-G are both essential links.

Working from right to left, as shown in the previous example, above, Kd = $84,000. However, it is necessary to determine how to calculate Kb in this example, where the absence of B would immediately eliminate D and reduce the contribution of E. The most obvious approach would be to take:

- Kb = KFdb * Kb + CFeb * Ce = ($84,000 *100%) + ($30,000 * 20%) = $90,000

Note the exception to the rule just given for remote contributions: that the essential contribution value should not diffuse back through the network. The exception is for situations as shown, and the value should diffuse back up to any incoming *essential* links.

However, this calculation for the value of Kb at $90,000 overstates the case for the following reason: If B is canceled, the outcome D (for which B is essential) will not occur, and so the financial outcome G (for which D is essential) will not be realized. All of the cost of losing G is already taken into account in the essential value of D and should not be counted a second time in the essential contribution to B that is due to E. Put another way, if G is not there because D is missing, then the contribution to E, and through that back to B, will be nonexistent. All of the nodes involved in this way are shown as an *essential network* in Figure 6.11. In a more complex example, there could be several such essential networks.

To understand and address the double-counting that led to Kb = $90,000, another approach for evaluating the essential value can be used:

- Assume that the essential node has been canceled.
- Since there is now no contribution across any of the links in the essential network, subtract the corresponding contribution shares from the destination nodes of each such link.
- The difference between the case with the essential node and the case without it gives the essential value of that essential node.

In detail, therefore, the first step is to reset to zero the value of the contributions—and the corresponding contribution shares—of all of the outcomes that would disappear (i.e., those within the essential network). The resulting BRM can then be recalculated as shown in Figure 6.12 with the reset and recalculated values highlighted (the values of the contributions of D and G are set to zero, and all of the other values such as A and B are then recalculated). The resulting diagram shows that the overall value of the portfolio left in the case in only $16,000 (the sum of the contributions from all of the initiatives), so that the loss from not realizing initiative B is $100,000 – $16,000 = $84,000 and not the $90,000 calculated in the earlier and simpler approach. The difference between $90,000 and $84,000, $6,000, comes from the fact that the simple approach counted the potential contribution of E (Ce = $30,000) to B (contribution fraction CFeb = 20% and contribution share CSe = CFeb * Ce = $6,000)

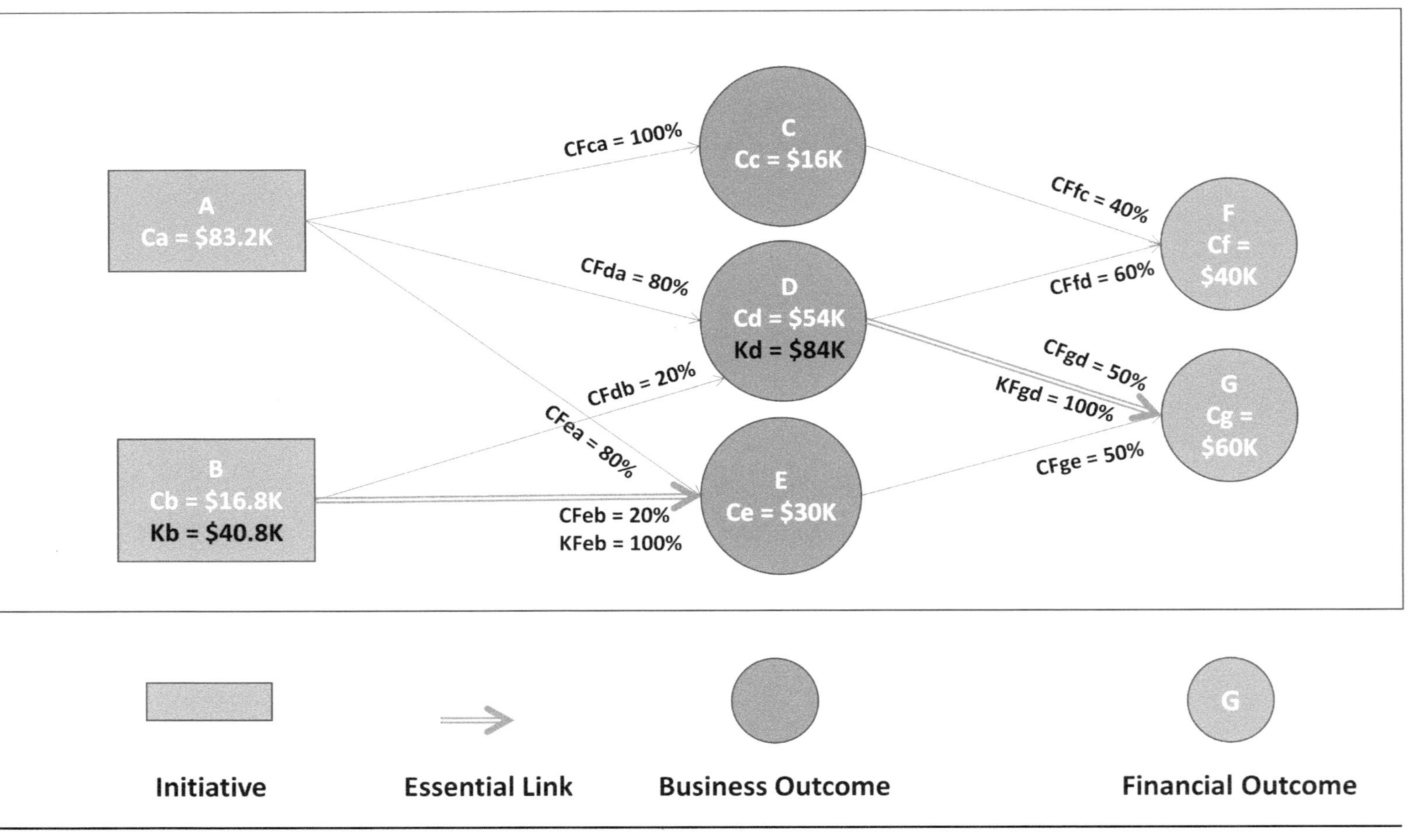

Figure 6.9 The Full Set of Values in a BRM with Separate Essential Nodes for Example 6-1

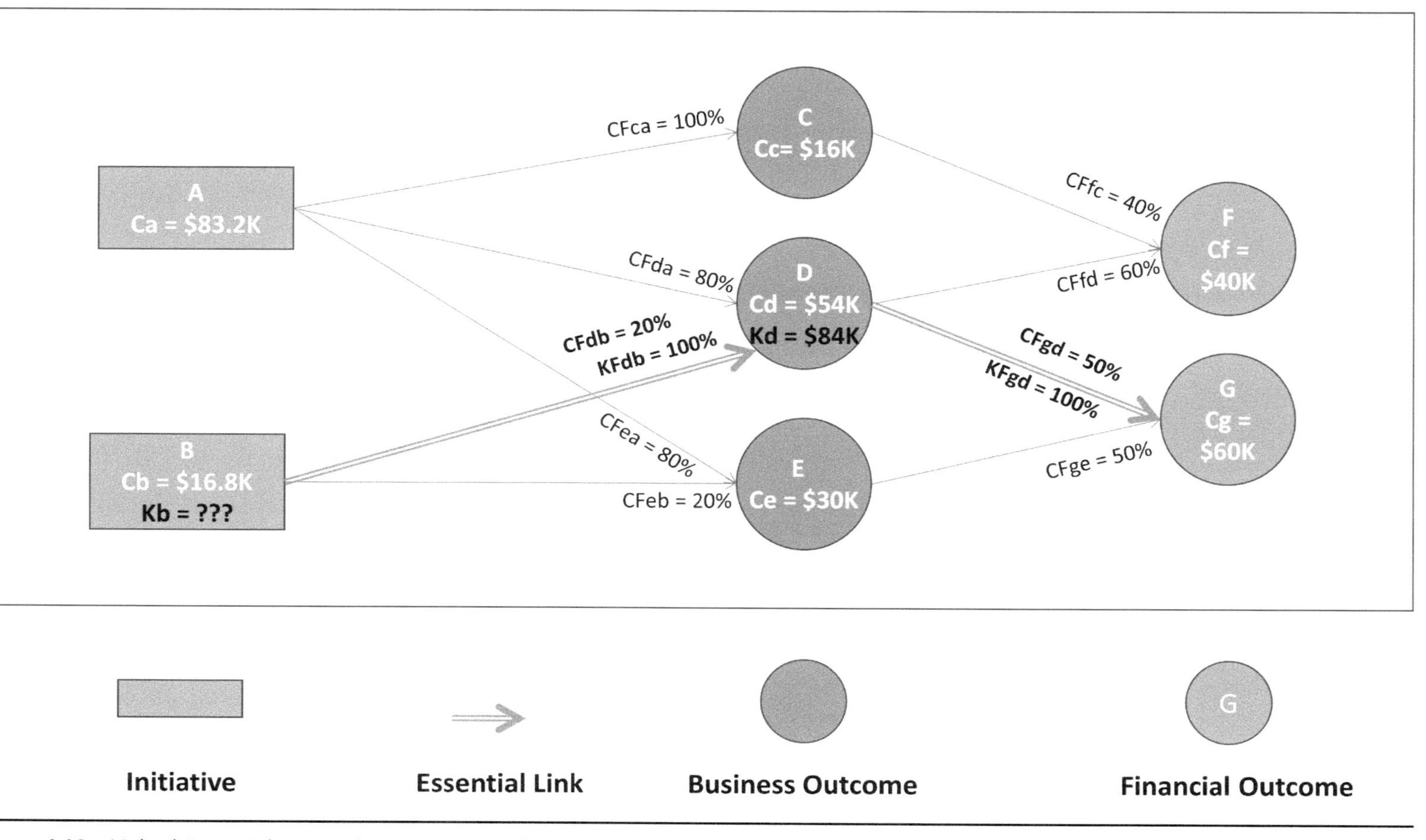

Figure 6.10 Linked Essential Dependencies (B-D-G) for Modified Example 6-1

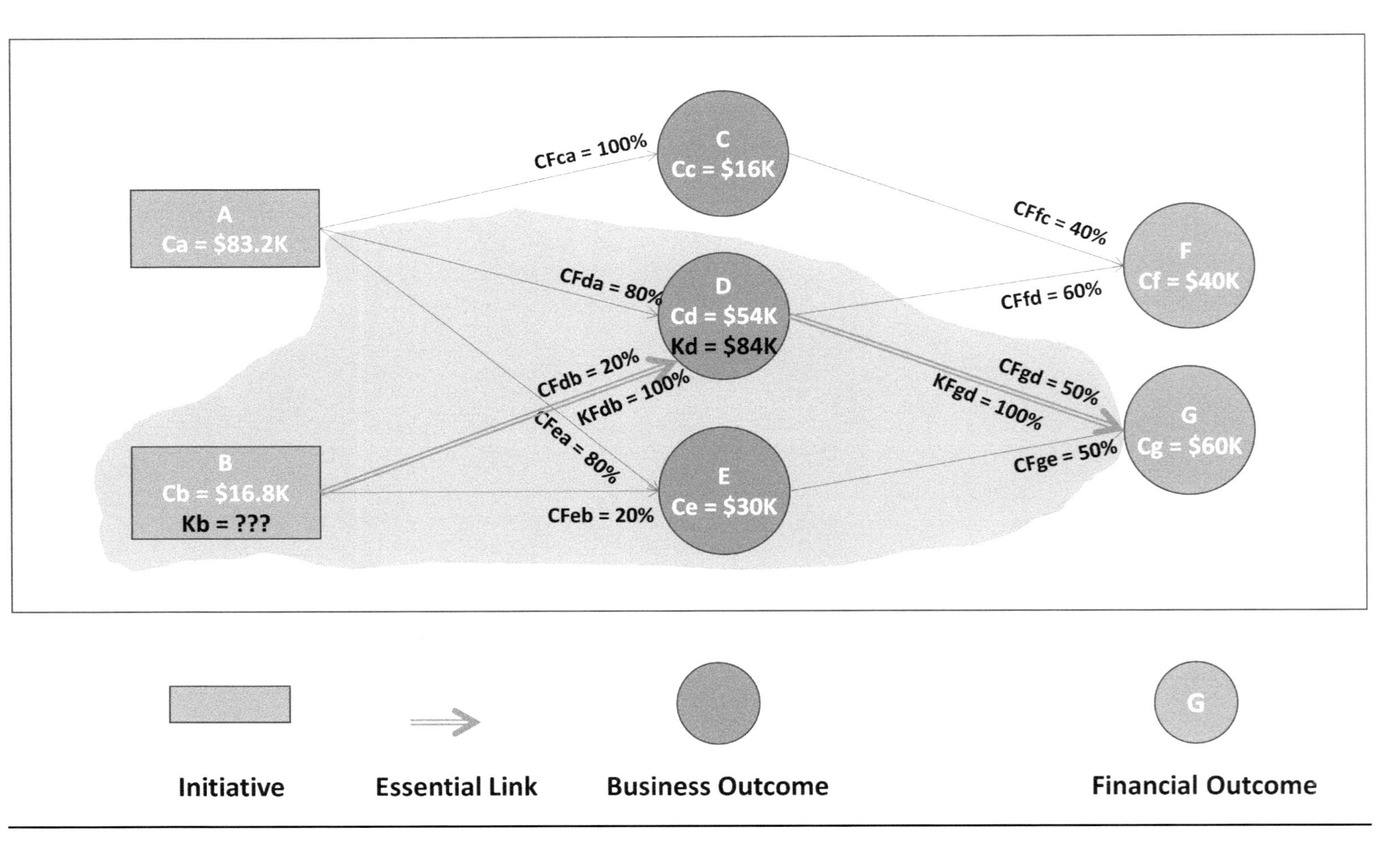

Figure 6.11 The Essential Network from B in Modified Example 6-1

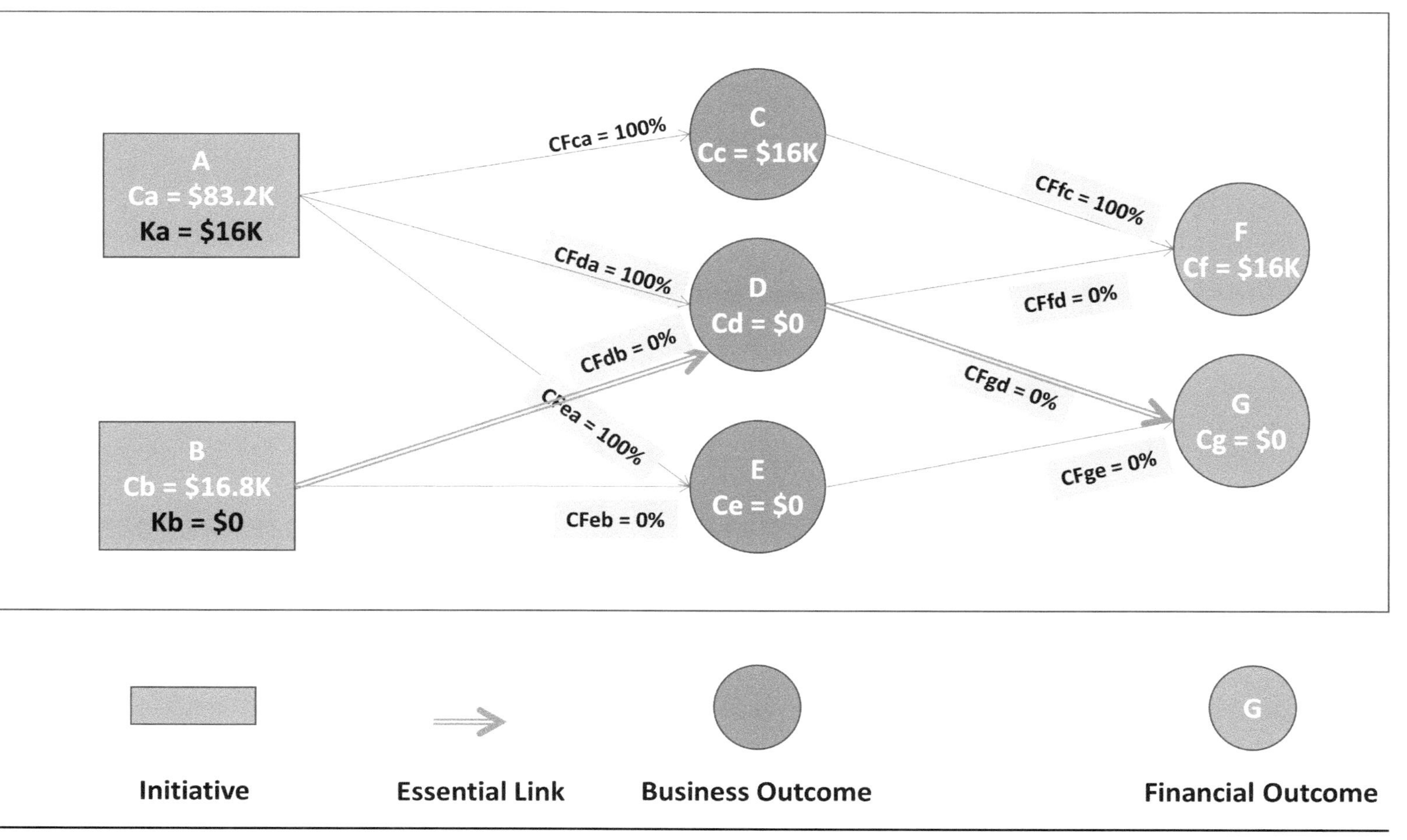

Figure 6.12 Setting Essential Effects to Zero and Recalculating for Modified Example 6-1

twice. The contribution of E should not be included because without B, D disappears, and without D, G disappears, and therefore its contribution drops to zero and the value of E (and CFei) also disappears.

Finally, the allocation fractions and allocation shares should be recalculated based on the recalculated contribution factions. The residual BRM after removal of all of the entities with a zero value is shown in Figure 6.13.

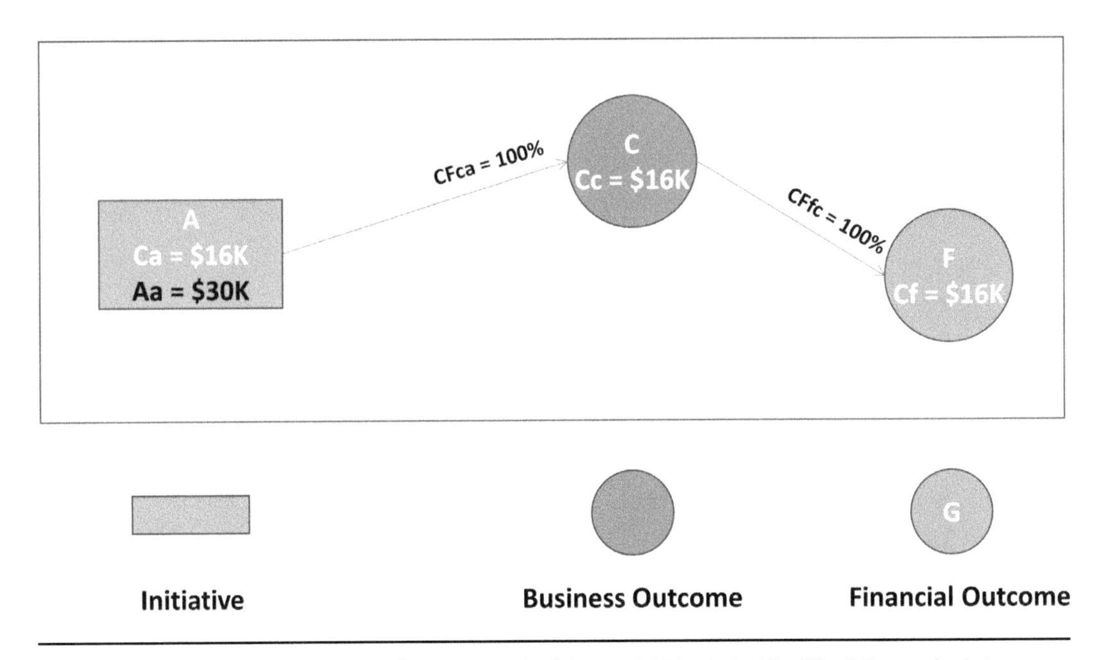

Figure 6.13 The Residual BRM after Removal of Essential Node in Modified Example 6-1

6.4.3 Conclusion: The Value of the Concept of Essential Contributions

The previous example shows the importance of this concept of *essential contributions*: Although initiative B only contributes 16.8 percent of the total benefit to the operational improvement (as shown in Figure 6.10), removing it would cost 84 percent of the benefit. This analysis now puts us in a good position to develop rational and justifiable business cases.

Even in this simple case, the analysis can seem fairly convoluted, so it is no surprise that so many mistakes are made in evaluating the potential effects of changes in major programs. To avoid mistakes, a more structured approach to carrying out the analysis is given next.

6.5 Structured Approach for Removing Nodes

This section provides the full algorithm for the removal of an initiative or another node, taking essential links into account where these are defined. The new example (Example 6-2) is used to explain the various steps.

Figure 6.14 shows the initial BRM for Example 6-2 with the essential links highlighted, evaluated to show the contributions and allocations. For the purpose of the example, we will evaluate the effect of cancelling initiative A.

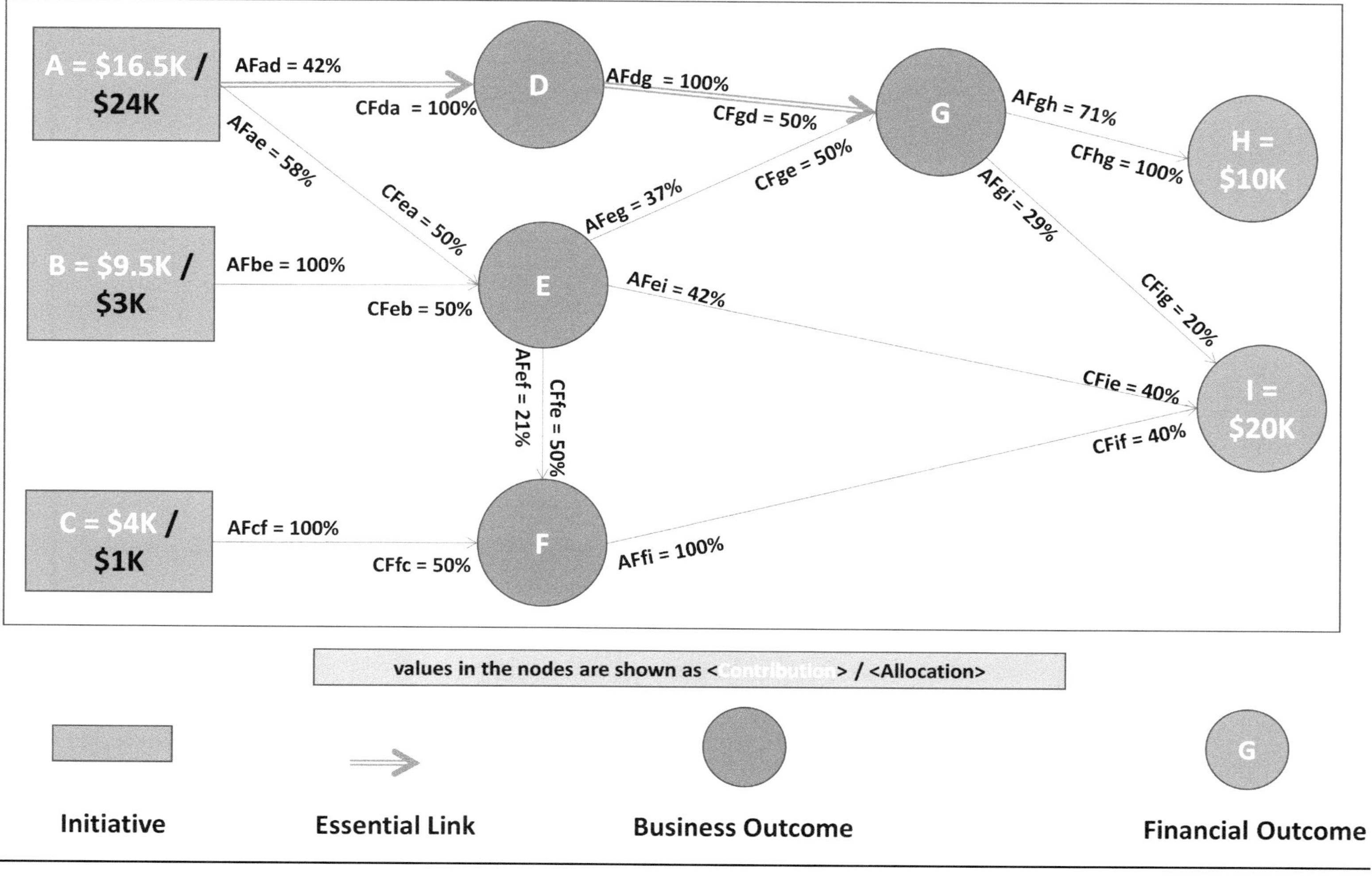

Figure 6.14 Example BRM showing Essential Links in Example 6-2

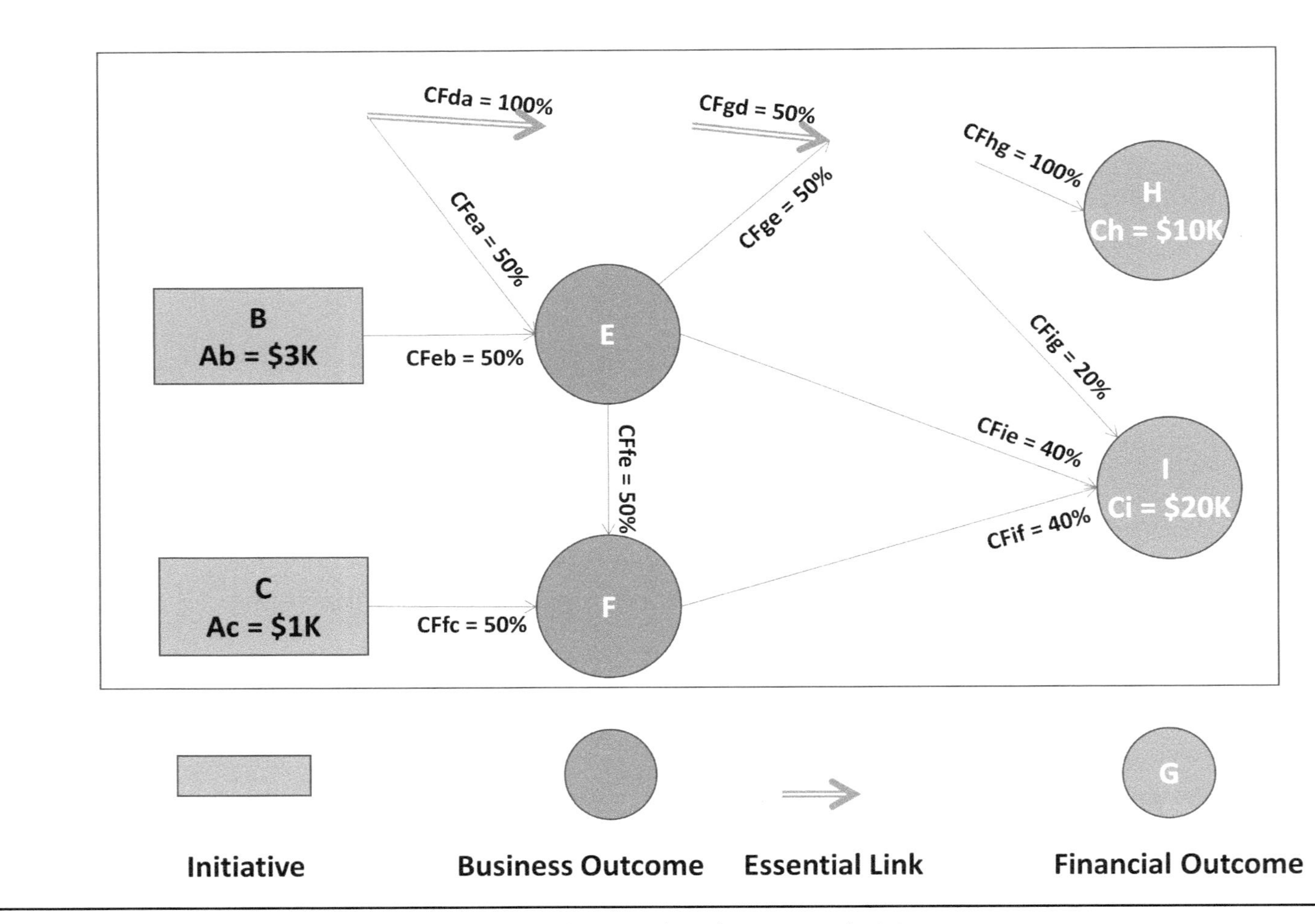

Figure 6.15 Example 6-2 after the Removal of the Linked Essential Nodes in Example 6-2

The algorithm for removing any essential node is as follows.
Start from the node to be removed; call it the *target node.*

1. Remove the target node.
2. Remove any nodes connected to it by essential links.
3. When a node has been removed, remove the corresponding links from it. For each of these links, if it is an essential link, set the corresponding destination node as a target node and go back to step 1.
4. If any of the remaining nodes has no incoming links, set each in turn to be a target node and go back to step 1.
5. You now have a network that shows only the remaining nodes and links.

For Example 6-2, on the first pass, this algorithm gives:

- Removal of (target node) A leads to the removal down link A to D of link D.
- Removal of (target node) D leads to the removal down essential link D to G of link G.

This first set of steps leads to the situation shown in Figure 6.15.

The next step is the removal of the dangling links shown in Figure 6.15. The result is shown in Figure 6.16.

The final step is the removal of the floating node (H) shown in Figure 6.16.

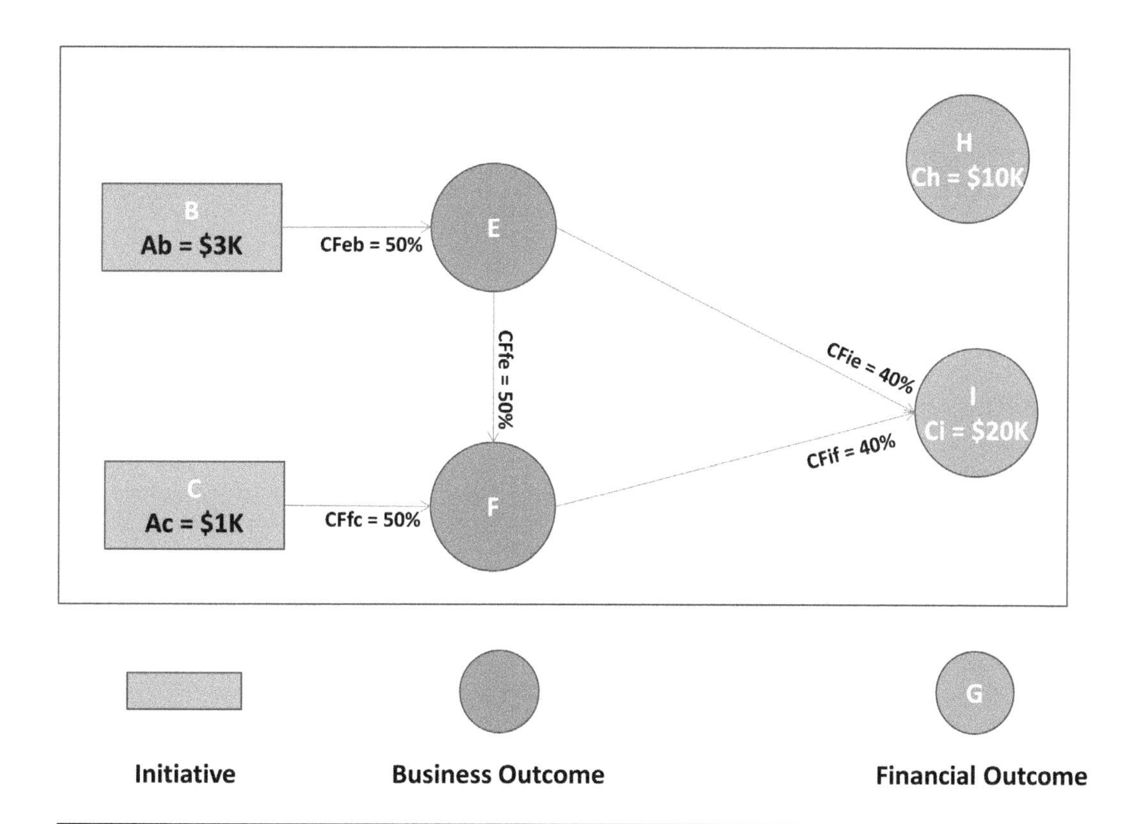

Figure 6.16 Example 6.2 after Removal of Dangling Links

The result of the algorithm will be known as the *pruned network* and is shown in Figure 6.17 with the removed nodes highlighted.

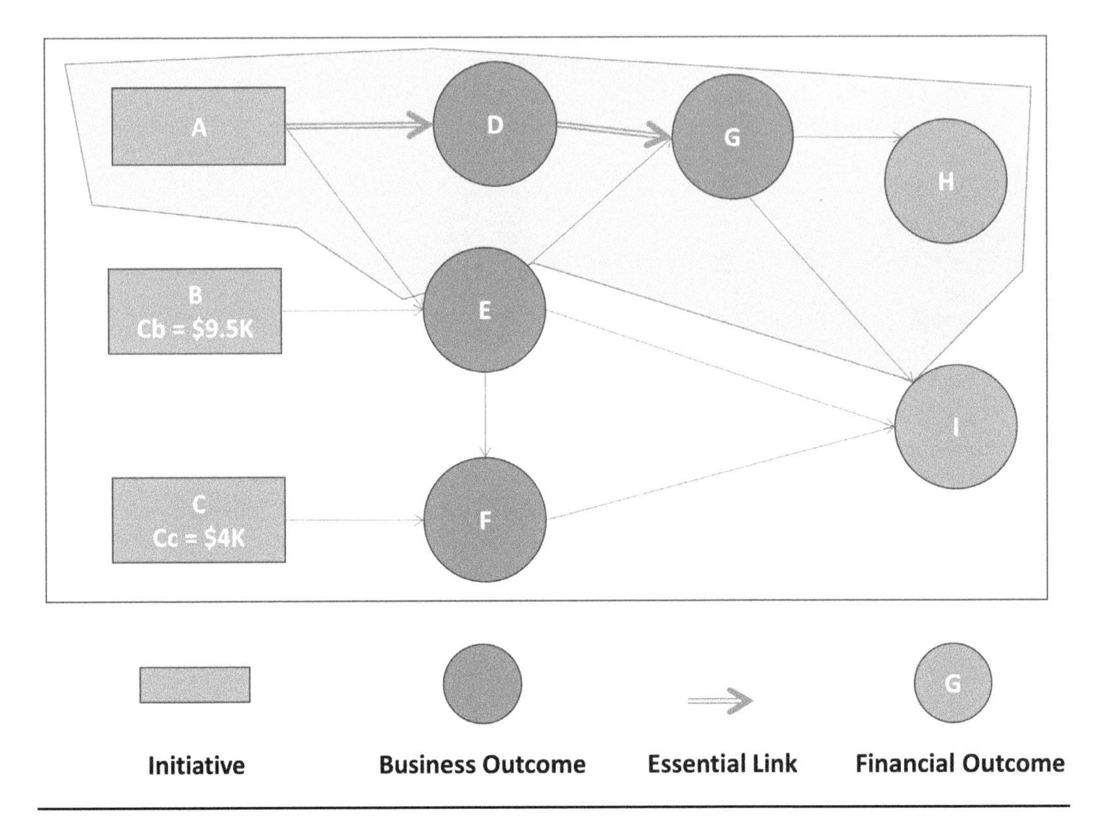

Figure 6.17 The Essential Network for Example 6-2

6.6 Recalculating the Pruned Network

Recalculating the pruned network entails carrying out five steps described below:

- Clear the redundant data.
- Calculate the Contribution Shares and Contributions.
- Calculate the Contribution Fractions.
- Reuse the Allocation values for the initiatives.
    - At this point, you have completed the residual BRM based on the pruned network.
- Recalculate the residual BRM.

The logic behind the second and third steps requires understanding the way in which the full BRM is calculated—that is, the BEER that links allocations and contributions.

6.6.1 What's in the BEER?

The key point here is to go back to the definition of the BEER. The algorithm for calculating the allocation fractions was designed so that when the initiatives are given allocation values equal to their contributions, each of the allocations calculated back through the network from left to right, using the allocation fractions, is equal to the corresponding contribution. By extension, therefore, if you want to know the values of the financial outcomes that would correspond to given values of contributions from the initiatives, you just need to set the values of the initiatives accordingly and carry out the left-to-right calculation. Similarly, if the contributions of the financial outcome nodes are set equal to their allocations as calculated from left to right, then, subsequent calculation of the contributions from right to left will produce contribution values identical to those of the allocations at every node. To state it in another way: The left-to-right calculation of the allocations is equivalent to the right-to-left calculation of the contributions. In its simplest form, the BEER leads to the fact that *a left-to-right allocation is equivalent to a right-to-left contribution and vice versa*. This concept is applied to the evaluation of the pruning, as explained next.

6.6.2 Step 1: Clear the Redundant Data

Some of the original data the business analyst evaluated will have to be restructured for the pruned network. It is important to note that the approach described next is made without the need for additional business assumptions or estimates to be made for the BRM as originally developed and approved. By avoiding any changes to the business assumptions, the original analysis is preserved—but in a different form, to take the pruning into account. This new analysis makes use of four sets of data in the original BRM:

- The topology of interconnections
- The allocation fractions on the remaining links
- The contribution values of the remaining initiatives
- The allocation values of the remaining initiatives
    - The allocation values of the initiatives are not used until step 4.

All of the information and data being used comes directly or by direct calculation from the initial business analysis. The reason this data integrity is fundamental is that it eliminates the possibility of any stakeholders taking advantage of changes to "adapt" the original BRM for their own purposes after it has been validated and approved.

The starting point is the full network from Example 6-2, shown in Figure 6.14. The result of removing node A, pruning the network and then removing all but the four sets of data defined above is shown in Figure 6.18.

6.6.3 Step 2: Recalculate the Pruned Network Contribution Shares and Contributions

We are now ready to calculate through the network in order to regenerate the contributions corresponding to the values (allocation = contribution) in the remaining initiatives.

Because of the BEER, the contribution shares in the pruned network are equal to the allocation shares—that is, the product of allocation fraction by node contribution. For example:

- CSeb = ASbe * Cb = 100% * $9,500 = $9,500

as the link from B to E is the only link to E, Ce = CSeb = $9,500

- CSfe = AFef * Ce = 21% * $9,500 = $2,000

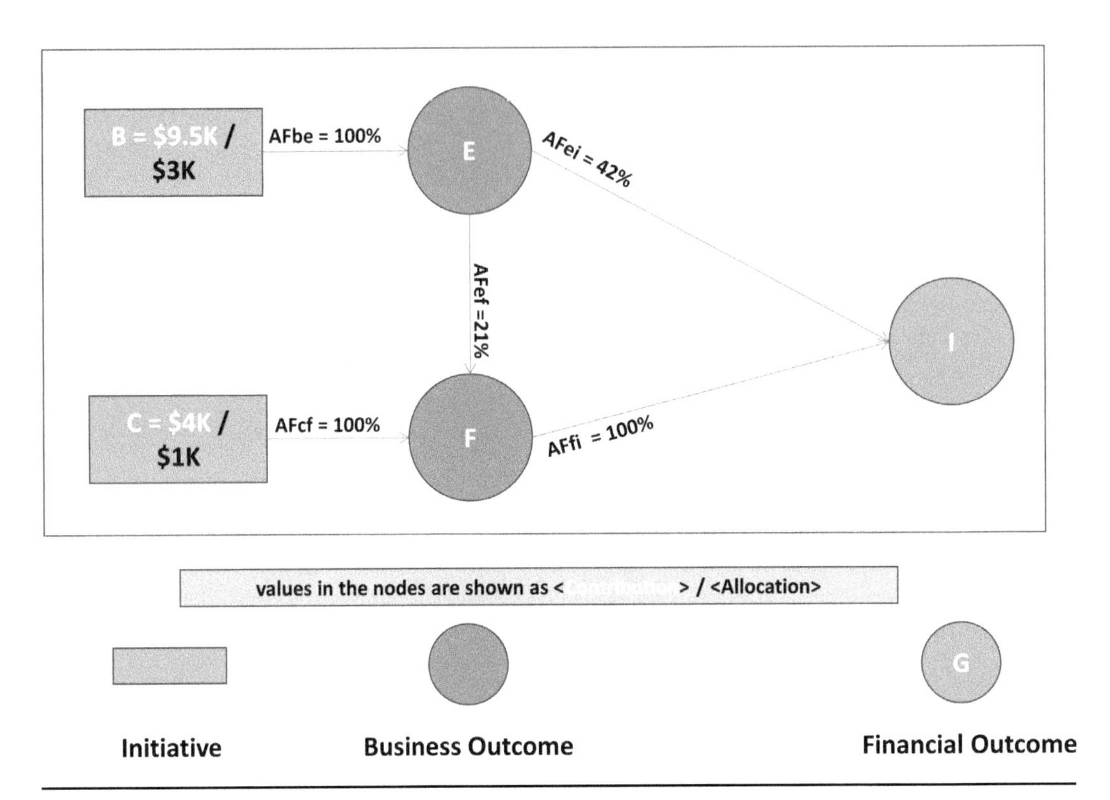

Figure 6.18 Example 6-2 Ready for Recalculation

The total contribution for each node is equal to the sum of its contribution shares. The result of the full set of calculations for this step is shown in Figure 6.19.

We now have the BEER values of the contributions based on the left-to-right calculation using the pruned allocation fractions and the contribution values of the remaining initiatives.

Next, we need to recalculate the contribution fractions from this rebuilt network.

6.6.4 Step 3: Calculate the Contribution Fractions

From the previous step, we know the contributions and the contribution shares as shown in Figure 6.19. These values allow us to complete the pruned diagram by calculating the contribution fractions from these values: For each node, take the ratio of each contribution share to the total contribution of the corresponding node.

The next step completes the regeneration of core data for the pruned BRM.

6.6.5 Step 4: Reuse the Initiative Allocations

The calculations up to this point were based on the left-to-right calculation, in which the allocation values of the initiatives were replaced with their calculated contributions. Now that the calculation has provided the values of the contributions and the contribution shares, the allocations for the initiatives need to be included in order to create a clean version of the residual BRM. The clean version of the BRM ready for this final set of calculations is shown in Figure 6.20.

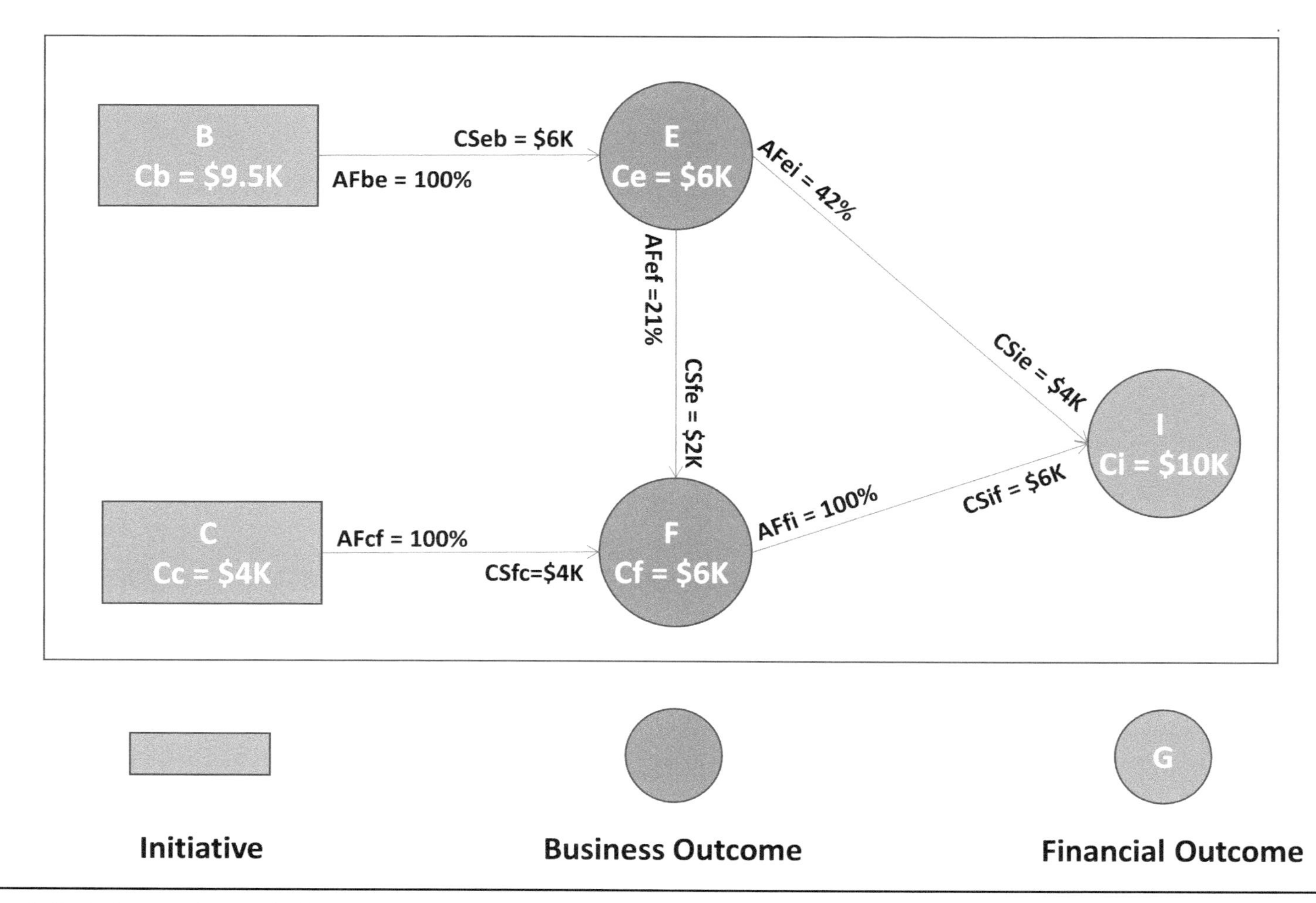

Figure 6.19 Left-to-Right Recalculation to Generate Remaining Contribution Shares and Contributions

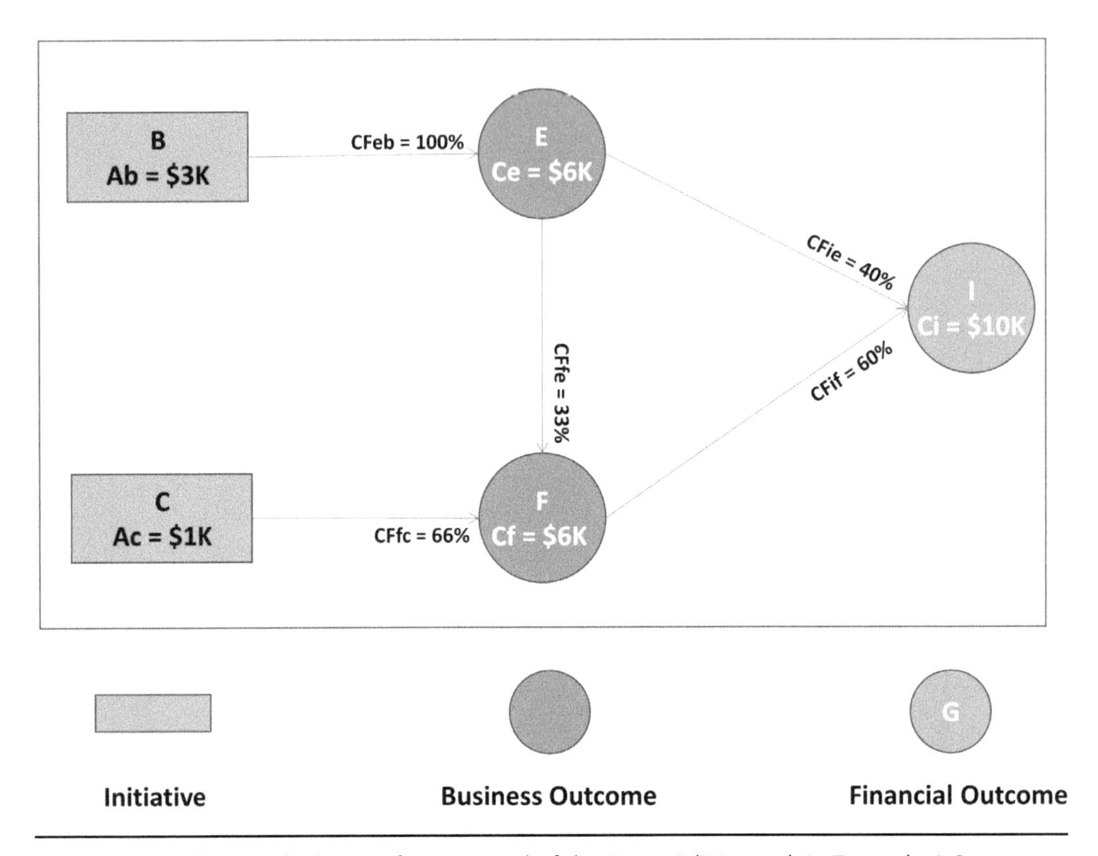

Figure 6.20 The Residual BRM after Removal of the Essential Network in Example 6-2

6.6.6 Step 5: Recalculate the Pruned Network

Now that all of the data has been regenerated as described above, the residual BRM can be recalculated using the BEER algorithm developed in the previous chapter. The result is shown in Figure 6.21.

That completes the development and explanation of all of the algorithms required for calculating and optimizing the BRM.

6.7 Summary

This chapter has taken the BRM concepts one step further by adding the concept of negative side effects (disbenefits) and nodes that are an absolute prerequisite for a successor (essential links). The additions to the algorithms for calculating the corresponding BRMs were developed. Based on this analysis, techniques were described for optimizing the business case by the removal of redundant initiatives.

6.8 Current Status

All of the principal concepts for translating the BRM into a consistent quantified plan are now in place:

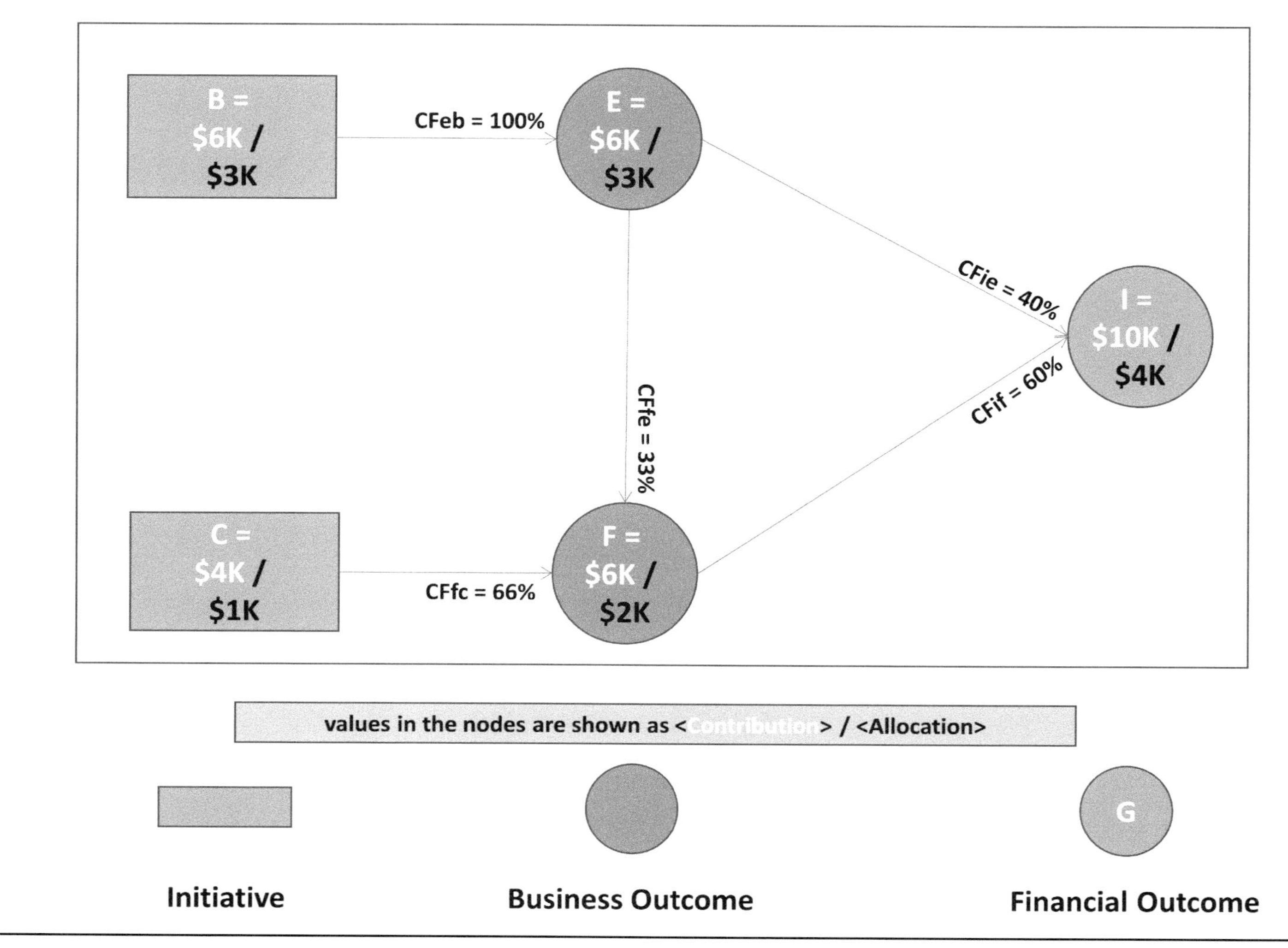

Figure 6.21 Fully Recalculated BRM for the Pruned Network in Example 6-2

- Calculating the contributions
- Calculating the allocations
- Including disbenefits
- Understanding essential links
- Removing nodes
- Recalculating the residual BRM

In the next chapter, the power of this technique will be applied to the QERTS case study introduced in Chapter 3. The analysis gives rise to interesting results that provide the basis for important insights in program management.

Chapter 7

Applying the BRM Approach to the QERTS Example

The concepts and tools developed in the preceding chapters are now applied to the case study example. Applying the algorithms and calculations to the full QERTS case study gives rise to some unexpected results, leading to a deeper understanding of the challenges and power of program management.

The numerical calculations in the following examples have been carried out using the Excel process described in the Appendix, implementing the algorithms defined in the preceding chapters.

7.1 Analyzing Outcomes and Options: The QERTS Example

Although it can be useful to calculate the individual contribution-allocation characteristics of each initiative separately, this somewhat defeats the object of a program, which is to make use of the synergies between the full set of components.

Consider the following example shown in Figure 7.1, based on merging two companies. The background was explained in detail in Chapters 3 and 4. Using the algorithms described earlier, the allocations can be calculated to give the values of allocations and contributions shown in Table 7.1.

This initial analysis gives management a dilemma. What action should be taken, given that that one initiative (B = Integrate Store Records) appears to cost more than it delivers and that one intermediate outcome (H = Integrated CRM) is similarly not justified from a cost–benefit point of view?

Two unfortunately popular approaches explain to some extent why people still find that most projects fail to deliver full satisfaction. The first approach is to rework the costs by applying pressure to the project managers and setting "stretch targets" or some such optimistic term that leads to a reduction in the cost *estimates*. The second unjustifiable approach that might also unfortunately be used is the optimistic inflating of the benefit figures.

The correct approach is to investigate options within the existing business case, as explained next.

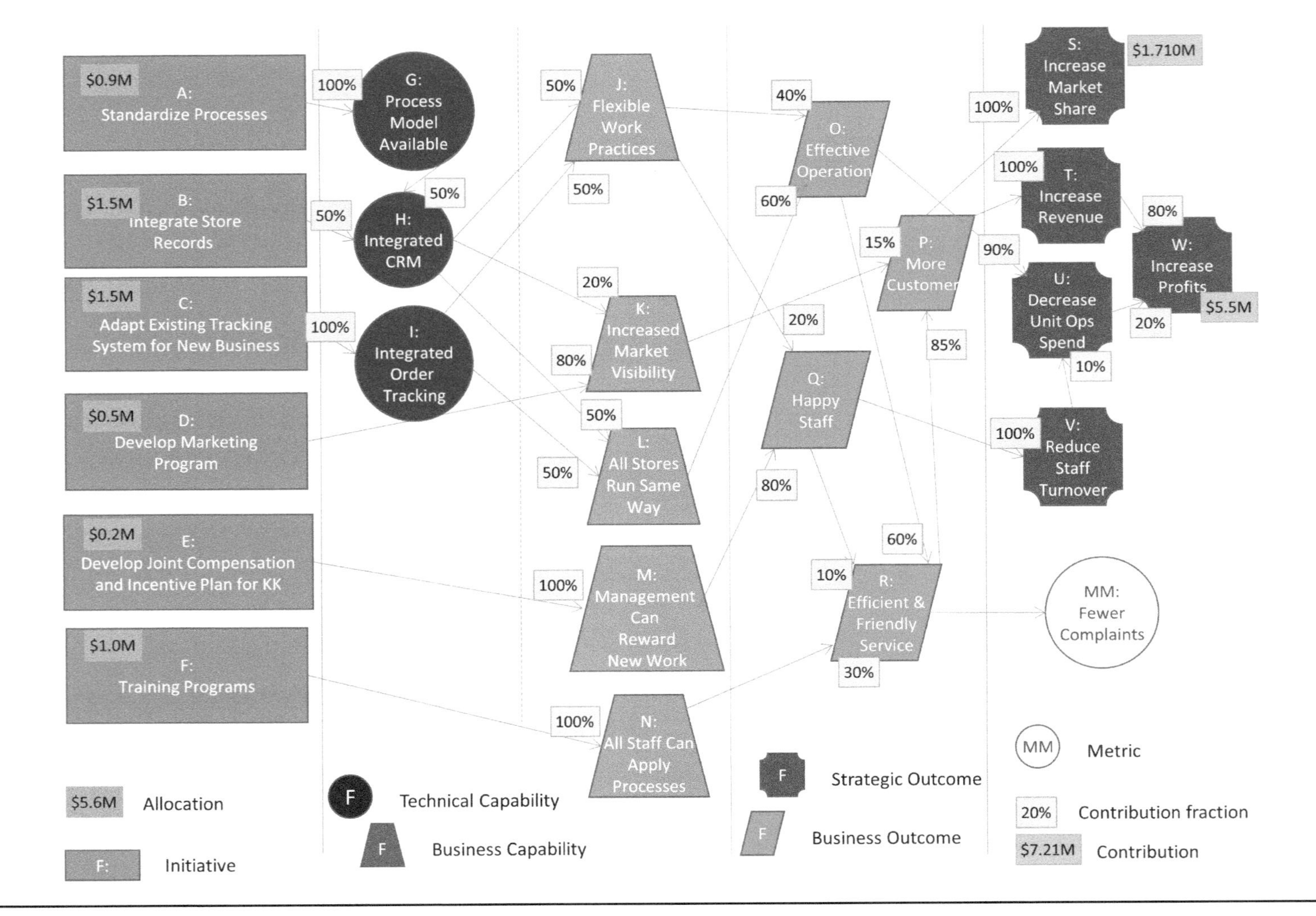

Figure 7.1 QERTS Service Merger Benefits Realization Map

Table 7.1 Allocations and Contributions for the Merger Example

		Allocation	Contribution	
A	Standardize Processes	$900,000	$1,149,643	
B	Integrate Store Records	$1,500,000	$1,149,643	Cost > Value
C	Adapt Tracking System	$1,500,000	$2,115,985	
D	Develop Joint Marketing	$500,000	$733,200	
E	Align Compensation	$200,000	$503,480	
F	Train Staff	$1,000,000	$1,558,050	
G	Process Model Available	$900,000	$1,149,643	
H	Integrated CRM	$2,400,000	$2,299,285	Cost > Value
I	Integrated Tracking	$1,500,000	$2,115,985	
J	Flexible Working Practices	$1,549,652	$1,768,310	
K	Increased Market Visibility	$691,329	$916,500	
L	All Stores Same	$2,159,019	$2,463,660	
M	Staff Rewards Defined	$200,000	$503,480	
N	Staff Trained and Capable	$1,000,000	$1,558,050	
O	Effective Operations	$3,598,365	$4,106,100	
P	More Customers	$4,678,181	$6,110,000	
Q	Happy Staff	$310,306	$629,350	
R	Great Service	$3,986,852	$5,193,500	
S	Higher Market Share	$1,309,278	$1,710,000	
T	Higher Revenue	$3,368,903	$4,400,000	
U	Operational Savings	$921,819	$1,100,000	
V	Less Turnover	$54,236	$110,000	
W	More Profit	$4,290,722	$5,500,000	

7.2 Revising the Program

7.2.1 Simple Initiative Removal

If the business pressure approach is either not applied or is ineffective, the most obvious action is to eliminate the unprofitable initiative—B in this case. [The benefits realization map (BRM) in this case is shown in Figure 7.2.]

As shown in Table 7.1, elimination of that node will "obviously" save more than it costs. The overall effect would be to reduce the investment by the amount allocated to integrating the store records ($1,500,000) and reduce the value of the program by the contribution of that initiative ($1,150,000). In this case, the return on investment (ROI) is predicted to increase from 129 percent (72/56) to 139 percent (61/41) and the absolute profit would appear to be raised from $1,610,000 to almost $2 million.

The full details of the effect on each node can be seen in Table 7.2 based on using the "pruning" algorithm described in Chapter 6 to eliminate initiative B.

However, none of these workarounds takes into account the real character of a program—that is, the interdependence and synergy between the initiatives and outcomes. There is a much more reliable approach based on the tools already described.

7.2.2 Accounting for Essential Links

For each node, start by asking the question: "Why do we need this node?" Whenever the answer is "because it is an absolute prerequisite for one of its successors," this will identify an essential link. In

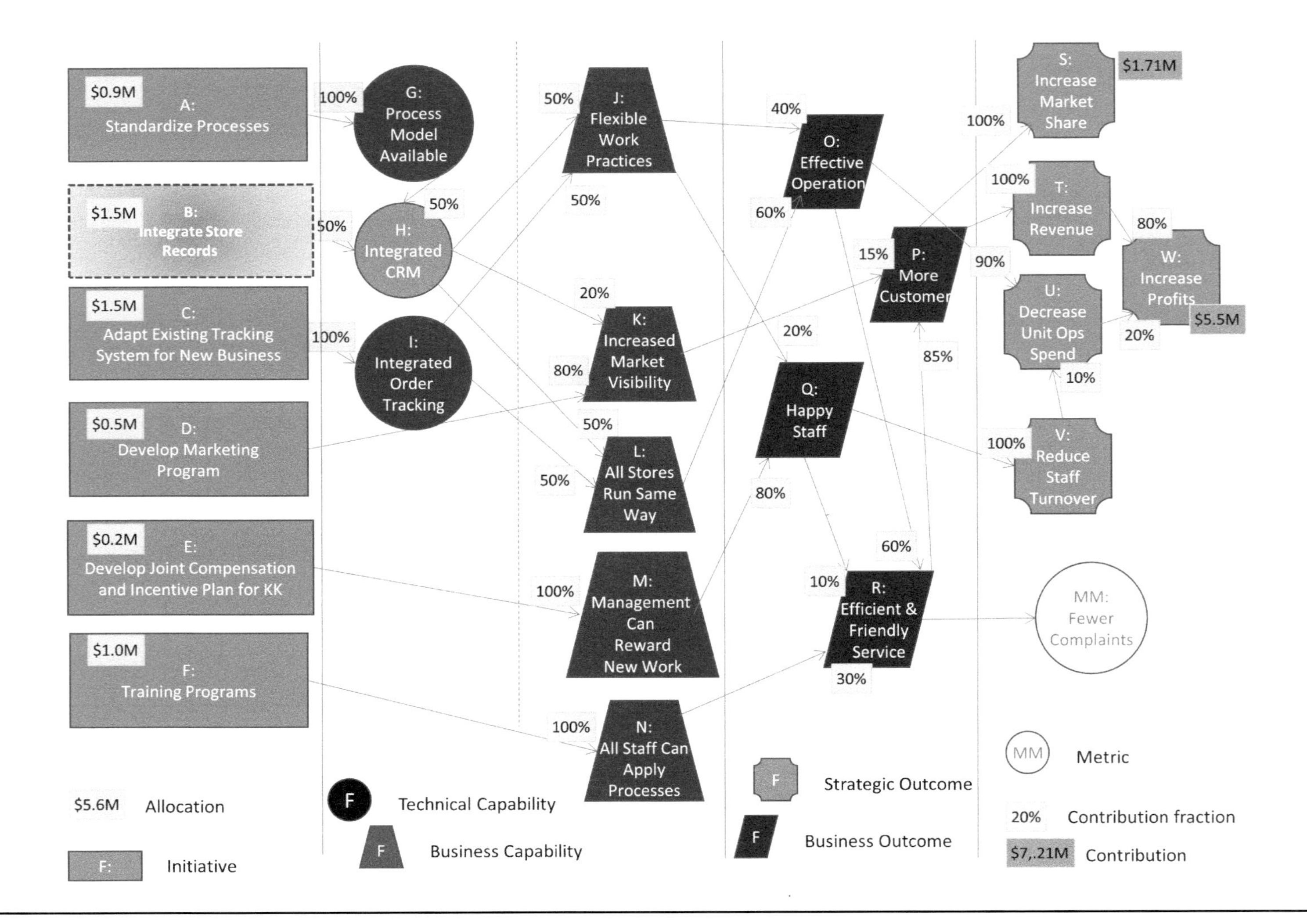

Figure 7.2 BRM after Removal of Node B

Table 7.2 Allocations and Contributions without Initiative B, if There Are No Essential Links

		Allocation	Contribution	
A	Standardize Processes	$900,000	$1,149,643	
B	Integrate Store Records	$0	$0	By Hypothesis
C	Adapt Tracking System	$1,500,000	$2,115,985	
D	Develop Joint Marketing	$500,000	$733,200	
E	Align Compensation	$200,000	$503,480	
F	Train Staff	$1,000,000	$1,558,050	
G	Process Model Available	$900,000	$1,149,643	
H	Integrated CRM	$900,000	$1,149,643	
I	Integrated Tracking	$1,500,000	$2,115,985	
J	Flexible Working Practices	$972,850	$1,326,233	
K	Increased Market Visibility	$571,748	$824,850	
L	All Stores Same	$1,355,402	$1,847,745	
M	Staff Rewards Defined	$200,000	$503,480	
N	Staff Trained and Capable	$1,000,000	$1,558,050	
O	Effective Operations	$2,259,003	$3,079,575	
P	More Customers	$3,508,284	$5,213,358	
Q	Happy Staff	$269,248	$597,883	
R	Great Service	$2,936,535	$4,388,508	
S	Higher Market Share	$981,860	$1,459,058	
T	Higher Revenue	$2,526,423	$3,754,300	
U	Operational Savings	$591,716	$847,000	
V	Less Turnover	$47,060	$104,500	
W	More Profit	$3,118,140	$4,601,300	
	TOTALS	$4,100,000	$6,060,358	

the given example (Figure 7.1), we will state that the link from B = Integrate Store Records to H = Integrated CRM is essential, as well as the link from H = Integrated CRM to K = Market Visibility. This additional information is shown in Figure 7.3.

Since the initial cost–benefit analysis suggested ruling out B = Integrate Store Records, we can analyze the effect of eliminating this node as before, but now taking into account the essential links we have just identified.

One key feature of this algorithm was described in Chapter 6: the elimination not only of destination nodes down essential links, but also, iteratively, of source nodes if the source contributes only to eliminated nodes. Let us see how this works in the example, if we decide to eliminate node B = Integrate Store Records.

In this case:

- Eliminating B = Integrate Store Records makes H = Integrated CRM impossible.
  - With H = Integrated CRM missing, G = Process Model Available becomes unnecessary.
    - Without G = Process Model Available, A = Standardize Processes can be left out.
  - Because of the essential link from H = Integrated CRM to K = Increased Market Visibility, K is eliminated.
    - Since D = Develop Joint Marketing only contributes to K, it is now unnecessary.

This gives rise to the new BRM shown in Figure 7.4.

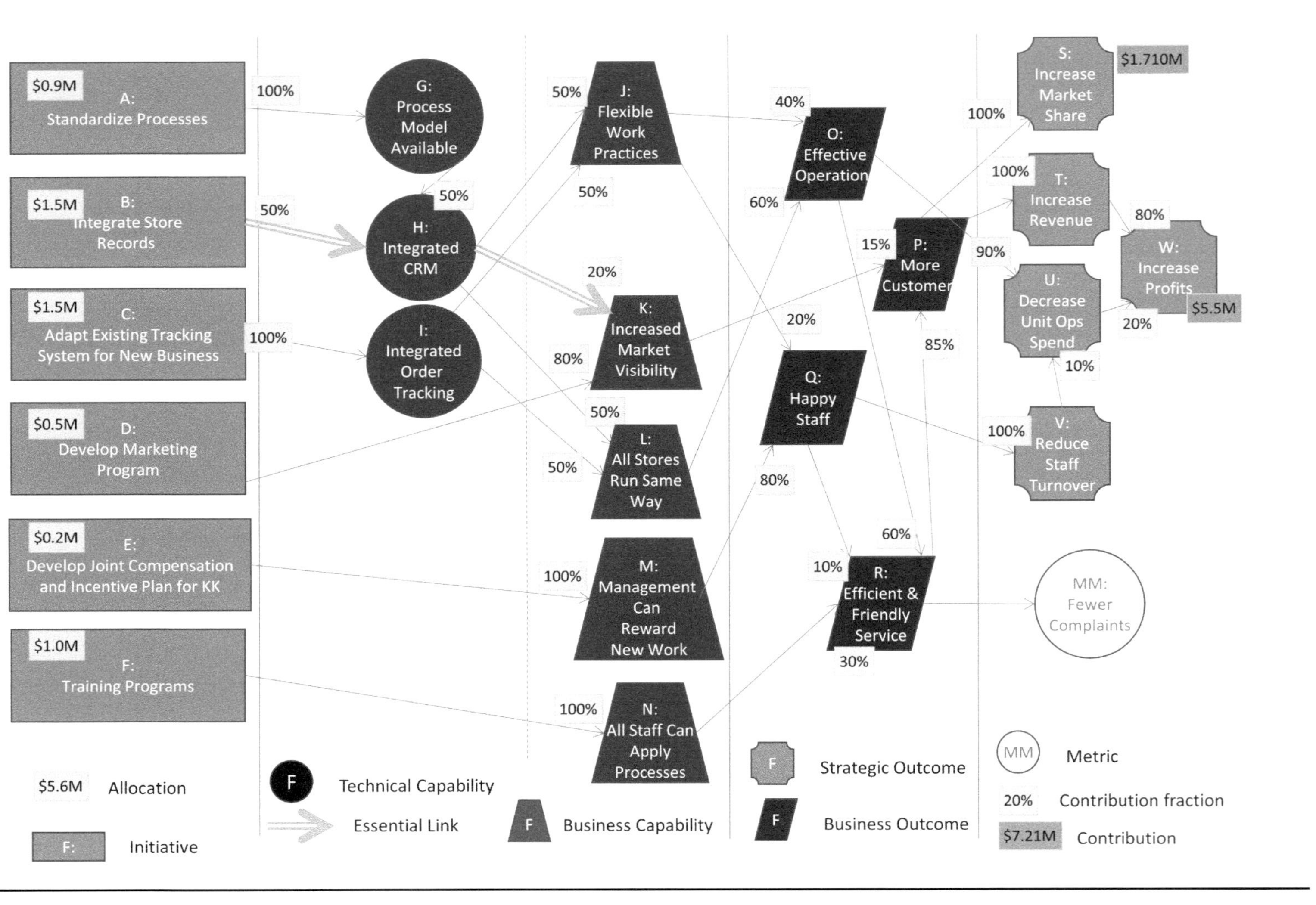

Figure 7.3 QERTS BRM Indicating the Essential Links

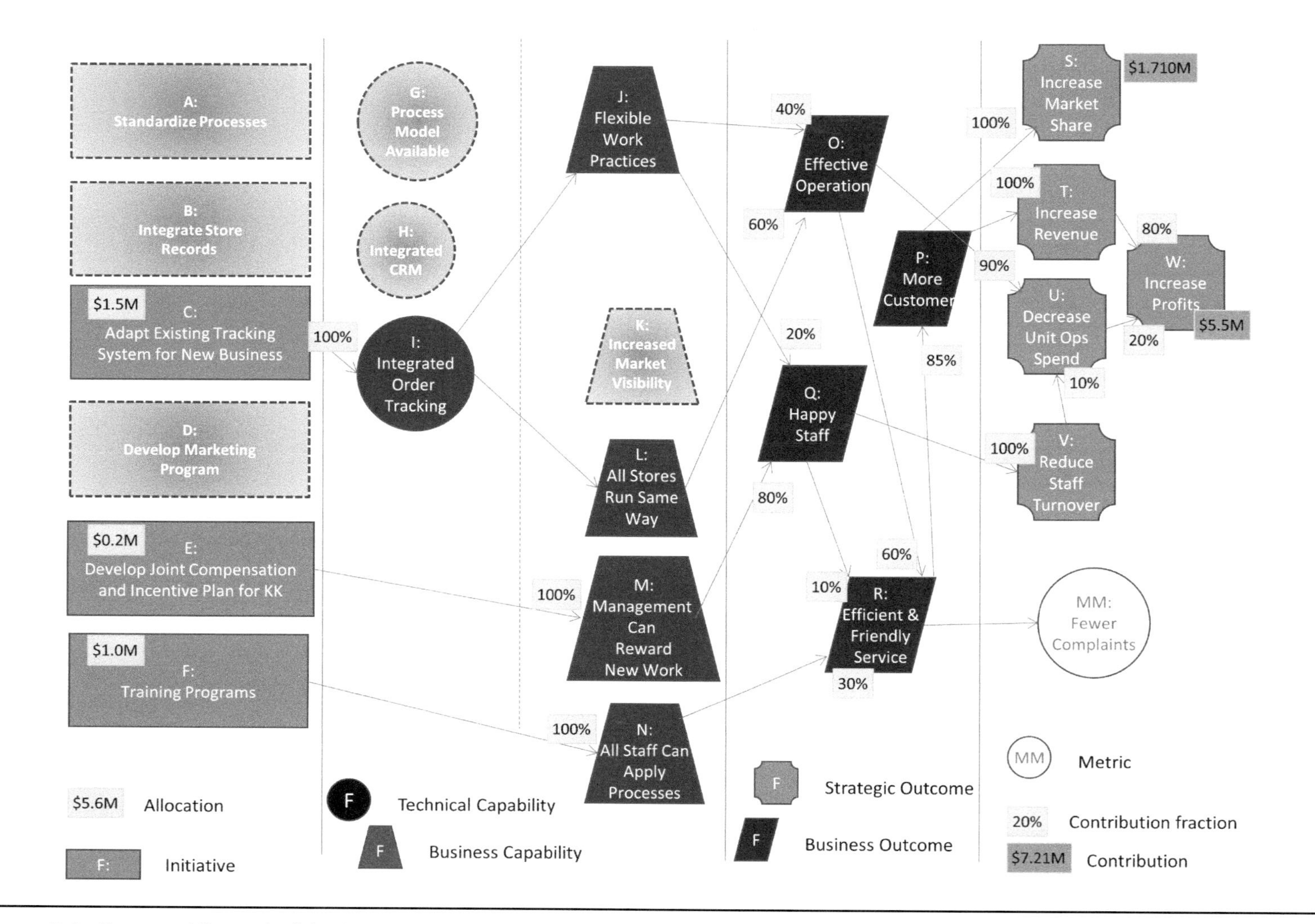

Figure 7.4 Proposed Stage 1 of the Merger Program

The pruning calculation algorithm from Chapter 6 can then be applied to recalculate this revised network.

In this example, after canceling initiative B = Integrate Store Records, the new calculations indicate the following, as shown in Table 7.3: The overall value of the program has dropped to $4,178,000 (from $7,210,000) for a total investment of $2,700,000 (from $5,600,000). The good news, however, is that all of the outcomes are financially justifiable based on the contribution–allocation figures.

Table 7.3 Contributions and Allocations after Eliminating B = Integrate Store Records

		Allocation	Contribution	
A	Standardize Processes	$0	$0	By Calculation
B	Integrate Store Records	$0	$0	By Hypothesis
C	Adapt Tracking System	$1,500,000	$2,115,985	
D	Develop Joint Marketing	$0	$0	By Calculation
E	Align Compensation	$200,000	$503,480	
F	Train Staff	$1,000,000	$1,558,050	
G	Process Model Available	$0	$0	By Calculation
H	Integrated CRM	$0	$0	By Calculation
I	Integrated Tracking	$1,500,000	$2,115,985	
J	Flexible Working Practices	$626,768	$884,155	
K	Increased Market Visibility	$0	$0	By Calculation
L	All Stores Same	$873,232	$1,231,830	
M	Staff Rewards Defined	$200,000	$503,480	
N	Staff Trained and Capable	$1,000,000	$1,558,050	
O	Effective Operations	$1,455,386	$2,053,050	
P	More Customers	$2,306,345	$3,583,515	
Q	Happy Staff	$244,614	$566,415	
R	Great Service	$2,306,345	$3,583,515	
S	Higher Market Share	$645,475	$1,002,915	
T	Higher Revenue	$1,660,870	$2,580,600	
U	Operational Savings	$393,655	$594,000	
V	Less Turnover	$42,754	$99,000	
W	More Profit	$2,054,525	$3,174,600	
	TOTALS	$2,700,000	$4,177,515	

This revised program provides a potentially interesting alternative for management: reduce the initial financial exposure by about $5,600,000 – $2,700,000 = $2,900,000 (52 percent) while reducing the forecast profit by about $7,210,000 – $4,178,000 = $3,032,000 (42 percent). Alternatively, this could be seen as increasing the ROI from 129 percent (72/56) to 155 percent (42/27).

Management might therefore choose to carry out the reduced program shown in Figure 7.4 as a first stage prior to standardization of working process, integration of store records, and developing a marketing program, as this second stage if that can be justified in the future.

However, if we had not taken the essential links into account, we would most probably have reverted to the approach shown above, in which only initiative B is canceled—so the cost of the program would be (A + C + D + E + F) = $4,100,000. The overall contribution would remain as calculated above: $4,178,000, *generating a program that only just breaks even!*

Another example is given next, with the aim of showing just how powerful and valuable this type of analysis can be. It is based on the same set of initiatives, but with a different set of contribution fractions.

7.3 Alternative QERTS Example

This example is based on the same overall scenario, except that a new analysis by the marketing department has led to a different set of contribution fractions for node P = More Customers. In the new case, K = Increased Market Visibility has increased in importance (from 15 percent to 40 percent) with respect to R = Efficient and Friendly Service (reduced from 85 percent to 60 percent), as shown in Figure 7.5.

This leads to the following Allocation–Contribution Table (Table 7.4).

Once again, given this initial analysis, management then needs to decide on the next steps, given that, as before, one initiative appears to cost more than it delivers and that, in addition, four intermediate outcomes are similarly not justified from a cost–benefit point of view.

The same analysis as above will now be carried out on this modified BRM.

7.3.1 Simple Initiative Removal

The effect of removing only the nonviable initiative B = Integrate Store Records is shown in Table 7.5.

In the case in which we only cancel initiative B = Integrate Store Records and do not take essential links into account, Table 7.5 shows that the overall effect would be to reduce the investment by the amount allocated to integrating the store records ($1,500,000) and reduce the value by the contribution of that initiative ($1,065,630, as shown in Table 7.4). In this case, the ROI is predicted to increase from 129 percent (72/56) to 139 percent (61/41), and the absolute profit would appear to be raised from $1,610,000 to just over $2 million. All of this is virtually identical to the situation in the previous example.

The situation changes considerably when essential links are taken into account.

7.3.2 Accounting for Essential Links

Once again, the essential links have to be taken into account. For the modified example, the effect of taking essential links into account is shown in Table 7.6.

By choosing this option, management would reduce the initial financial exposure by about $5,600,000 – $2,700,000 = $2,900,000 (52 percent). This choice does, however, reduce the forecast profit by about $7,210,000 – $3,124,000 = $4,086,000 (57 percent) and reducing the ROI from 129 percent (72/56) to 116 percent (31/27).

Whereas, in the previous example, management might still choose to carry out the reduced program as shown in Figure 7.4 as a first stage, in the present case, the ROI of this reduced option may be well below the organizational threshold for program approval. The choice in this case would therefore be to abandon the merger or to commit to the full program—despite the "loss-making" integration of store records. The option of going ahead with the merger might alternatively be deemed to be preferable from a long-term "strategic" point of view.

As in the previous example, if we had not understood the impact of essential links, we would most probably have decided simply the cancel initiative B, so the cost of the program would be (A + C + D + E + F) = $4,100,000 because we would not realize that initiatives A and D would have become redundant

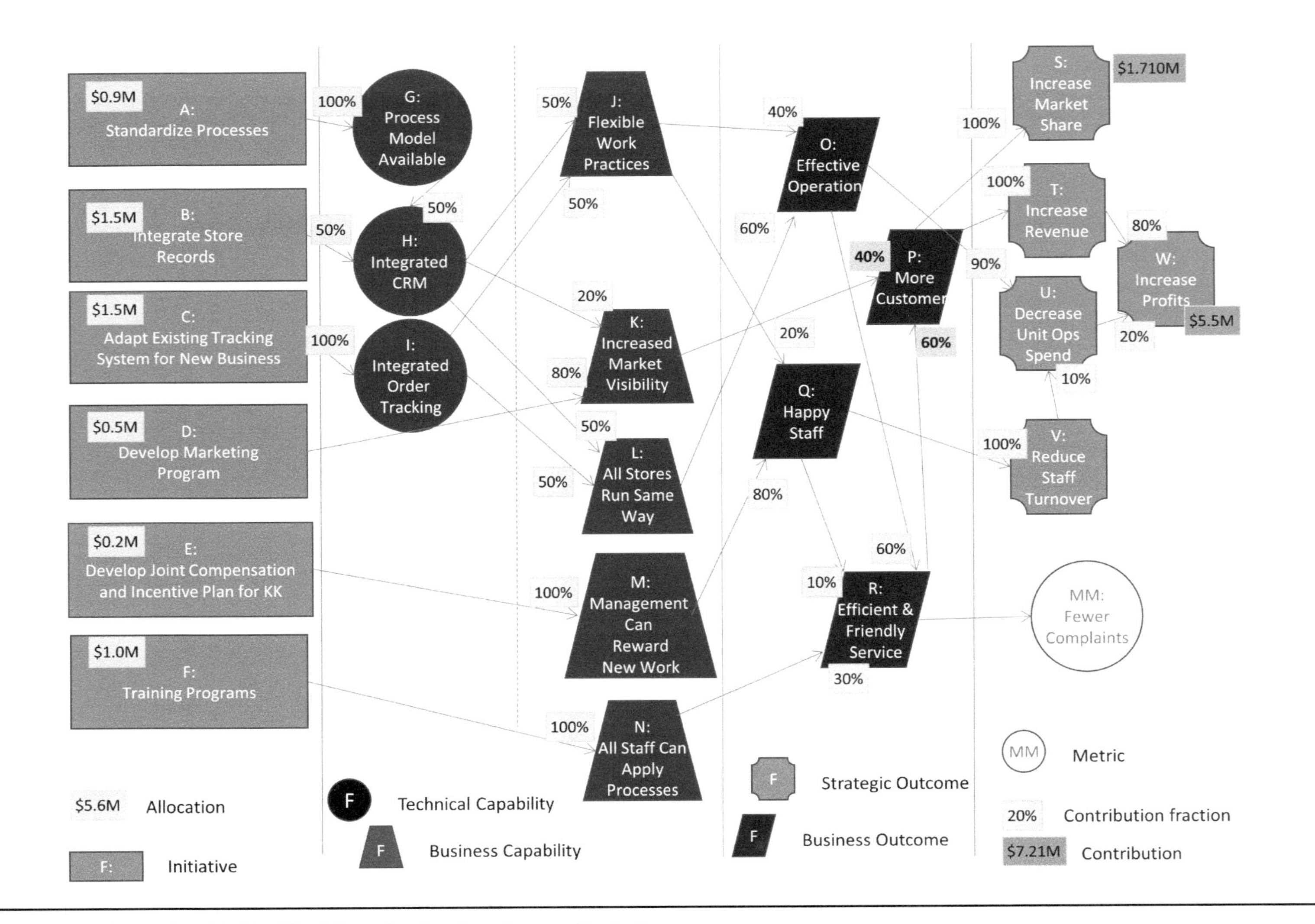

Figure 7.5 Example with Modified Contribution Fractions to Node P

Table 7.4 Allocation and Contributions for the Modified QERTS Example

		Allocation	Contribution	
A	Standardize Processes	$900,000	$1,065,630	
B	Integrate Store Records	$1,500,000	$1,065,630	Cost > Value
C	Adapt Tracking System	$1,500,000	$1,642,460	
D	Develop Joint Marketing	$500,000	$1,955,200	
E	Align Compensation	$200,000	$381,280	
F	Train Staff	$1,000,000	$1,099,800	
G	Process Model Available	$900,000	$1,065,630	
H	Integrated CRM	$2,400,000	$2,131,260	Cost > Value
I	Integrated Tracking	$1,500,000	$1,642,460	
J	Flexible Working Practices	$1,398,144	$1,371,160	Cost > Value
K	Increased Market Visibility	$1,050,435	$2,444,000	
L	All Stores Same	$1,951,422	$1,913,760	Cost > Value
M	Staff Rewards Defined	$200,000	$381,280	
N	Staff Trained and Capable	$1,000,000	$1,099,800	
O	Effective Operations	$3,252,369	$3,189,600	Cost > Value
P	More Customers	$4,521,924	$6,110,000	
Q	Happy Staff	$297,196	$476,600	
R	Great Service	$3,471,489	$3,666,000	
S	Higher Market Share	$1,265,547	$1,710,000	
T	Higher Revenue	$3,256,377	$4,400,000	
U	Operational Savings	$1,078,076	$1,100,000	
V	Less Turnover	$68,593	$110,000	
W	More Profit	$4,334,453	$5,500,000	

as shown in Figure 7.4. However, because of the effect of essential links, the overall contribution would remain as calculated above: $3,124,000, *generating a LOSS of almost $1 million!*

As with the previous example, this analysis underlines still further the importance of determining and understanding essential links and underlines the lesson: *If essential links are overlooked, the numerical basis for decision making is likely to be fatally flawed.*

7.3.3 An Alternative Remedy

An alternative to removing node B would be to extend the scope of the program in such a way as to add components that benefit from the integration of store records. The negative risk in doing this is that you may be indulging in scope creep to create an unwarranted justification for a given initiative. The potential upside, however, is that you identify beneficial outcomes and potentially contributory initiatives that you may have missed in your original analysis. In the current example, a brainstorming session on the value of integrated store records could lead to a number of creative ideas such as analyzing business trends per district in order to enhance the marketing plans.

This reworking of the overall BRM must take place before the final business case has been approved and any definitive expectations have been set. If the need for reworking the approach and the BRM appears later in the lifecycle, the amount of rework may exceed the value of the redesign.

Table 7.5 Allocations and Contributions when Only Node B is Removed

		Allocation	Contribution	
A	Standardize Processes	$900,000	$1,065,630	
B	Integrate Store Records	$0	$0	By Hypothesis
C	Adapt Tracking System	$1,500,000	$1,642,460	
D	Develop Joint Marketing	$500,000	$1,955,200	
E	Align Compensation	$200,000	$381,280	
F	Train Staff	$1,000,000	$1,099,800	
G	Process Model Available	$900,000	$1,065,630	
H	Integrated CRM	$900,000	$1,065,630	
I	Integrated Tracking	$1,500,000	$1,642,460	
J	Flexible Working Practices	$915,626	$1,028,370	
K	Increased Market Visibility	$706,413	$2,199,600	
L	All Stores Same	$1,277,961	$1,435,320	
M	Staff Rewards Defined	$200,000	$381,280	
N	Staff Trained and Capable	$1,000,000	$1,099,800	
O	Effective Operations	$2,129,935	$2,392,200	
P	More Customers	$3,378,052	$5,297,370	
Q	Happy Staff	$263,652	$452,770	
R	Great Service	$2,671,638	$3,097,770	
S	Higher Market Share	$945,412	$1,482,570	
T	Higher Revenue	$2,432,639	$3,814,800	
U	Operational Savings	$721,948	$847,000	
V	Less Turnover	$60,851	$104,500	
W	More Profit	$3,154,588	$4,661,800	
	TOTALS	$4,100,000	$6,144,370	

7.3.4 Insights for Program Management

One point that is important that this example underlines is that initiative B = Integrate Store Records cannot be considered in isolation within the program:

- On its own, it does not financially justify its inclusion in the program since it represents a loss of $434,000 (cost of $1,500,000 for a return of $1,066,000).
- If it is removed, it leads to a drop in value of the program of $4,086,000.
    - For a reduction in program costs of $2,900,000
- Its removal, however, ensures that each remaining initiative has a positive business justification when considered on its own, *but that reduced, the program as a whole is at best marginally interesting from a business point of view.*

It should be understood that the set of three initiatives (A = Standardize Processes, B = Integrate Store Records, plus D = Develop Joint Marketing) form an *emergent subprogram* that costs $2,900,000 and delivers $4,086,000 of value. This profit-making result can only be achieved if this complete A-B-D subprogram is carried out within the context of the rest of the program. This interdependency effect is shown in the modified diagram below (Figure 7.6). The diagram is created by incorporating the relevant

Table 7.6 Result on the Modified BRM of Taking the Essential Links into Account

		Allocation	Contribution	
A	Standardize Processes	$0	$0	By Calculation
B	Integrate Store Records	$0	$0	By Hypothesis
C	Adapt Tracking System	$1,500,000	$1,642,460	
D	Develop Joint Marketing	$0	$0	By Calculation
E	Align Compensation	$200,000	$381,280	
F	Train Staff	$1,000,000	$1,099,800	
G	Process Model Available	$0	$0	By Calculation
H	Integrated CRM	$0	$0	By Calculation
I	Integrated Tracking	$1,500,000	$1,642,460	
J	Flexible Working Practices	$626,116	$685,580	
K	Increased Market Visibility	$0	$0	By Calculation
L	All Stores Same	$873,884	$956,880	
M	Staff Rewards Defined	$200,000	$381,280	
N	Staff Trained and Capable	$1,000,000	$1,099,800	
O	Effective Operations	$1,456,474	$1,594,800	
P	More Customers	$2,191,728	$2,529,540	
Q	Happy Staff	$243,526	$428,940	
R	Great Service	$2,191,728	$2,529,540	
S	Higher Market Share	$613,397	$707,940	
T	Higher Revenue	$1,578,331	$1,821,600	
U	Operational Savings	$508,272	$594,000	
V	Less Turnover	$56,206	$99,000	
W	More Profit	$2,086,603	$2,415,600	
	TOTALS	$2,700,000	$3,123,540	

nodes (A, B, and D) into a subprogram (called ABD in Figure 7.6), and including the other nodes that were identified for pruning. The new BRM for the total program is created by linking subprogram ABD back into the pruned network shown in Figure 7.4. The corresponding links from subprogram ABD to the nodes in the pruned network, with their corresponding percentages, are then reconnected back to this subprogram.

The strategic objectives of the subprogram considered on its own are the contribution shares corresponding to the subprogram's links into the overall program. These subprogram strategic objectives are shown in Figure 7.7.

7.3.5 Resolving Subprograms

As shown in this example, a subprogram could be formally defined as "a separable subset of a program." The term *separable* indicates the following:

- The initiatives of the subprogram connect only to the nodes of the subprogram, with no links to other nodes (in Figure 7.7, these initiatives are A = Standardize Processes, B = Integrate Store Records, and D = Develop Marketing Program).

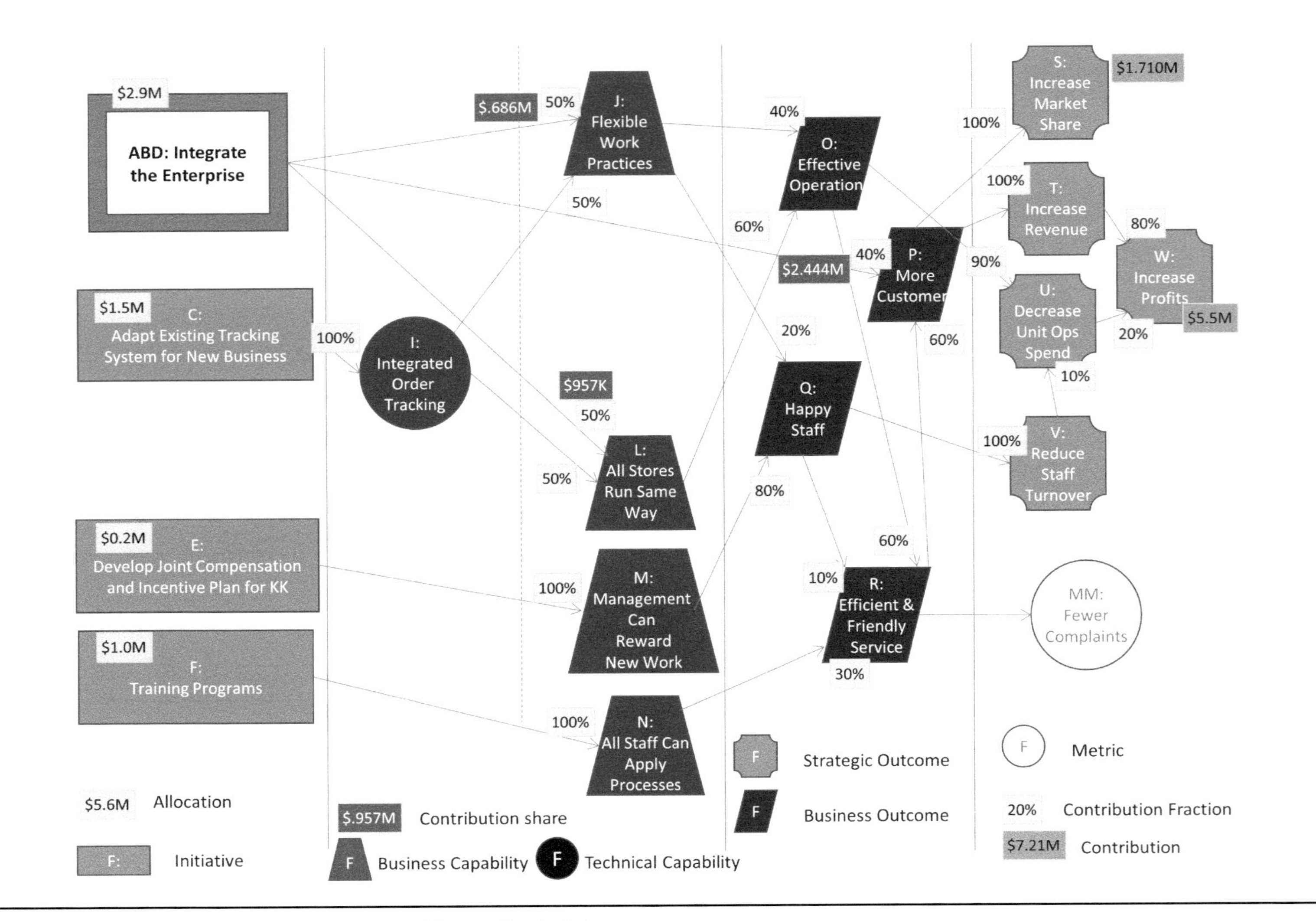

Figure 7.6 BRM Showing Initiatives A, B, and D as a Single Subprogram

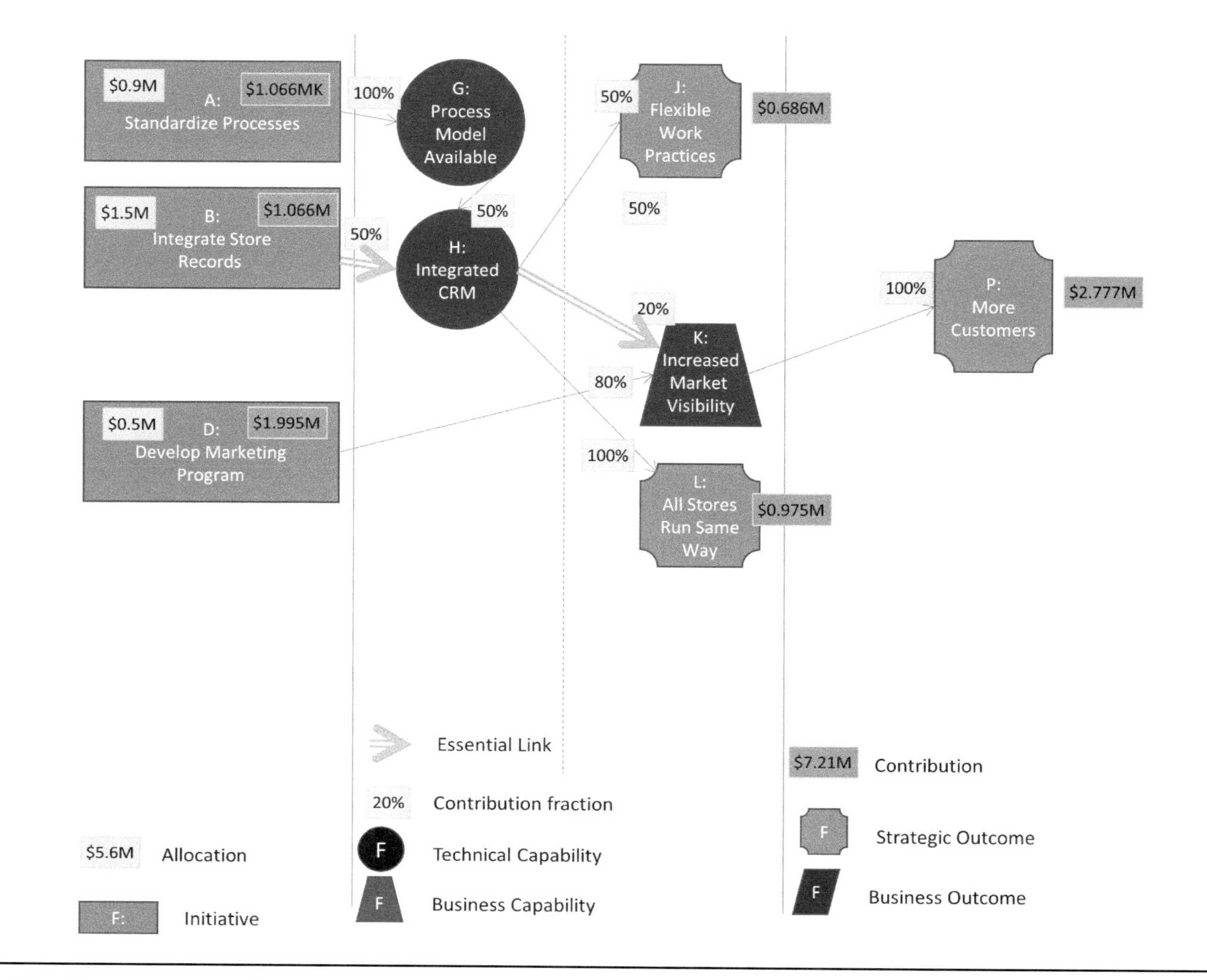

Figure 7.7 The ABD = "Integrate the Enterprise" Subprogram

- There are no links from the overall program into the nodes of the subprogram other than to its terminal nodes—that is, the nodes that represent the strategic outcomes of the subprogram (in Figure 7.7, these are J = Flexible Work Practices, P = More Customers, and L = All Stores Run the Same Way). For example in the full program, P = More Customers is also enhanced by R = Friendly and Efficient Service as shown in Figure 7.6.
- The internal nodes of the subprogram do not link directly to any nodes outside the subprogram. In Figure 7.7, the internal nodes are G = Process Model Available, H = Integrated CRM, and K = Increased Market Visibility.

The BRM for the subprogram can be developed directly from the BRM for the whole program as follows:

- All of the links between nodes are maintained.
- The contributions are calculated as follows:
  - For the initiatives and the internal nodes, the contribution is identical to the contribution in the program map.
  - All of the contribution shares are unchanged.
  - For each terminal node, the contribution is the sum of the contribution shares of incoming links from the subprogram.
  - For each link, the contribution fraction is (node contribution) / (contribution share of that link).
- The allocations for the resulting BRM can then be calculated using the algorithm explained in Chapters 5 and 6.

7.3.6 Insights from Subprograms

Subprogram ABD = Integrate the Enterprise as shown in Figure 7.7 provides a typical example of the synergies and interdependencies that can and should occur due to the program management approach. This insight into program synergy links back to the discussion on the difference between projects and programs in Chapter 1.

This program synergy underlines the different roles and responsibilities between project managers and program managers. If project managers were required to sign off the business case for the project to which they are assigned, in the current example, they would be most likely to veto B = Integrate Store Records, the "loss-making" project, on financial grounds. In the current situation, that would lead to an overall loss to the program of $1 million. To repeat the earlier message: Project managers should be accountable for the cost and capabilities of what the project delivers; it is up to the project sponsor—in this case, the program manager—to ensure that this delivery is justified based on the broader business perspective.

The other dangerous blind spot that can occur in developing the business cases for programs is the topic of disbenefits—that is, the potential downside or negative side effect of certain initiatives or outcomes. The way in which disbenefits affect this merger example is addressed next.

7.4 Including Disbenefits in the QERTS Example

The initial concepts associated with disbenefits were presented in Chapter 6. In that example, one specific outcome made a negative contribution to the result. In the modification of the QERTS BRM presented below, we consider the more habitual—and more troublesome—case where a given outcome has a negative effect on one node (i.e., is a disbenefit from that point of view) but, at the same time, has positive effects on other nodes (i.e., contributes to a benefit).

The change to the QERTS BRM in Figure 7.1 is that we postulate that we have discovered that the plan to standardize all of the stores (Node L) will upset a proportion of the staff who came from the QE (Quick and Easy) part of the merger. In fact, the negative impact is estimated as halving staff happiness. As shown in Figure 7.8, two nodes contribute positively to node L = Happy Staff. The connected nodes are J = Flexible Work Practices (which makes a 20 percent contribution to staff happiness) and M = Management Can Reward New Work (80 percent).

It is important to note an addition to the rules governing contribution fractions in the case of disbenefits. The *benefits* percentages must still add up to 100 percent as in the case without disbenefits. The disbenefits percentages are simply added into the BRM without any change to the values of the contribution fractions of the benefits. The addition of the specified disbenefits is shown in Figure 7.8. Including the negative contribution fractions in this way means that the total of all contribution fractions will be less than 100 percent for any node affected by disbenefits.

The algorithm described in Chapter 5 can be applied to this diagram that includes the disbenefit. However, due to the fact that the sum of percentages at the node affected by the disbenefit is less than 100 percent, the reverse calculation from left to right no longer satisfies the BEER rule, and, as a consequence, the recalculated values of the contributions (calculated left to right using the allocation percentages) are different from the values initially calculated right-to-left based on the contribution percentages. The difference between the two numbers reflects the effect of the disbenefit on the original assumptions of the value of the strategic outcome. The corresponding list of contributions and allocations is shown in Table 7.7 and discussed in detail in the following section.

An analysis of Table 7.7 shows a number of interesting features of the situation caused by including the disbenefit—that is, the reduction of staff happiness due to store standardization.

In numbers, the net financial benefit of the original BRM is $7,210,000 – $5,600,000 = $1,610,000. The disbenefit has reduced this to $6,971,700 – $5,600,000 = $1,371,700—that is, a decrease of almost a quarter of a million dollars. The ROI has dropped from 129 percent to 124 percent, which does not seem so dramatic.

A more detailed analysis provides additional insights:

- It is no surprise that the value of none of the nodes has increased.
- Because the disbenefit is due to reduced flexibility, as would be expected, the contributions of the nodes directly associated with the standardization (A = Standardize Working Practices, B = Integrate Store Records, and C = Adapt Tracking System) have been reduced.
    - The same can be said for the outcomes from these nodes (G = Process Model Available, H = Integrated CRM, I = Integrated Tracking, and, of course, L = All Stores the Same).
- The indirect effect from this downside from standardization is to reduce the contribution (i.e., the value derived from) all of the downstream outcomes.
- The only nodes that are not affected are those associated with marketing, compensation, and training:
    - D = Develop Joint Marketing, E = Align Compensation, and F = Train Staff, which generate:
        - J = Flexible Working Practices, K = Increased Market Visibility, M = Staff Rewards Defined, N = Staff Trained and Capable, and O = Effective Operations.
- However, to preserve this downstream value, the cost (i.e., the allocation) attributed to some of the nodes has increased. This increase in allocation is shown in Table 7.7 for:
    - J = Flexible Working Practices, K = Increased Market Visibility, O = Effective Operations.

7.4.1 Mitigating the Disbenefits in the QERTS Example

It is not sufficient to understand and analyze a problem. A positive response needs to be developed. Ideally, the management reaction to this type of analysis should be to examine the options for

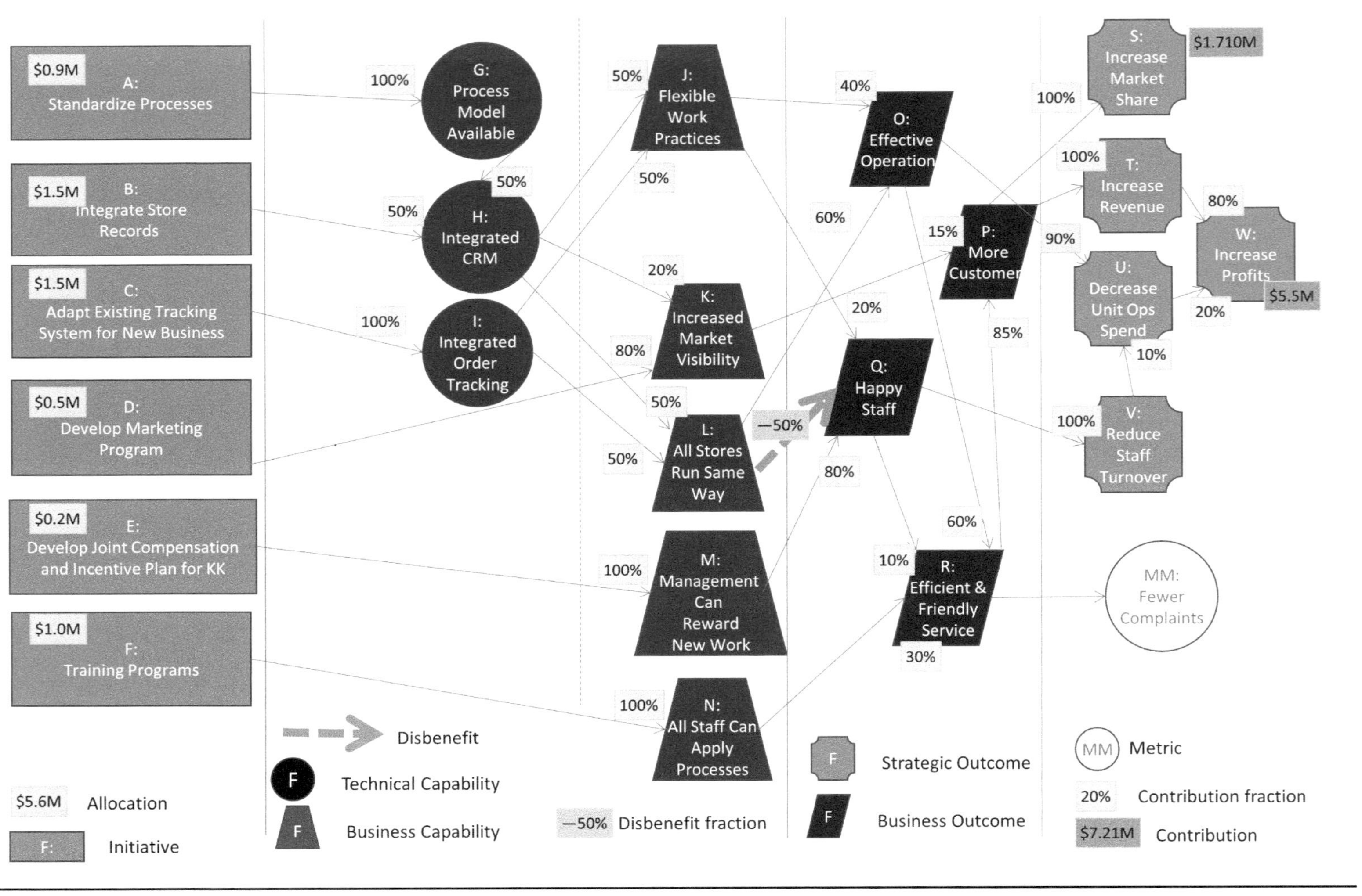

Figure 7.8 BRM with the Addition of the Disbenefit from L to Q

Table 7.7 Allocations and Contributions in the Disbenefit Case

		Allocation	Contribution		Change in Allocation	Change in Contribution
A	Standardize Processes	$900,000	$1,006,055		$0	-$59,575
B	Integrate Store Records	$1,500,000	$1,006,055	Cost>Value	$0	-$59,575
C	Adapt Tracking System	$1,500,000	$1,523,310		$0	-$119,150
D	Develop Marketing Program	$500,000	$1,955,200		$0	$0
E	Deliver Compensation Program	$200,000	$381,280		$0	$0
F	Train Staff	$1,000,000	$1,099,800		$0	$0
G	Process Model Available	$900,000	$1,006,055		$0	-$59,575
H	Integrated CRM	$2,400,000	$2,012,110	Cost>Value	$0	-$119,150
I	Integrated Order Tracking	$1,500,000	$1,523,310		$0	-$119,150
J	Flexible Practices	$1,492,834	$1,371,160	Cost>Value	$94,690	$0
K	Increased Market Visibility	$1,083,030	$2,444,000		$32,595	$0
L	All Stores Run the Same Way	$1,824,137	$1,675,460	Cost>Value	-$127,285	-$238,300
M	Management Can Reward	$200,000	$381,280		$0	$0
N	All Staff Can Apply Processes	$1,000,000	$1,099,800		$0	$0
O	Effective Operations	$3,472,638	$3,189,600	Cost>Value	$220,269	$0
P	More Customers	$4,511,918	$5,926,700		-$10,007	-$183,300
Q	Happy Staff	$44,332	$238,300		-$252,864	-$238,300
R	Efficient & Friendly Service	$3,428,888	$3,482,700		-$42,601	-$183,300
S	Increased Market Share	$1,262,746	$1,658,700		-$2,801	-$51,300
T	Increased Revenue	$3,249,171	$4,268,000		-$7,206	-$132,000
U	Decreased Operational Spend	$1,088,082	$1,045,000	Cost>Value	$10,007	-$55,000
V	Reduced Staff Turnover	$10,232	$55,000		-$58,361	-$55,000
W	Increased Profits	$4,337,254	$5,313,000		$2,801	-$187,000
	Totals	$5,600,000	$6,971,700		$0	-$238,300

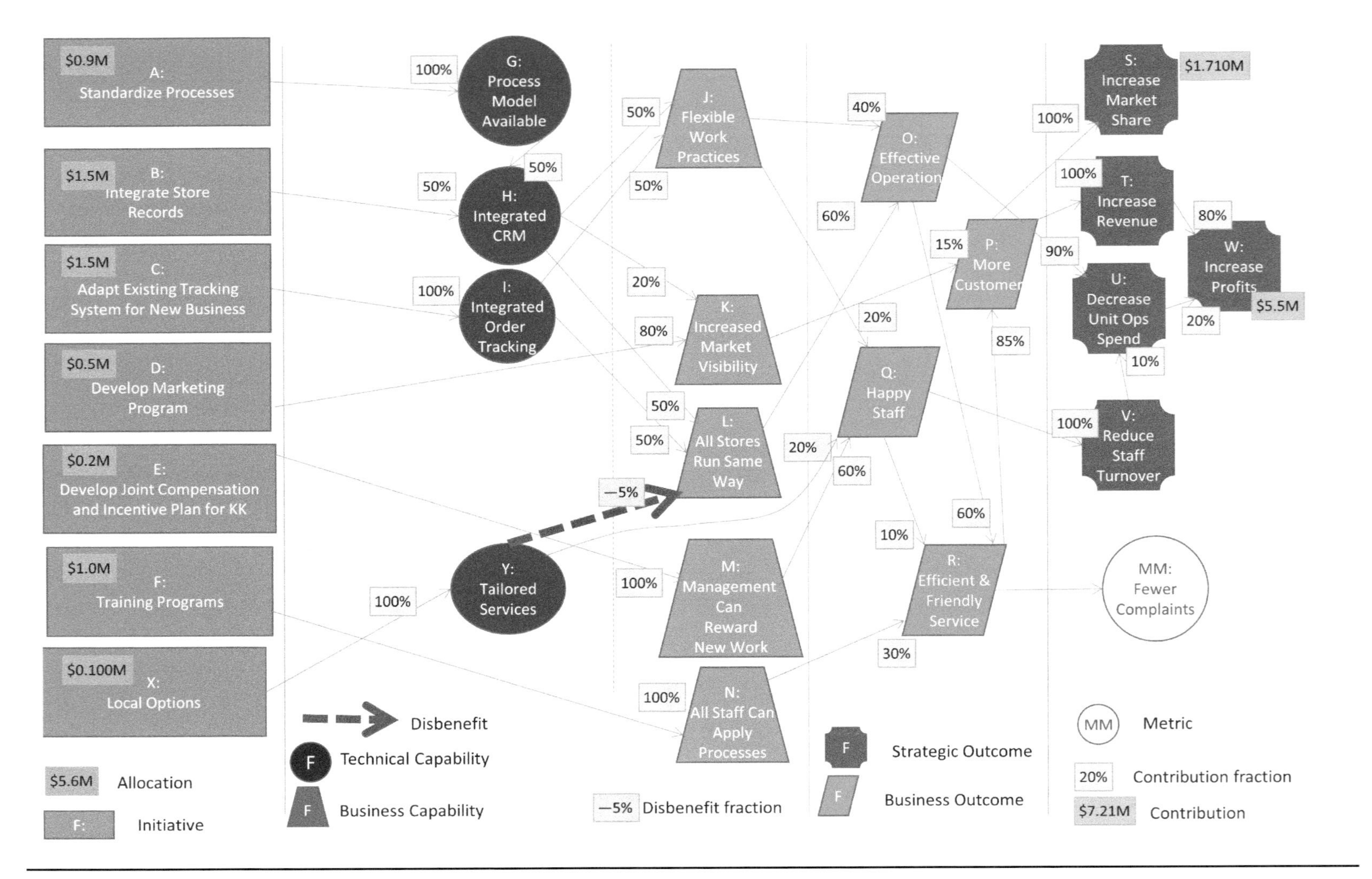

Figure 7.9 Expanded Program to Mitigate Standardization Disbenefit

responding to these staff morale issues. The current example will analyze the option of adding a specific initiative aiming to provide some of the flexibility and creativity that the ex-QE staff require. In this case, the disbenefit due to standardization will then decrease, but an additional cost in time and effort (at least) would be added to the overall cost of the program. This additional cost would help to mitigate the loss of market share and the corresponding impact on profit for the business. This option would allow management to address a problem that, if allowed to occur, might become an intractable issue.

The modified BRM to incorporate this mitigation would look as shown in Figure 7.9, with the addition of a new initiative, X = Local Options, at an additional cost of $100,000. This additional initiative delivers the capability Y = Tailored Services. Y has three effects, two of which are positive and one, negative:

- It eliminates the disbenefit caused by all stores being the same.
- In addition, it contributes directly to the level of happiness by 20 percent.
- However, it adds a new disbenefit in that it reduces the benefit of standardization by 5 percent (link Y to L).
  - The importance of identifying potential disbenefits in any mitigation strategy should always be borne in mind!

The numerical calculations on the effect of the expanded program are shown in Table 7.8. This table now shows that the total cost of the program has increased by $100,000 to $5,700,000, but that the values as shown by the sum of the contributions of nodes S and W are now $7,114,000, giving an overall increase in value with respect to the unmitigated case of $143,000—a net return on this investment of 43 percent.

Some remarks can also be made on the effect of the changes to the business situation based on the figures in Table 7.8:

- The contribution of E = Align Compensation has dropped by over 20 percent with the corresponding reduction in M = Staff Rewards Defined. This reduction occurs because of the additional contribution of Y = Tailored Services to staff happiness (node Q).
- The contribution of O = Effective Operations has been reduced due to the fact that not all stores are now run the same way due to the tailored services (node Y).

This example also calls for a comment on organizational management and the importance of people for program success. The mitigation approach adopted to address the issue—in this case, taking staff issues to heart and acting on them—also has the unquantified benefit of showing that the joint management of the integrated QERTS company is willing to take the needs and expectations of the previous QE staff into account and invest in supporting their working conditions. The fact that this actually has a positive ROI makes it a real win–win situation.

The result of this analysis on the original BRM are shown in Table 7.9, with the following results:

- The overall cost has of course increased by the $100,000 cost of the mitigation action (X = Local Options).
- The total value has dropped by $95,688.
  - Due to reduction in the value of both market share and profit

Analysis of the reductions in the contributions of the intermediate nodes shows the following:

- Although the Y = Tailored Services do contribute to Q = Staff Satisfaction, the mitigation approach has reduced this by 25 percent (the contribution fraction in the original case was 80 percent and, with mitigation, it has dropped to 60 percent).

Table 7.8 Allocations and Contributions for the Disbenefits Mitigation Case Relative to the Unmitigated Situation

		Allocation With Mitigation	Contribution with Mitigation		Change in Allocation	Change in Contribution
A	Standardize Processes	$900,000	$1,065,630		$0	$59,575
B	Integrate Store Records	$1,500,000	$1,065,630	Cost>Value	$0	$59,575
C	Adapt Tracking System	$1,500,000	$1,642,460		$0	$119,150
D	Develop Marketing Program	$500,000	$1,955,200		$0	$0
E	Deliver Compensation Program	$200,000	$285,960		$0	-$95,320
F	Train Staff	$1,000,000	$1,099,800		$0	$0
G	Process Model Available	$900,000	$1,065,630		$0	$59,575
H	Integrated CRM	$2,400,000	$2,131,260	Cost>Value	$0	$119,150
I	Integrated Order Tracking	$1,500,000	$1,642,460		$0	$119,150
J	Flexible Practices	$1,398,144	$1,371,160	Cost>Value	-$94,690	$0
K	Increased Market Visibility	$1,050,435	$2,444,000		-$32,595	$0
L	All Stores Run the Same Way	$27,953,595	$1,818,072	Cost>Value	$26,129,459	$142,612
M	Management Can Reward	$200,000	$285,960		$0	-$95,320
N	All Staff Can Apply Processes	$1,000,000	$1,099,800		$0	$0
O	Effective Operations	$29,254,543	$3,093,912	Cost>Value	$25,781,905	-$95,688
P	More Customers	$2,529,533	$6,044,012		-$1,982,385	$117,312
Q	Happy Staff	-$25,604,978	$476,600	**Negative Allocation**	-$25,649,310	$238,300
R	Efficient & Friendly Service	$1,479,098	$3,600,012		-$1,949,790	$117,312
S	Increased Market Share	$707,938	$1,691,532		-$554,808	$32,832
T	Increased Revenue	$1,821,595	$4,352,480		-$1,427,577	$84,480
U	Decreased Operational Spend	$3,170,467	$1,070,300	Cost>Value	$2,082,385	$25,300
V	Reduced Staff Turnover	-$5,909,668	$110,000	**Negative Allocation**	-$5,919,900	$55,000
W	Increased Profits	$4,992,062	$5,422,780		$654,808	$109,780
X	Local Options	$100,000	-$368	Cost>Value	$100,000	-$368
Y	Tailored Services	$100,000	-$368	Cost>Value	$100,000	-$368
		$5,700,000	$7,114,312		$100,000	$142,612

Table 7.9 Comparing the Original BRM with the BRM after Disbenefits Mitigation

		Allocation With Mitigation	Contribution with Mitigation		Change in Allocation	Change in Contribution
A	Standardize Processes	$900,000	$1,065,630		$0	$188,934
B	Integrate Store Records	$1,500,000	$1,065,630	Cost>Value	$0	$188,934
C	Adapt Tracking System	$1,500,000	$1,642,460		$0	$72,368
D	Develop Marketing Program	$500,000	$1,955,200		$0	$1,222,000
E	Deliver Compensation Program	$200,000	$285,960		$0	-$633,000
F	Train Staff	$1,000,000	$1,099,800		$0	-$1,134,556
G	Process Model Available	$900,000	$1,065,630		$0	$188,934
H	Integrated CRM	$2,400,000	$2,131,260	Cost>Value	$0	$377,868
I	Integrated Order Tracking	$1,500,000	$1,642,460		$0	$72,368
J	Flexible Practices	$1,398,144	$1,371,160	Cost>Value	-$509,916	-$501,020
K	Increased Market Visibility	$1,050,435	$2,444,000		$299,538	$1,527,500
L	All Stores Run the Same Way	$27,953,595	$1,818,072	Cost>Value	$26,044,971	$175,632
M	Management Can Reward	$200,000	$285,960		$0	-$633,000
N	All Staff Can Apply Processes	$1,000,000	$1,099,800		$0	-$1,134,556
O	Effective Operations	$29,254,543	$3,093,912	Cost>Value	$25,304,459	-$1,012,188
P	More Customers	$2,529,533	$6,044,012		-$2,076,510	-$65,988
Q	Happy Staff	-$25,604,978	$476,600	Negative Allocation	-$26,039,121	-$672,100
R	Efficient & Friendly Service	$1,479,098	$3,600,012		-$2,376,048	-$1,593,488
S	Increased Market Share	$707,938	$1,691,532		-$581,151	-$18,468
T	Increased Revenue	$1,821,595	$4,352,480		-$1,495,359	-$47,520
U	Decreased Operational Spend	$3,170,467	$1,070,300	Cost>Value	$2,176,510	-$29,700
V	Reduced Staff Turnover	-$5,909,668	$110,000	Negative Allocation	-$5,951,241	$0
W	Increased Profits	$4,992,062	$5,422,780		$681,151	-$77,220
X	Local Options	$100,000	-$368	Cost>Value	$100,000	-$368
Y	Tailored Services	$100,000	-$368	Cost>Value	$100,000	-$368
	Total	$5,700,000	$7,114,312		$100,000	-$95,688

- This is not compensated directly by the 20 percent of contribution from Y = Tailored Services, as the disbenefit caused to L = All Stores Run the Same Way (which loses nearly $100,000 in value) actually removes all its direct value.
    - Such is the cost of mitigation!

This type of analysis shows the potential danger there would have been of only comparing the mitigated and the unmitigated cases. If that had been the only information provided, all we could have shown to the decision-makers was the idea of spending an extra $100,000 that would lead to a reduction in value of almost $100,000. That "business case" that would be rejected immediately: "What makes you think it is worth losing almost $200,000 just to satisfy the whims of newly acquired employees? What is the worst they can do?" What they will do has already been shown in Table 7.7: Their loss of happiness (and therefore of motivation) generates a BRM indicating a *consequential loss of $238,000—almost 20 percent higher than this $200,000 figure.*

These calculations allow management to make decisions based on all of the facts and impacts of any scenarios they may wish to investigate. The power of this approach is that the scenarios are all based on the original business analysis and therefore should be impervious to unwarranted pressure. The decisions and their impacts are also traceable from the start to the end of the program.

Despite this analysis of the disbenefit risk, for the purpose of the QERTS case study, it turns out that management does not agree to the additional investment of $100,000 because it would invalidate their initial budget expectations and create a "dangerous precedent" according to the finance department. The unmitigated case is therefore used in subsequent chapters.

Although these analyses and calculations provide a measure of what the plan will deliver, the management team also needs to have information about delivery dates. Scheduling concepts are addressed in the next chapter.

7.5 Summary

The worked example has shown a number of powerful features of the analysis that allow a program's business case to be optimized in a number of ways:

- Identifying *subsidized* initiatives—that is, initiatives that cost more than they contribute, although the overall program is financially justifiable
- Including the concept of *essential links*—that is, links to contributing nodes that must be available before the destination node can have any value
- Defining the way of removing subsidized initiatives even in the case of essential links
- Explaining the concept of "true" subprograms
- Addressing the concept and challenge of disbenefits:
    - The method for calculating the overall impact
    - A case-study example of an approach for mitigating the effects of a disbenefit

Chapter 8

A Generalized Approach to Scheduling and Cash Flow

To support the specific characteristics of programs, this chapter proposes a major conceptual change to the way in which the standard scheduling models are viewed and applied. It aims to rationalize and integrate the currently used approaches and structures for developing schedule models.

This approach is applied to the QERTS case study, and the way in which it can be used to forecast cash flow is explained.

The benefits realization map (BRM) provides a view of how the situation is expected to have changed once the program is complete and all of the benefits are accruing. However, the ramp-up period needs to be considered not only while the initiatives are being implemented but also beyond that date, since results do not necessarily appear the moment the previous entity completes.

It is important to model this ramp-up period in order to set the expectations of the various stakeholders that the time required to recover the initial investment can be considerable. Including the details of this slow build-up in the business case can help to ensure consistent stakeholder support throughout the lifetime of the program.

These delays, therefore, need to be evaluated and factored into the model.

Once these values are available, the standard scheduling tools can be used to calculate the corresponding dates. Although program scheduling can be carried out using any of the standard schedule planning methodologies and tools, a variation on the scheduling models is proposed below. The proposed approach is a generalization of current techniques as each known model is a special case of this method.

8.1 Background

The precedence diagramming method (PDM)—also called the "activity on node" method—for creating a scheduling model represents the activities by nodes and indicates the relationships between any two activities by an arrow between the corresponding nodes. By definition, the PDM builds up a model from predecessor and successor nodes. The concepts of predecessor and successor are logical when the relationship is *finish to start*—that is, the successor cannot start until the predecessor has finished, as shown below (Figure 8.1).

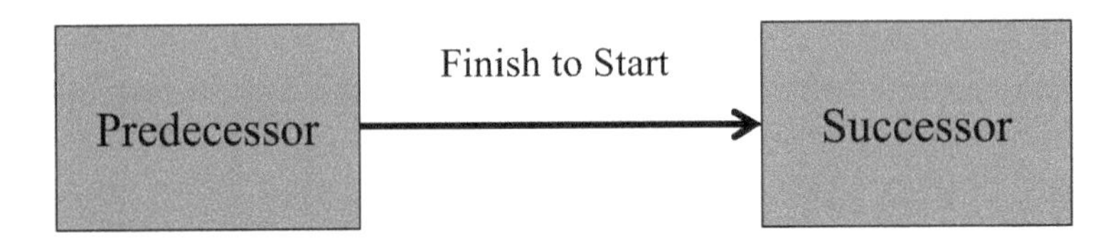

Figure 8.1 Finish to Start: A Respecter of Chronology

The definition becomes much less applicable, however, for other types of dependency. Consider, for example, *start to finish*, where the successor cannot finish until the predecessor has started (Figure 8.2). In this case, it is confusing, to say the least, to state that the logical relationship shows the sequence in which the activities are to be performed, since it linguistically reverses the arrow of time.

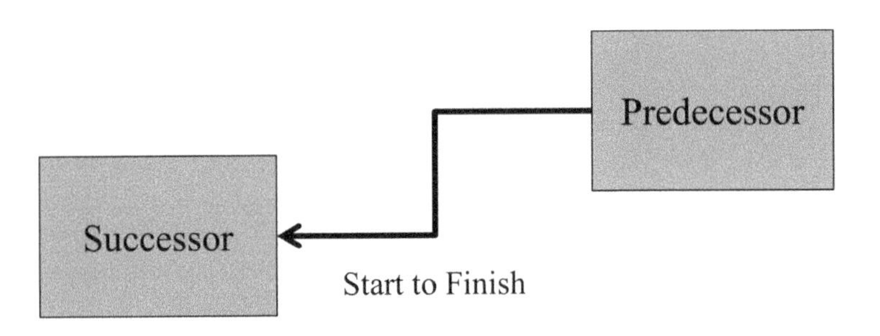

Figure 8.2 Start to Finish: Just What Does *Predecessor* Mean Here?

This predecessor–successor concern was the initial insight that led, through a few steps, to the development of a more consistent and simpler approach to schedule modeling. This chapter follows the same set of steps toward the final conclusion. The technique is then demonstrated to be well suited for program schedule management.

The problem comes from confusing activities with events, as discussed in the next section.

8.2 New View

So what is the answer? The answer is that it is not the activity itself that is the predecessor or successor; it is one of the two possible activity *events*—that is, an *activity start* event or *activity finish* event. The four possible relationships (FS = finish-to-start, SS = start-to-start, SF = start-to-finish, FF = finish-to-finish) correspond to the set of combinations of these two events.

This event-based view is also totally compatible with the major deliverable of the PDM: the calculation of the possible dates at which these events can occur—that is, the *early* and *late* values of the start and finish dates.

The only other conceptual step is to realize that not only dependency relationships but also activities can be represented by an arrow. This arrow links the activity start event to the finish event of the same activity. The duration value of this arrow is the duration of the activity.

This new approach makes explaining and understanding the algorithm for the critical path method (CPM) (Project Management Institute, 2011) much more straightforward, as it only needs to address the events and the links. In addition, the technique integrates the concepts of *activity on arrow* and *activity on node* representations in a natural manner to provide a consistent and generalized scheduling model.

8.3 The New Approach: Characterizing the Components

The new model has only two types of components:

- Events
- Arrows

These are explained in more detail below.

8.3.1 Events

Events are linked to activities; as such, there are two main types of events, with some additional uses, as follows:

1. Activity-related event types:
   a. *Activity start*, with two scheduling values
      i. Early start
      ii. Late start
   b. *Activity finish*, with two scheduling values
      i. Early finish
      ii. Late finish
2. Additional uses:
   a. *Project start*
   b. *Project finish*
   c. *Project milestone*, with two scheduling values:
      i. Earliest possible milestone date
      ii. Latest possible milestone date

8.3.2 Arrows as Vectors

The arrows link events and also separate them in time by representing the duration to be applied from the start to the end on the arrow. Since the arrows in this model have a direction and a length, the term *arrow* will be replaced in all that follows by the more technically correct term *vector*.

Vector Length

A vector can be used to separate any two events in the schedule and can represent different scheduling constructs, as follows:

1. *Activity*: This vector represents the duration of an activity.
   a. The ends correspond to the start and the end of the activity.
   b. The length corresponds to the estimated duration.
2. *Time buffer*: This vector represents a contingency buffer in general, or a critical chain buffer—a project buffer or feeding buffer. More details for critical chain project management (CCPM; Newbold, 1998) are given later.
   a. The length of the vector defines the contingency time that has been made available for all of the preceding activities in the project or in the chain.
3. *Dependency*: The dependency vector constrains and defines the next event in the path (there can be several such vectors from and/or to any given event). It can have a positive, negative or zero length.
   a. The length of the vector defines the lead (positive vector) or the lag (negative length) that applies a delay or an overlap between the events. For example, in Figure 8.3, there is a lag of 3 between the start event of Act 2 and the start event of Act 3, as well as a lead (–3) between the finish event of Act 3 and the start event of Act 4.

Vector Characteristics

All vectors impose a constraint on the events that they connect: The event at the far end of the vector cannot arise until the event at the origin has occurred and the duration implied by the vector has elapsed.

It is useful to add characteristics to each vector such as shown in the following non-exhaustive list:

1. The type of duration. Durations can be:
   a. Fixed (zero or non-zero)
   b. Resource-dependent (i.e., based on specified values of effort and assigned resources)
   c. Inherited (e.g., in a "hammock" where the duration is evaluated from the start and finish events, which are calculated from the rest of the network)
   d. Probabilistic (based a range of values)
   e. Correlated (dependent on other events)
2. Effort required (for effort-related activities)
3. The assigned resources:
   a. Name or category
   b. Percentage available of each name or category
   c. Budgetary information
4. Type of dependency:
   a. Mandatory dependency: A physical, unbreakable rule (e.g., undercoat prior to top coat when painting)
   b. Discretionary dependency: Applied tactically to "improve" the model (e.g., risk reduction, house rules, serving cheese before the fruit course, etc.)
   c. Resource exclusion (based on priority use of scarce resources)
5. Others, such as:
   a. Tracking rules (e.g., the way of defining and using "percent complete" for earned value calculations)
   b. Splitability (Can the activity be interrupted once it has started or does it have to be an uninterrupted duration?)

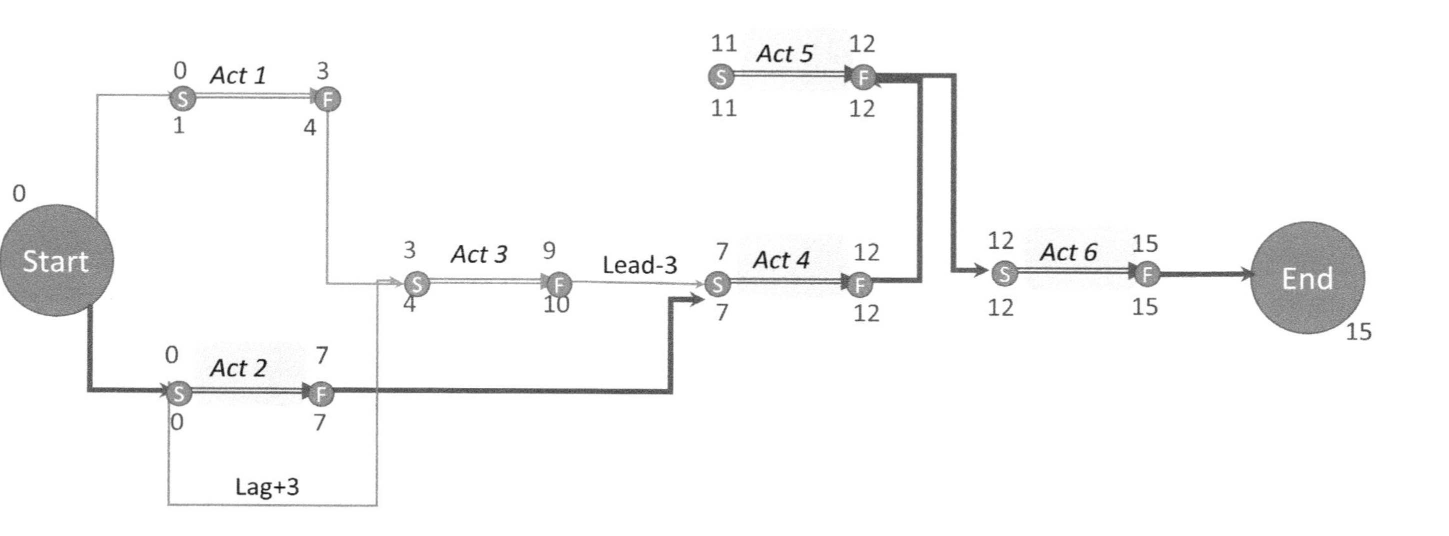

Figure 8.3 Example Applying the VEST Representation to the CPM

8.4 Toward a Simplified Scheduling Model

If you take this new model to its logical conclusion, you can effectively remove the activity node from the model by considering just the activity start and activity finish events. These events are linked by a start–finish dependency vector specifying the activity duration. Considering an activity as a vector in this way provides a new scheduling method as follows:

> *Vectorized event scheduling technique (VEST):* A technique used for constructing a schedule model, in which activities are represented by their start and finish events. The events are graphically linked by one or more vectors indicating relationships (activities and interevent constraints) along with their characteristics such as duration, resource impacts, and the like to specify the sequence in which the events—and, consequentially, the activities—are able to occur.

The diagram in Figure 8.3 applies this technique to the CPM on a simple example. The critical path is highlighted and shown in red. To make the correspondence clearer, activity vectors are drawn in a different format from dependency vectors, although this is not conceptually necessary.

This approach can then be applied not only to the CPM—on which the initial analysis above was based—but also to other classical scheduling methods.

8.5 Extensions to Other Scheduling Methods

The following methods can be integrated seamlessly into the VEST:

- Program evaluation and review technique (PERT)
- CCPM
- Milestone planning

8.5.1 PERT

The approach works directly with the PERT method once you understand that the PERT uses event-on-node but allows only finish-to-start dependencies, and restricts the types of vector to:

1. Activity.
2. Dummy (a zero-duration activity vector to allow path convergence to be represented in the PERT diagram): This is one of the well-known complicating drawbacks of the PERT model itself and is avoided by using the VEST, as shown in Figure 8.4.

8.5.2 CCPM

The feeding buffers, resource buffers, and program buffer required by the CCPM approach can also be incorporated into the VEST without any change to the overall concept. Each type of CCPM buffer is represented by a vector that defines the duration assigned to the buffer. The CPM example has been "converted" to a CCPM approach (ignoring resource constraints) by:

- Halving the activity estimates.
- Leaving the leads and lags unchanged.

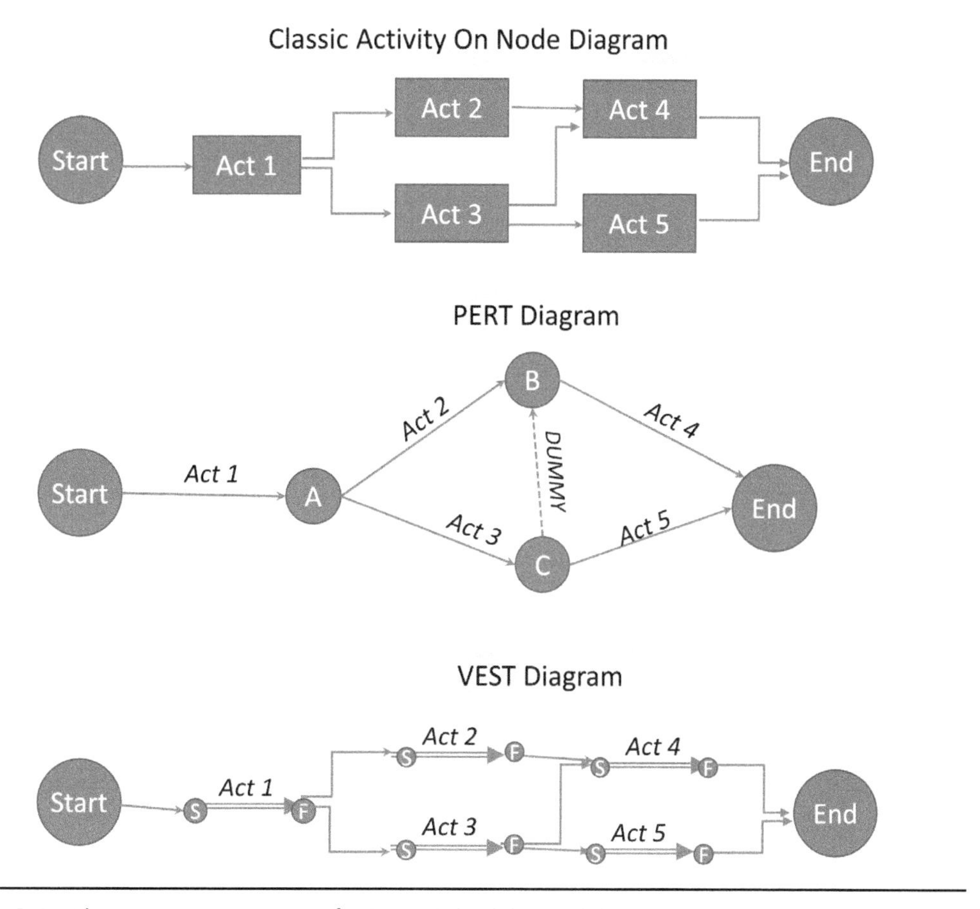

Figure 8.4 Three Representations of a Same Schedule Model

- Adding a feeding buffer at the point where the non-critical chain (Act 1 + Act 3) joins the critical chain:
  - The length of the feeding buffer is set to half the total duration of the preceding activities on this chain.
- Adding a program buffer just before the End node:
  - The length of the program buffer is set to half the total duration of the preceding activities on this chain.
- No resource buffers have been included in this example, as shown in Figure 8.5.

With the addition of the various buffers in the CCPM, the CPM concept of "float" is no longer relevant for critical chain evaluation, and all activities should start at the calculated Late Start date.

8.5.3 Milestone Planning

A milestone chart is in fact a classical event-on-node diagram of the PERT variety. The QERTS BRM, with the addition of dates, is an example of a milestone chart (see Figure 8.6).

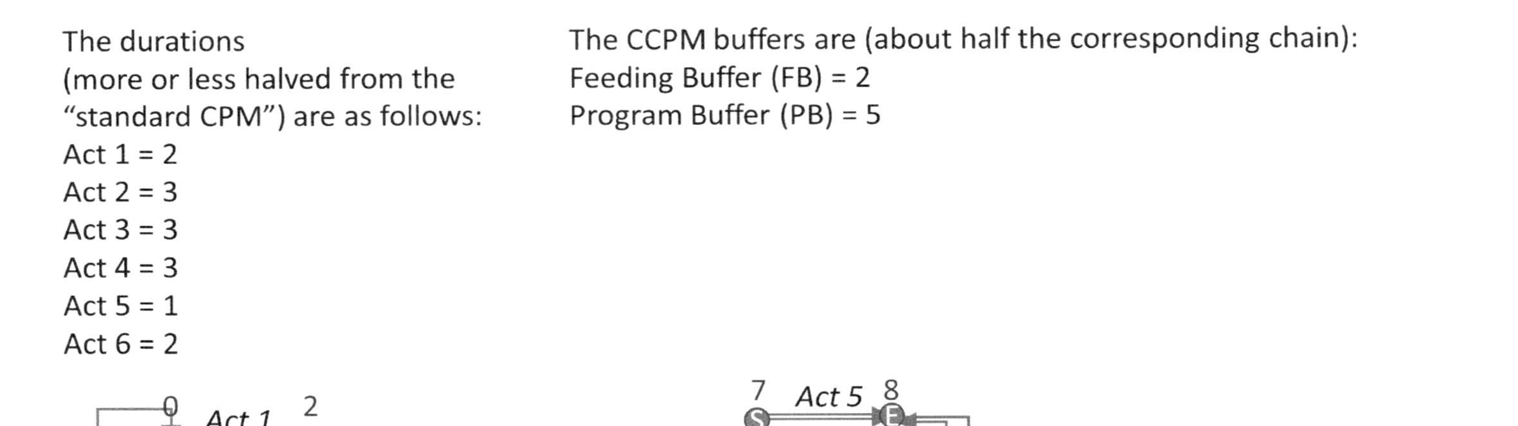

Figure 8.5 Applying CCPM to the Example, Using the VEST Representation

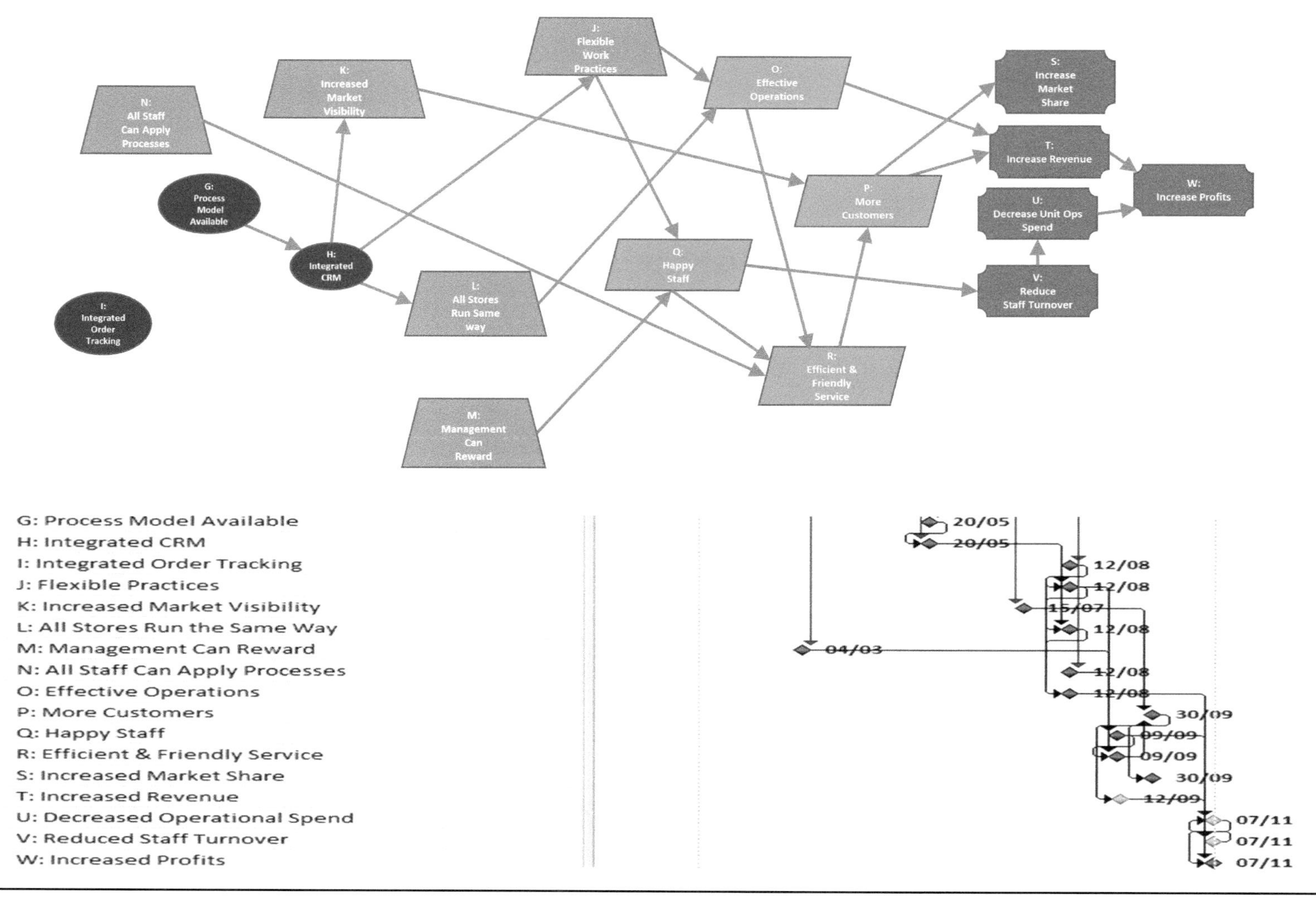

Figure 8.6 The QERTS BRM Milestone Chart as a PERT Diagram

8.6 Applying the VEST Approach to Programs

Programs show a number of characteristics for which the application the VEST approach is well-suited. The BRM shows not only components but also logical dependencies between the various categories of program components—initiatives, capabilities, outcomes, and benefits. Program schedule analysis does, however, require taking into account the fact that not only do the initiatives take time, but also that the time required for a deliverable to become a capability, for a capability to become an outcome, and for an outcome to deliver a given level of benefit has to be modeled.

With the VEST approach, each of these features can be represented as a vector and calculated in a consistent manner. The endpoint of each vector represents a potentially significant milestone toward achieving the strategic objectives of the program. As such, the VEST information provides the basis for effective communication for all categories of stakeholders in the organization. In addition, due to its general applicability, it allows the organization to use the most appropriate approach—waterfall, critical chain, agile, and the like—to manage each of the initiatives.

8.7 Additional Definitions

For completeness, the elements of this method can be defined and integrated with some of the PDM definitions as follows:

Event: A significant point in the schedule model.

Activity event: A significant point in the execution of an activity. There are two types: the *start event* and the *finish event.*

Start event: The point at which the execution of an activity begins.

Finish event: The point at which the execution of an activity completes.

Milestone event: The point at which a specific status is achieved.

Successor event (replaces the PDM concept of *successor activity*): An event that logically depends on the occurrence of a given event. It should more correctly be called a *dependent event. Note:* It is used to mean *direct successor*, rather than any of the subsequent events (that is to say that the successors to successors are not to be considered successor events in the context of this definition).

Predecessor event (replaces *predecessor activity*): An event that logically leads to a given event in the schedule. A better term would be a *driving event.* Note: A similar comment to the note on successor events applies.

Schedule network diagram: A graphical representation of the events and the logical relationships between them.

Relationship vector: An arrow indicating an event-to-event dependency constraint, along with the delay that this relationship imposes, such as an activity duration, a lag, a lead, a contingency buffer, and the like.

Interevent relationship: The logical relationship between the driving event and the dependent event. For activities, there are four such relationships: start-to-start (SS), finish-to-start (FS), start-to-finish (SF) and finish-to-finish (FF).

Activity: An element of work that takes some time and during which effort is expended in order to create an output.

Activity duration: The time separating the start and finish events of an activity.

Late date: In the critical path method, the latest point in time at which the corresponding (start or finish) event could occur without entailing a modification to the end date based on the existing schedule model. For activity scheduling, the late start and late finish dates are calculated.

Early date: In the critical path method, the earliest point in time at which the corresponding (start or finish) event could logically occur based on the existing schedule model. For activity scheduling, the early start and early finish dates are calculated.

8.8 Applying the VEST to the QERTS Example

8.8.1 Scheduling the Initiative Activities

The first step in developing the program schedule is to schedule the initiatives. Although the BRM seems to indicate that the initiatives in our example are independent of each other, this is not actually the case, as is shown in the following extract from the corresponding WBS (work breakdown structure) dictionary established by the QERTS technical team (Table 8.1).

Table 8.1 Extract from QERTS Transition Program WBS Dictionary

Task #	Name	
	Initiatives	**Relationship**
1	**A: Standardize Processes**	
2	Define QERTS Commercial Services	
3	Define Set of Processes	
4	Define Data Requirements for Processes and Tracking	2FS
5	Plan and Implement New Processes	3FS
6	Validate Processes	4FS
7	**B: Integrate Store Records**	8FS,5FS
8	Plan and Implement Records Integration	
9	**C: Adapt Tracking System**	4FS
10	Plan and Implement Tracking	
11	Validate Tracking	4FS
12	**D: Develop Marketing Program**	8FS,10FS
13	Plan and Implement Marketing Program	
14	Launch Marketing Program	2FS
15	**E: Deliver Compensation Program**	6FS,13FS
16	Plan and Implement Compensation	
17	**F: Train Staff**	2FS
18	Develop Training Program	
19	Roll Out Training	3FS,5FF

Table 8.2 QERTS Transition Program WBS Dictionary, Including Capabilities and Outcomes

Task #	Name	Start	Duration
	Initiatives		
1	**A: Standardize Processes**	**Mon 04/01/16**	
2	Define QERTS Commercial Services	Mon 04/01/16	3 wks
3	Define Set of Processes	Mon 25/01/16	3 wks
4	Define Data Requirements for Processes and Tracking	Mon 15/02/16	4 wks
5	Plan and Implement New Processes	Mon 14/03/16	2 mons
6	Validate Processes	Mon 09/05/16	2 wks
7	**B: Integrate Store Records**	**Mon 14/03/16**	
8	Plan and Implement Records Integration	Mon 14/03/16	6 wks
9	**C: Adapt Tracking System**	**Mon 20/06/16**	
10	Plan and Implement Tracking	Mon 20/06/16	6 wks
11	Validate Tracking	Mon 01/08/16	2 wks
12	**D: Develop Marketing Program**	**Mon 25/01/16**	
13	Plan and Implement Marketing Program	Mon 25/01/16	1 mon
14	Launch Marketing Program	Mon 23/05/16	1 mon
15	**E: Deliver Compensation Program**	**Mon 25/01/16**	
16	Plan and Implement Compensation	Mon 25/01/16	6 wks
17	**F: Train Staff**	**Mon 11/04/16**	
18	Develop Training Program	Mon 11/04/16	4 wks
19	Roll Out Training	Mon 23/05/16	3 mons
	Technical Capabilities		
20	G: Process Model Available	Fri 20/05/16	
21	H: Integrated CRM	Fri 20/05/16	
22	I: Integrated Order Tracking	Fri 12/08/16	
	Business Capabilities		
23	J: Flexible Practices	Fri 12/08/16	
24	K: Increased Market Visibility	Fri 15/07/16	
25	L: All Stores Run the Same Way	Fri 12/08/16	
26	M: Management Can Reward	Fri 04/03/16	
27	N: All Staff Can Apply Processes	Fri 12/08/16	
	Business Outcomes		
28	O: Effective Operations	Fri 12/08/16	
29	P: More Customers	Fri 30/09/16	
30	Q: Happy Staff	Fri 09/09/16	
31	R: Efficient & Friendly Service	Fri 09/09/16	
	Strategic Outcomes		
32	S: Increased Market Share	Fri 30/09/16	
33	T: Increased Revenue	Mon 07/11/16	
34	U: Decreased Operational Spend	Mon 07/11/16	
35	V: Reduced Staff Turnover	Mon 07/11/16	
36	W: Increased Profits	Mon 07/11/16	

As shown in Table 8.1, records integration (task 7) depends of the existence of the data requirements for processes and tracking (the result of task 4), as does the tracking system adaptation initiative (task 9); the marketing program (task number 12) and the compensation program (task number 15) depend on the results of defining the commercial services (task 2, which is carried out within the process standardization project). Similarly, the staff training initiative needs the defined set of processes (from task 3) and is also constrained by the availability of the processes (task 5) for the completion of the training material (see the FF relationship for task 18).

These relationships underline once again the interdependencies and synergies that are one of the key characteristics of programs that distinguish them from projects and from generic portfolios.

The full schedule for the program requires addressing not only the initiatives but also the capabilities and the outcomes generated by these initiatives. The schedule calculations will be based on the expanded WBS dictionary containing the capabilities and outcomes in the form of milestones. The relevant information from the full WBS dictionary is shown in Table 8.2.

The use of milestones to indicate capabilities and outcomes takes advantage of the potential characteristics of the VEST relationship vectors, as follows. Any delay between the finish event of an initiative and the milestone can be characterized to represent the evolution of the situation from the point of view of benefits realization. For example, the one-month delay between the launch of the marketing program (task 14) and the increased market visibility (task 24) is the time it will take for the information to be broadcast and enter into people's consciousness. Flexible practices (23) will take a month to be appreciated by the staff (30) and the actual effect of job-based compensation (26) will take a good two months before the effect can be seen; in each case, the build-up can be represented by a simple graph rather than an "all-or-nothing" step function.

8.8.2 Creating an Integrated Schedule

Creation of an integrated schedule includes the following steps:

1. Using the VEST, develop the schedule up to the end of the initiatives (Figure 8.7).
2. Include the capabilities that are directly linked to these initiatives (Figure 8.8).
3. Link the rest of the BRM as-is to the benefits milestones—with changes to the layout, if required, to increase readability (Figure 8.9).

8.9 Linking Cash-Flow to the Schedule

Whereas the schedule and the cost estimates for the initiatives can be used to evaluate the spending required for the program, the financial benefits from these initiatives only start to appear a certain time after the end of the initiatives. For the QERTS example, the various waiting times are shown as "lags" in the WBS dictionary (Table 8.2). The build-up from zero to the full contribution can be modeled, and the cumulative contribution curve as explained below can then be evaluated.

The approach for calculating the cumulative contribution curve is explained by reference to the QERTS example. The information required on each link is as follows:

- The lag and the start and end dates of the vector
- The ramp-up curve over that time
- The contribution fraction corresponding to the link
- The type of link: contributory or essential

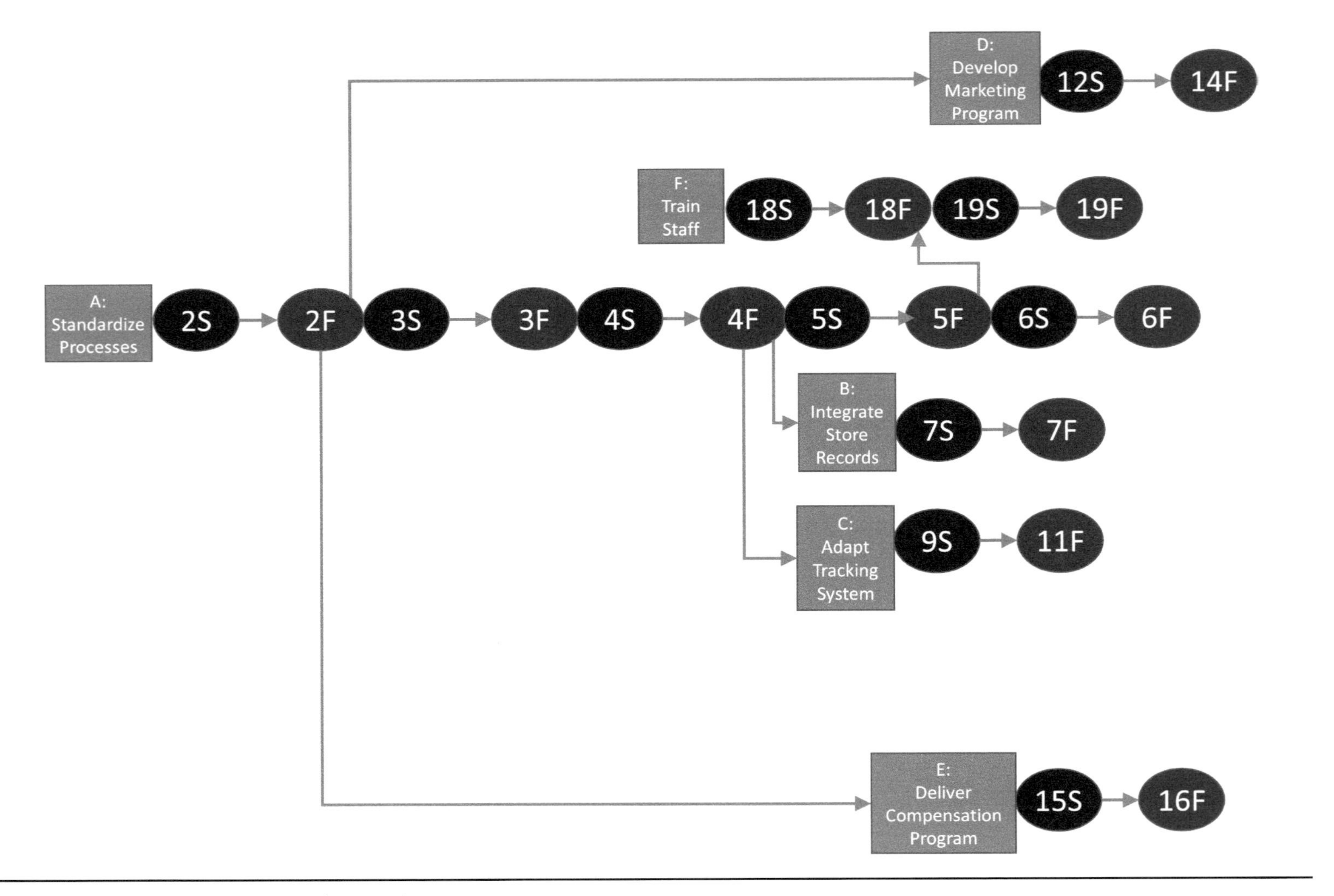

Figure 8.7 Applying the VEST to the Set of Initiatives

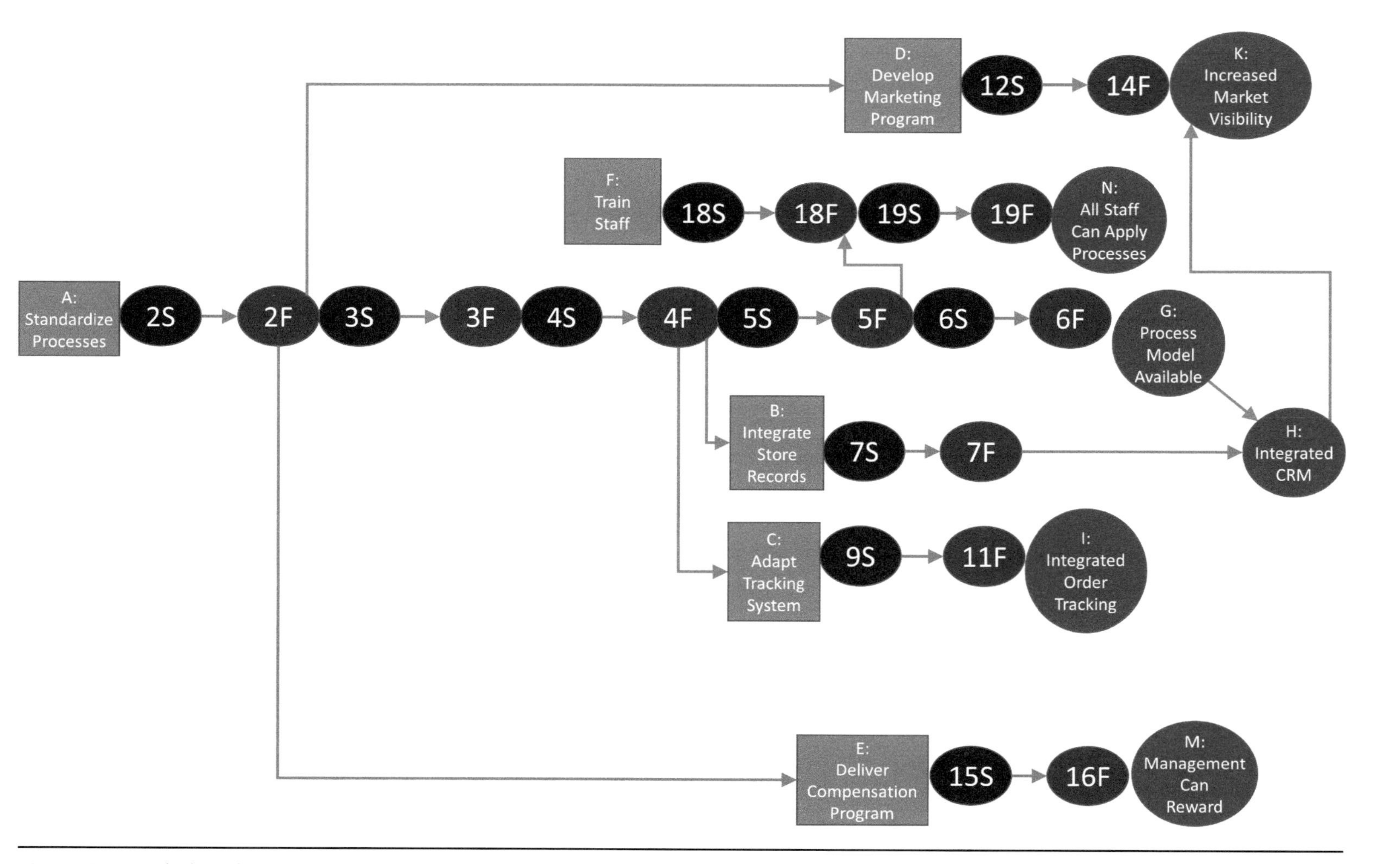

Figure 8.8 Including the Immediate Capabilities

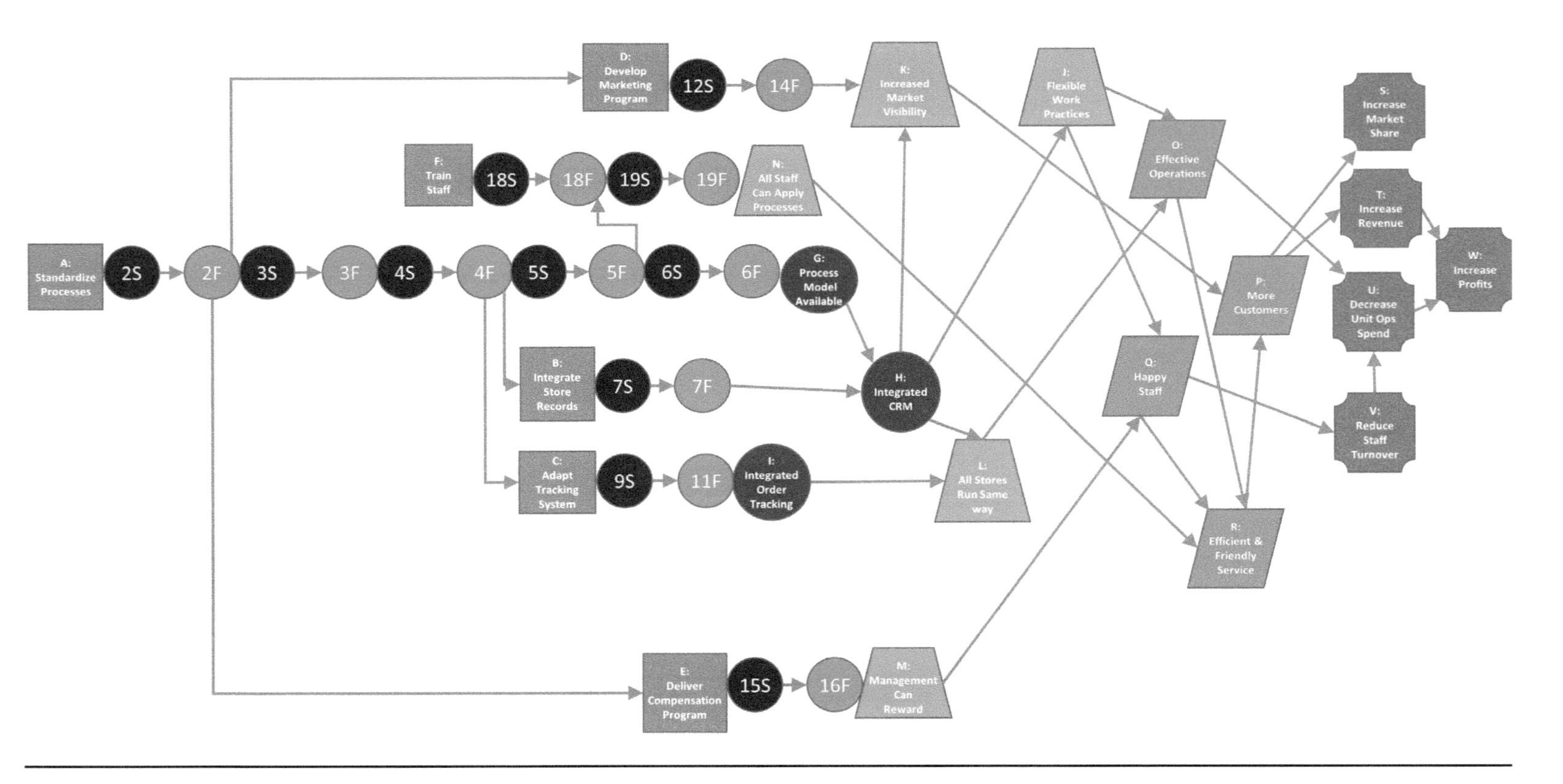

Figure 8.9 Appending the Rest of the BRM

The contributions to the benefits diffuse from left to right through the BRM, subject to the delay imposed by the ramp-up process. The calculations therefore have to be carried out from left to right based on chosen time intervals from the completion of each initiative.

8.9.1 Effective Dates

Since the effects are delayed across the BRM from left to right, all of the calculations have to be aligned to a set of *effective dates*. The effective date of a driving event with respect to a given dependent event at a given point in time is the date that defines the applicable value of the driving event. The effective date of each link (or, more formally, of the driving event of the corresponding dependency vector) is calculated from right to left by subtracting the lag (delay) associated with each vector from the date of the specific dependent event. For example, if the effect of the marketing program takes one month to fully increase market visibility, the degree of visibility on July 1 will depend on the status of the marketing program on June 1. This date determines the *effective contribution share* that should be used. The lags in the QERTS example are shown in Figure 8.10.

The calculation can be carried out as follows:

- Create a *calculation* table with the VEST events as rows and the calculation dates as columns. The cells should contain the contribution of the given event at the corresponding date. The contributions of the initiatives should be set to zero until the initiative is complete.
- Create an *effective date* table with the VEST links as rows and the calculation dates as columns. The cells should contain the effective date of that link. The effective dates can be used to identify the effective contribution shares when required.

The limit dates for the tables should be based on the calculated schedule from the start of the program until the date at which all of the benefits are delivering their full value.

For the QERTS example, assuming a start date of January 1, 2016, the schedule is as shown in Table 8.3. The range of dates for this example will be taken as January 1, 2016 until December 31, 2016 to cover the whole ramp-up period.

The starting point for calculating the values of the contributions at any point is, as before, based on the BEER approach of using the contribution values of the initiatives at a given date and calculating from left to right using the allocation fractions to find the equivalent contribution for each node. However, the calculation has to be adapted to take the *effective date* of each event into account. This provides the time for the result of one event to affect its dependent events in the case where a lag has been defined, as will be shown in the following example.

Based on the schedule dates and assuming calculating for performance at monthly intervals, the initial calculation table for the initiatives would be as shown in Table 8.4. This table is calculated from the solution to the BRM as evaluated in Chapter 7. It provides the value to be assigned to each initiative for the BEER calculations in the phase during which the initiatives are being implemented.

From January 1 until the of end of March, there are no completed initiatives, which means no possibility of accruing benefits. However, E = Deliver Compensation Program should finish on April 1, so we can calculate the effect of having the corresponding capability available. If we now calculate from left to right starting with the situation on April 1 shown in Table 8.4, we find non-zero values for:

- M = Staff Rewards Defined: $918,960
- P = More Customers: $830,960
- Q = Happy Staff: $918,960
- R = Great Service: $830,960

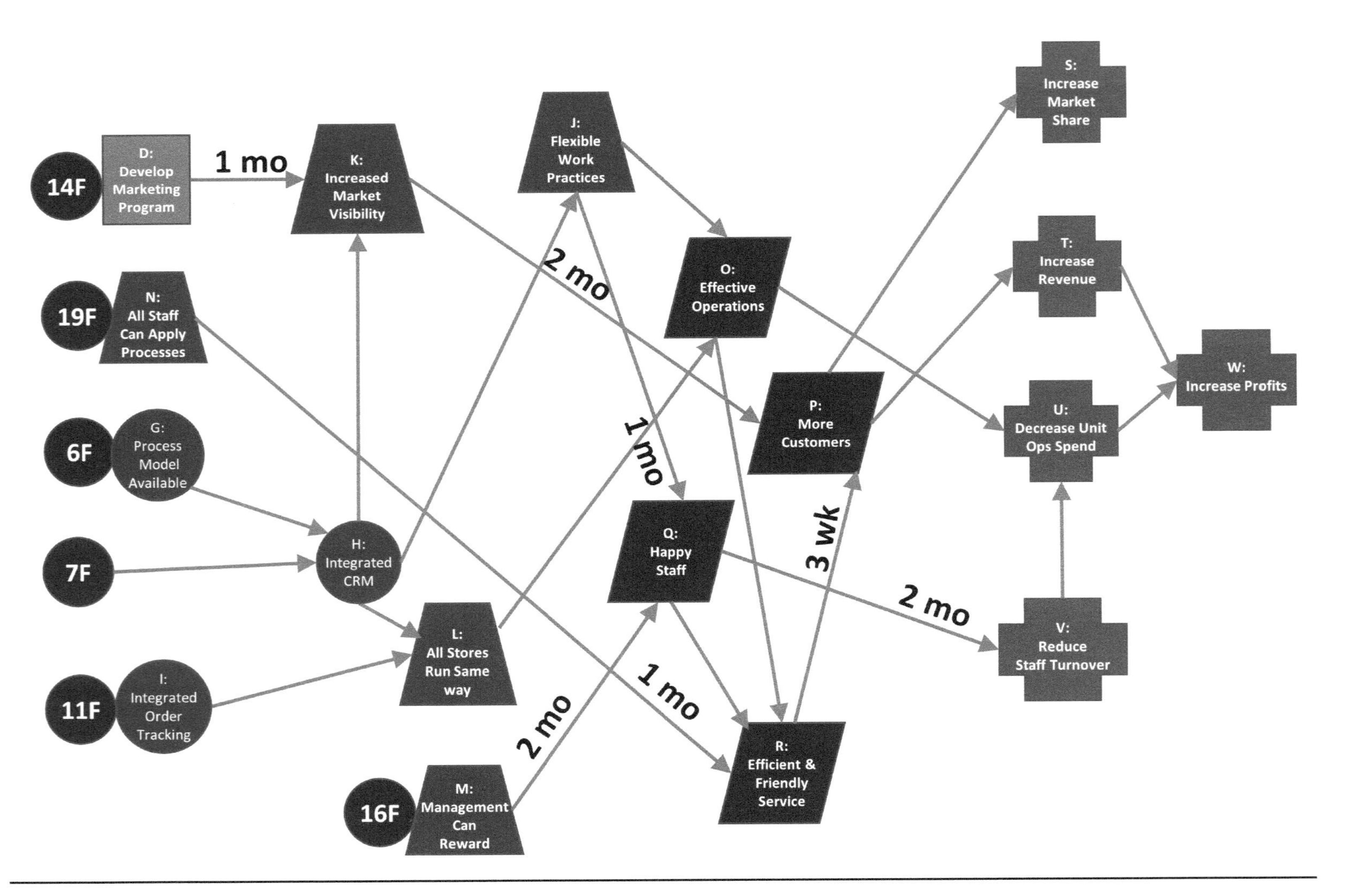

Figure 8.10 QERTS Example, Including Lags

Table 8.3 Scheduled Dates for the QERTS Example

Task #	Name	Start	Finish
	Initiatives		
1	**A: Standardize Processes**	**Mon 04/01/16**	**Fri 20/05/16**
2	Define QERTS Commercial Services	Mon 04/01/16	Fri 22/01/16
3	Define Set of Processes	Mon 25/01/16	Fri 12/02/16
4	Define Data Requirements for Processes and Tracking	Mon 15/02/16	Fri 11/03/16
5	Plan and Implement New Processes	Mon 14/03/16	Fri 06/05/16
6	Validate Processes	Mon 09/05/16	Fri 20/05/16
7	**B: Integrate Store Records**	**Mon 14/03/16**	**Fri 22/04/16**
8	Plan and Implement Records Integration	Mon 14/03/16	Fri 22/04/16
9	**C: Adapt Tracking System**	**Mon 20/06/16**	**Fri 12/08/16**
10	Plan and Implement Tracking	Mon 20/06/16	Fri 29/07/16
11	Validate Tracking	Mon 01/08/16	Fri 12/08/16
12	**D: Develop Marketing Program**	**Mon 25/01/16**	**Fri 17/06/16**
13	Plan and Implement Marketing Program	Mon 25/01/16	Fri 19/02/16
14	Launch Marketing Program	Mon 23/05/16	Fri 17/06/16
15	**E: Deliver Compensation Program**	**Mon 25/01/16**	**Fri 04/03/16**
16	Plan and Implement Compensation	Mon 25/01/16	Fri 04/03/16
17	**F: Train Staff**	**Mon 11/04/16**	**Fri 12/08/16**
18	Develop Training Program	Mon 11/04/16	Fri 06/05/16
19	Roll Out Training	Mon 23/05/16	Fri 12/08/16
	Technical Capabilities		
20	G: Process Model Available	Fri 20/05/16	Fri 20/05/16
21	H: Integrated CRM	Fri 20/05/16	Fri 20/05/16
22	I: Integrated Order Tracking	Fri 12/08/16	Fri 12/08/16
	Business Capabilities		
23	J: Flexible Practices	Fri 12/08/16	Fri 12/08/16
24	K: Increased Market Visibility	Fri 15/07/16	Fri 15/07/16
25	L: All Stores Run the Same Way	Fri 12/08/16	Fri 12/08/16
26	M: Management Can Reward	Fri 04/03/16	Fri 04/03/16
27	N: All Staff Can Apply Processes	Fri 12/08/16	Fri 12/08/16
	Business Outcomes		
28	O: Effective Operations	Fri 12/08/16	Fri 12/08/16
29	P: More Customers	Fri 30/09/16	Fri 30/09/16
30	Q: Happy Staff	Fri 09/09/16	Fri 09/09/16
31	R: Efficient & Friendly Service	Fri 09/09/16	Fri 09/09/16
	Strategic Outcomes		
32	S: Increased Market Share	Fri 30/09/16	Fri 30/09/16
33	T: Increased Revenue	Mon 07/11/16	Mon 12/09/16
34	U: Decreased Operational Spend	Mon 07/11/16	Mon 07/11/16
35	V: Reduced Staff Turnover	Mon 07/11/16	Mon 07/11/16
36	W: Increased Profits	Mon 07/11/16	Mon 07/11/16

Table 8.4 Calculation Table Showing the Initiatives

				Calculation Table									
				01-Jan	01-Feb	01-Mar	01-Apr	01-May	01-Jun	01-Jul	01-Aug	01-Sep	01-Oct
	Initiatives												
1	A: Standardize Processes			0	0	0	0	0	$970,305	$970,305	$970,305	$970,305	$970,305
7	B: Integrate Store Records			0	0	0	0	$970,305	$970,305	$970,305	$970,305	$970,305	$970,305
9	C: Adapt Tracking System			0	0	0	0	0	0	0	0	$1,757,310	$1,757,310
12	D: Develop Marketing Program	K	1 mo	0	0	0	0	0	0	$733,200	$733,200	$733,200	$733,200
15	E: Deliver Compensation Program			0	0	0	$918,960	$918,960	$918,960	$918,960	$918,960	$918,960	$918,960
17	F: Train Staff			0	0	0	0	0	0	0	0	$1,859,920	$1,859,920

- S = Higher Market Share: $232,560
- T = Higher Revenue: $598,400
- U = Operational Savings: $598,400
- V = Less Turnover: $88,000
- W = More Profit: $686,400

This set of values therefore provides the non-zero effective contribution shares and the effective contribution (EC) of each node, based on the state of the program at the chosen data date—April 1, in this case.

Repeating the exercise for the next calculation date of May 1 will bring in the contribution of B = Integrate Store Records, as that initiative will by then be ready to contribute. If we carry out the direct left to right calculation as for April 1, but including as well B = Integrate Store Records, we obtain a result for the contributions on May 1. However, that set of calculations does not take the lags into account. The uncorrected values are shown in Table 8.5.

Table 8.5 Uncorrected Contributions on May 1 Showing Nodes Affected by Lags

A	Standardize Processes	$970,305
B	Integrate Store Records	$970,305
C	Adapt Tracking System	$0
D	Develop Marketing Program	$0
E	Deliver Compensation Program	$918,960
F	Train Staff	$0
G	Process Model Available	$970,305
H	Integrated CRM	$1,940,610
I	Integrated Order Tracking	$0
J	Flexible Practices	$936,090
K	Increased Market Visibility	$183,300
L	All Stores Run the Same Way	$821,220
M	Management Can Reward	$918,960
N	All Staff Can Apply Processes	$0
O	Effective Operations	$1,642,440
P	More Customers	$2,364,570
Q	Happy Staff	$1,033,830
R	Efficient & Friendly Service	$2,181,270
S	Increased Market Share	$661,770
T	Increased Revenue	$1,702,800
U	Decreased Operational Spend	$495,000
V	Reduced Staff Turnover	$99,000
W	Increased Profits	$2,197,800

The lags need to be applied to the highlighted nodes. For initiatives, the lags were allowed for directly by setting the initiative allocation to be equal to the contribution of that initiative at the equivalent date and then applying the BEER. However, for intermediate nodes, some other means of allowing for the lags is required. This entails defining and applying an additional set of rules (the Lag Evaluation Rules, or LAGER) in conjunction with the BEER.

8.9.2 The Lag Evaluation Rules

The LAG Evaluation Rules (LAGER) are based on the following considerations:

- For each link on which a lag is defined, the standard BEER approach will deliver an incorrect value until the ramp-up process is complete, because the predecessor node value should be based on an effective date prior to the evaluation date.
- Each node may be the source of links with different lags.

For every link that has a lag, allow for it as follows: Adapt the corresponding allocation fraction by the ratio between a) the value of the source node calculated earlier for the effective date and b) the calculated value of the source node based on the latest calculation. If there is no lag, the two terms in the ratio are, by definition, equal, and no change is made to the allocation fraction applied in the BEER.

To apply the LAGER, start the calculation date from the start date of the program—that is, when all of the allocations and contributions are zero. Then go forward one evaluation period at a time as explained above recording all of the contribution values corresponding to each evaluation date.

The approach is therefore to address the shares on each link by modifying the allocation fraction that is used for the left-to-right evaluation of the contribution. The detailed iterative algorithm below implements the LAGER for a given date:

LAGER Step 1

Start by ignoring the lags and calculating the model at the chosen date:

- Set the contribution value of each initiative pro rata to the percentage completed at the calculation date of the final contribution value.
- Start with the full BEER matrix. This defines the initial values of the *current allocation fractions* (CAFs).
- Calculate (left to right) the values of the contributions based on the contributions of the initiatives at the given date.
  - These are the *current date contributions* (CDCs).
- Set *new contributions* (NCs) to be equal to the CDCs.

LAGER Step 2

Now to take the lags into account:

1. Calculate the *sum of the NC* (SONC).
2. For each link with a lag (logically you can do this for every link, but for doing it by hand, limiting it to links with lags reduces the amount of work required), calculate the *effective date factor* (EDF) from the effective contribution and the current date contribution as EDF = EC/CDC.
   a. Modify the corresponding allocation fraction in the BEER matrix to be CAF = CAF * EDF.
3. Recalculate (left to right) the BRM using the modified BEER matrix to deliver a new set of contribution values (NC).

Repeat from 1 if the sum of NC is different from SONC (you can just compare the overall totals because any change that is due to lags will reduce the contribution of at least one node).

This will provide the values of the contributions across the entire BRM for the chosen date. However, once the lags have been taken into account, there is still one additional consideration to be addressed: How to deal with essential links.

8.9.3 Dealing with Essential Links

As explained in Chapter 6, the existence of the node at the destination end of an essential link depends on the existence of the source node, regardless of the state of the other sources for this node.

Partial Contributions

An additional convention needs to be defined to deal with the case in which a precursor node on an essential link is only partially complete. The rule to be applied is that the contribution of the successor node is multiplied by the Effective Date Factor (EDF) of the essential precursor node. If there is more than one essential link, the minimum of the EDFs is taken as the value by which to multiply the contribution. The normal rule for essential links is satisfied in this expanded convention because, if any essential precursor node has not been started or is otherwise unavailable, the percent complete is zero, leading to a zero EDF.

This convention can be applied in the additional approach for dealing with essential links—the Algorithm for Link Evaluation (ALE), as follows.

The Algorithm for Link Evaluation

For a given node, start by calculating the essential links factor (ELF) to be applied to its contribution value. The ELF is 1 for nodes with no essential links into them. For a node with one or more incoming essential links, the essential link factor is the minimum of the EDF values of all of the essential source nodes. This value is called *essential node adjustment* (ENA) for the destination node.

The contribution of a node is then calculated by modifying step 3 of the LAGER algorithm to include the ENA as follows:

a. Recalculate (left to right) the BRM using the full BEER matrix of CAFs.
b. Calculate the ENA for each node.
c. Multiply the calculated value of the contribution by its ENA. This will be the new contribution value (NC).

The rest of the iteration is unchanged.

The same approach for dealing with essential links will be applied when calculating the earned benefit as explained in Chapter 12 and for validating the model based on forecast progress (Chapter 13).

8.9.4 Reviewing the Effect of Multiple Lags

To understand the effect of the lags on the ramp up of benefits, the dates from Table 8.3 have been added into the dependency diagram in Figure 8.10. This is shown in Figure 8.11 and illustrates in chronological order how the benefits diffuse, as follows:

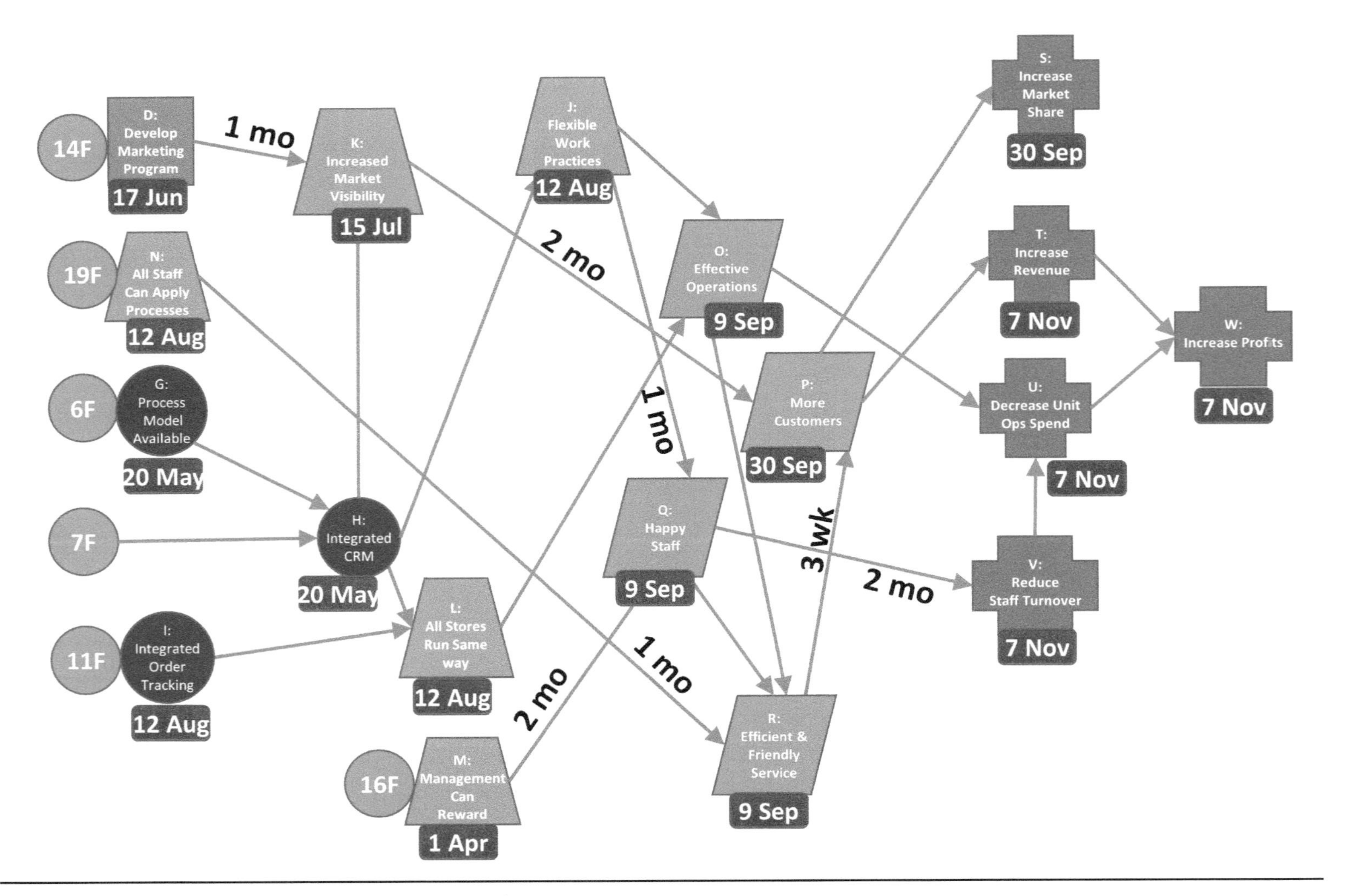

Figure 8.11 Dependency Diagram Showing Lags and Dates

- M = Staff Rewards Defined has an effective date of April 1.
- The contribution from M to Q = Happy Staff ramps up from zero to full value over two months (April and May), but the total outcome of Q = Happy Staff is only achieved once J = Flexible Work Practices have been in place for one month—that is, September 9.
- This full effect of Q = Happy Staff on R = Great Service is therefore delayed until September 9. It will have started once M = Management Can Reward Staff (April 1) and built up over two months (to the end of May). The final ramp up will come once J = Flexible Working Practices becomes available (August 12) and be complete two months later (September 9th).
- The ramp up of R = Great Service serves to support the ramp up of P = More Customers over an additional three weeks (September 9–30).
- Q = Happy Staff supports the ramp up of V = Reduced Staff Turnover over an additional two months (September 9 to November 7).
- P = More Customers and V = Reduced Staff Turnover then contribute to the rest of the BRM, giving ramp-up values from April to November 7.

These effects of lags are further quantified by developing a cumulative contribution curve.

8.9.5 The Cumulative Contribution Curve for the QERTS Example

The BEER, LAGER, and ALE can be applied along with the VEST to forecast the contribution values at each review date in the program. These values can then be used to create the planned cumulative contribution curve.

The contribution curve shows the evolution of the value of the contributions. Since the calculated amount measures the value of the benefit level that the situation would reach during a given period or interval, the cumulative contribution has to be calculated as the area under the contribution curve. This calculation will be approximated by representing the contribution curve by a step function. The evolution of the contributions, the cumulative contribution and the cumulative spend from the initiatives across 18 monthly intervals for the QERTS example is shown in Figure 8.12. This form of cash flow analysis is a valuable addition to the business case. In the example, it shows that break-even is reached within 18 months from the start of the program.

8.9.6 Optimizing the Cash Flow

The example was calculated based on launching each initiative at its earliest possible start date. However, other timing options could be considered and compared in order to optimize the cash flow of the program with respect to the financial model chosen (e.g., net present value, payback period, etc.) or to minimize work in process by taking essential links into account as discussed in Chapter 6.

This set of forecasts can then be used as a set of key performance indicators (KPIs) to be used in order to track performance and check the accuracy of the model once the initiatives have been completed. KPIs are discussed in more detail in Chapter 13.

8.10 Summary

This chapter has presented an integrated scheduling technique (VEST) that combines the various scheduling models, and applied it to the case study. The implementation schedule has been extended by the use of the Lag Evaluation Rules (LAGER) and the Algorithm for Link Evaluation (ALE) to forecast

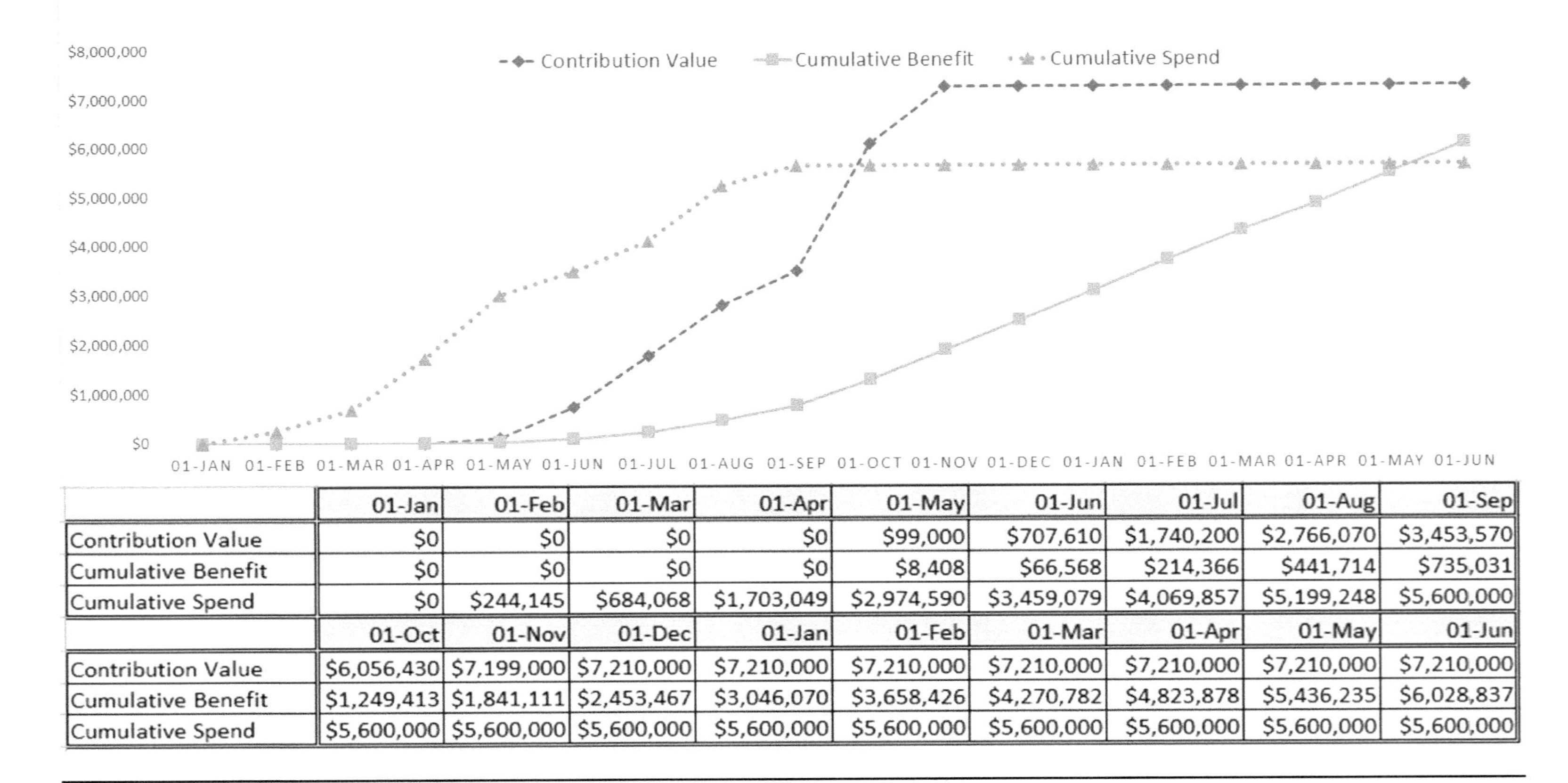

	01-Jan	01-Feb	01-Mar	01-Apr	01-May	01-Jun	01-Jul	01-Aug	01-Sep
Contribution Value	$0	$0	$0	$0	$99,000	$707,610	$1,740,200	$2,766,070	$3,453,570
Cumulative Benefit	$0	$0	$0	$0	$8,408	$66,568	$214,366	$441,714	$735,031
Cumulative Spend	$0	$244,145	$684,068	$1,703,049	$2,974,590	$3,459,079	$4,069,857	$5,199,248	$5,600,000
	01-Oct	**01-Nov**	**01-Dec**	**01-Jan**	**01-Feb**	**01-Mar**	**01-Apr**	**01-May**	**01-Jun**
Contribution Value	$6,056,430	$7,199,000	$7,210,000	$7,210,000	$7,210,000	$7,210,000	$7,210,000	$7,210,000	$7,210,000
Cumulative Benefit	$1,249,413	$1,841,111	$2,453,467	$3,046,070	$3,658,426	$4,270,782	$4,823,878	$5,436,235	$6,028,837
Cumulative Spend	$5,600,000	$5,600,000	$5,600,000	$5,600,000	$5,600,000	$5,600,000	$5,600,000	$5,600,000	$5,600,000

Figure 8.12 Cash Flow Calculations and Curves

the ramp up of benefits over time. The results of applying this technique to the case study have been used to develop a cash-flow forecast up to the break-even point between spend and actual benefit.

8.11 From Schedule to Risk

This modeling activity should be considered only the starting point, because in reality, the situation is not deterministic, and the uncertainty of the links between capabilities and outcomes, for example, needs to be taken into account. In addition, one major constraint in all programs and portfolios is based on the availability of resources: the people or technology required to support and carry out the planned activities. Some challenging features of resource management will be addressed in Chapter 10 once the uncertainty topic has been addressed in the next chapter.

8.12 References

Newbold, R. *Project Management in the Fast Lane: Applying the Theory of Constraints.* Boca Raton, FL, USA: The CRC Press Series on Constraints Management, CRC Press, Taylor and Francis, 1998.

Chapter 9

Total Risk and Issue Management

Programs and portfolios are, because of their multiple interdependencies, considerably more complex than projects. In particular, the management of program and portfolio risks—and especially the planning and execution of response strategies—requires a complete and structured approach to predict and avoid unwelcome secondary effects as well as the techniques to capture opportunities that might otherwise have been missed. In addition, in programs, the way in which issues are anticipated and dealt with is a major contributor to program success. This chapter proposes a consistent process for addressing the identification, analysis, planning, and execution of responses covering both risks and issues, to minimize the downside and accentuate the positive possibilities.

This chapter started out as a presentation to the PMI® EMEA Congress 2012 in Marseille, France. Many of the illustrations are taken from the corresponding presentation.

9.1 Background

The trigger for developing an integrated risk and issue approach was a statement that is often made: "Once a risk occurs, it becomes an issue." I was initially happy with that statement and even repeated it in the *Practice Standard for Project Risk Management* (Project Management Institute 2009) by David Hillson, David Hulett, and myself. But I then started thinking about it.

The first shortcoming came from the fact that it is now generally agreed in the field of risk management that risks can be positive (opportunities) or, as is more usual in everyday use, negative (threats). The statement about issues being risks that have occurred should therefore imply that issues can also be positive or negative. I had not seen anything supporting that concept in mainstream publications, but I realized that it was not only a logical extension of the idea for risk management, but a valuable addition.

The second point was that, although most organizations include risk management in their formal project and program management methodology, the corresponding outcomes are not as successful as you might anticipate, because the risk management process is frequently not carried through effectively or completely to its conclusion. The commitment to the process seems to weaken progressively as one moves through the successive stages from *identification*, with a large number of risks identified, to *definition*, where the risk descriptions can start to become somewhat incomplete, to *analysis*, where assumptions, attitudes, and biases are ignored, *prioritization*, using unreliable or inconsistent models, *response planning*, which is rarely verified for practical feasibility and secondary effects, *risk treatment*, which may not even be included in an updated action plan, and *risk monitoring*, which should assess the results of response actions and invoke, as necessary, fallback plans—in the unlikely event that any have been developed—so that, by the time the risk occurs, there are few resources or little motivation available to deal with the corresponding issue. My proposal for ensuring end-to-end risk and issue management is to turn the process on its head and start with what really matters: ensuring that you will be capable of dealing with the issues that could occur.

The third point was that, although an issue is a risk that has occurred, many of the definitions of "risk" in the standards were not easy to adapt to defining an issue. This challenge led me to rework the definition of risk from the first principles that are generally agreed: Risks are something the occurrence of which, though uncertain, would have an effect on the success of the endeavor in question. A few intermediate definitions were required to make this statement more precise, but basically, it was clear that risks have three dimensions: the situation, the uncertainty about whether the situation would exist, and the potential impact of that situation on a set of agreed success metrics. An issue, then, simply becomes a risk without the uncertainty as to its occurrence—or, since the approach is to start with issues and not with risks: A risk is a potential issue.

At the original presentation, I started out with what I hoped to be a challenging claim to people interested in risk management: "Risks don't matter." I then began to explain that it is the corresponding issues that matter, because risks, as potential issues, are not reality. At the end of the presentation, when I thought I had made all of this approach clear, one of the first questions I received was "When you said 'risks don't matter,' were you joking?" I tried to explain that it was a challenging statement, based on reality, by which I was hoping to create a change of perspective in people's minds. This explanation did not seem to satisfy the person who had asked the question. Since the event, I have now realized that it might have helped to explain that even though risks do not matter, *risk management* matters a great deal. In fact, the back-to-front analysis of issue and risk management later on shows that risk management can, and should, be considered as *preemptive issue management*—or, as J. R. Ewing recommended in the TV series *Dallas*: "Do unto others before they do unto you."

This discussion led me to appreciate the benefit of creating an integrated risk and issue management approach and to broaden the scope of the model to make it more generally applicable, including:

- Moving away from pure statistics for the analysis of the uncertainty and trying to understand the meaning of uncertainty or indeterminacy when applied to a single occurrence of a situation
- Developing an *integrated* risk and issue management model to address total program management and thereby to create a *total* risk and issue management (TRIM) model

9.2 Overview

9.2.1 Contentious Introductory Assertion

Risks do not matter because what has not yet happened cannot hurt you. Therefore: risks only matter when they are no longer uncertain (i.e., when they are no longer risks). Since it is generally agreed that once a risk occurs, it becomes an *issue*, it is issue management and not risk management that should be the central concern.

9.2.2 Background

Although there are a number of risk management standards (ISO 31000, etc.), plus focused risk management chapters in other standards (The Project Management Institute's *PMBOK® Guide*, Chapter 11; The Office of Government Commerce [OGC 2011] standard *Managing Successful Programmes* [MSP®], "Risk and Issue Management;") and the PMI® *Practice Standard for Project Risk Management*, referred to earlier, the position of issue management is less well documented, and the actual definition of *issue*—even within the set of PMI Standards—is far from standardized. As an example, Max Wideman's "Comparative Glossary of Project Management Terms" provides the following collection of definitions:

> Issue
>
> - Something in dispute or to be decided. (Derived from the glossary of terms used by the Defense Systems Management College, Virginia, USA)
> - An immediate problem requiring resolution. (Abstracted from Association of Project Management (UK) APMP Syllabus 2nd Edition, January 2000, Abridged Glossary of Project Management Terms (Rev.4))
> - A major problem that will impede the progress of the project and cannot be resolved by the project manager and project team without outside help (Mochal T, the TenStep Project Management Process Glossary)
> - A concern raised by any stakeholder that needs to be addressed, either immediately or during the project. As issues are reviewed during the project, they may become a threat and hence a risk to the project. (Tasmanian Government Project Management Glossary, Version 3.0, 2002. See Tasmanian Government Project Management Home Page.)
> - An unexpected risk event whose time has come for management decision.
>
> (*Source:* Wideman, R. M., Composite additions from various sources, 1998–2001.)

The glossary in the *PMBOK® Guide* gives a definition that is far removed from the concept of a risk that has occurred: "A point or matter that is in dispute," although the term *issue* is used as a synonym for *problem* in most of the standard and is explicitly described in Chapter 10 as a negative risk that has occurred.

None of these definitions provides a satisfactory basis for understanding where issues fit into the practice of total program management or, more importantly, for developing an effective approach for managing them. In addition, even where issues and risks are considered together, as in the MSP® document, the approaches for dealing with issues and risks are not integrated into a single, consistent process. However, the concepts of risks and issues overlap in many areas (see Figure 9.1) and the corresponding management techniques should accommodate this.

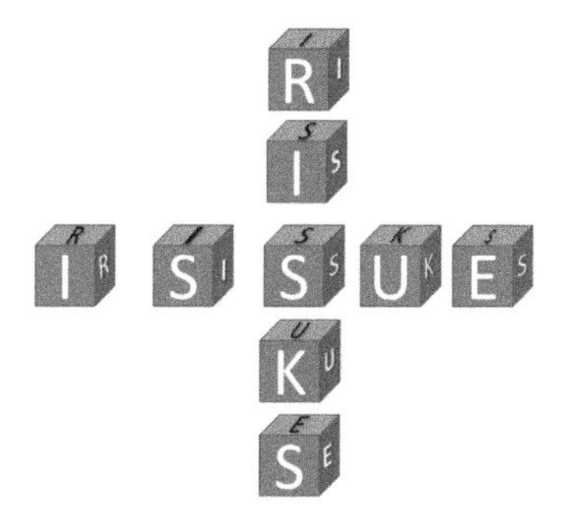

Figure 9.1 Risks and Issues Are Inseparable

These shortcomings will be addressed, starting with:

- Developing a new definition of *issue*, which will serve as the basis for a definition of risk
- Defining a process for managing issues and showing how, for completeness, this issue management process has to include all of the components of risk management

9.3 Issues

The development of the definition in this section is based on the thinking that has been used in defining risk in most of the risk publications.

Put simply, an issue is something that matters enough for you to feel the need to take some action. To be usable, this definition needs to be made more precise: what is the *something* and how do you specify *matters*?

The *something* is an event, set of events or a condition and will be identified by the term *situation*.

Why does it *matter*? It can only matter with respect to a specific set of objectives that are relevant to the corresponding endeavor (e.g., project, program, etc.)—that is, it only matters if it can affect success, positively or negatively. The precise definition of *success* depends on the environment; this definition for programs is developed below.

But first, we need to examine these two components—the situation, and the concept of success—in more detail.

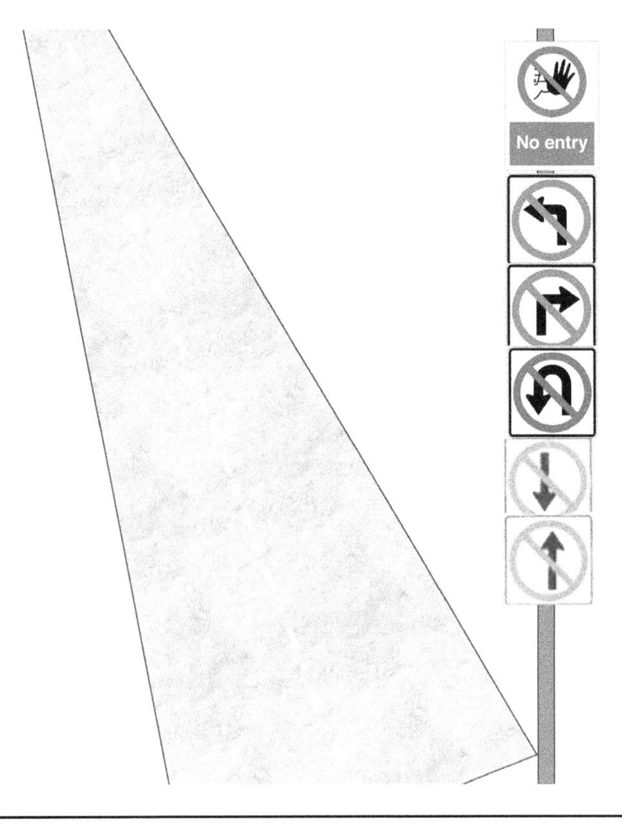

Figure 9.2 We Have a Situation

9.3.1 The Situation

The use of the expression *an event or set of conditions* rather just *an event* has been chosen to cover the full set of eventualities. Take, for example, Figure 9.2, in which there would be no issue as such for a driver if even one of the signs were missing: The situation is only significant because of the full set of road signs, plus, potentially, some associated condition or constraint such as the need to obey them.

9.3.2 Success

Success is measured by delivery in accordance with the objectives and achievement of specified benefits. However, associated constraints have also to be taken into account. There are in fact two major categories of constraints on success in project and program management: internal constraints and external constraints.

Measurable Objectives

The objectives of a project or program are a formalization of the reason for which the work was instigated. The goal in effective definition of objectives is to develop an unambiguous description of what is required, and of how you will know whether you have achieved it. Tracking progress toward success in this way is addressed in detail in Chapter 15 (Key Performance Indicators).

Internal Constraints

As a general rule, for example, success in projects is defined as complying with each of the components of the "triple constraint" triangle made up of schedule, cost and performance. As explained in Chapter 2, each domain (project, program, or portfolio) has its own view of what constitutes performance, providing criteria to measure the quality of the management of the endeavor. Internal constraints are therefore derived from the objectives and define the conditions that need to be obeyed for the endeavor to be considered a success.

The set of internal constraints is the main focus of performance monitoring because these constraints are the ones that most directly mirror the decisive parameters of the business justification for the endeavor. For programs, time, cost, and beneficial outcomes form a generic triple constraint. They have mutual dependencies, in that it is normally impossible to modify one of these constraints without having an immediate impact on at least one of the others:

- If the allowed time is reduced, you can either do less or will have to acquire additional resources (cost increase).
- If the budget is reduced, you can either do less, or you may be able to use slower, less costly resources.
- If more is required from the performance, it is likely that the endeavor will take longer and cost more.

These constraints in turn result in constraints on the component projects, and the program benefits realization map (BRM) is the basis for creating the business justification for each of these components.

The internal program constraints are examined below.

- Schedule
  - The schedule-related objectives normally include more than the finish date. As explained in the previous chapter, the effect on the cash flow may modify the value of the overall program. In addition, intermediate milestones may be linked to contractual conditions or to business imperatives for the customer or the performing organization.
- Cost
  - Cost in this context means more than money: It should take into account all of the budgeting considerations for resources (people, equipment, space, management involvement, etc.) as well as for money. For this reason, the term budget should be used instead of cost where it seems more appropriate.
  - The BRM for a program, will, for example, compare the estimated costs against the value forecast for the completed work. To be successful, the program must complete in accordance with the BRM.
- Performance
  - For a project, performance is related to its deliverables; for a program, to its outcomes and for a portfolio, to its strategic value.

External Constraints

Although the internal constraints are a necessary condition for success, they are not normally sufficient. There are other important constraints which affect all programs:

- Organizational environment
- Regulatory environment
- Key stakeholders
- Contractual considerations

So, how do these influence success?

The Organizational Environment

Every organization has internal rules that must be obeyed, such as tools and templates that must be used, reporting formats, and the like. Some of these tools may be mandated by other considerations, such as ISO 9000 (2009) or CMMI (2010) compliance. In addition, the organization may impose specific constraints on the program, such as the use of a preferred supplier list, use of specific tools, and so forth.

A program will not be considered fully successful if it is not carried out in a way that the organization considers in compliance with its *best practice*.

The Regulatory Environment

Formal rules and regulations must be taken into account in the way the work is carried out as well as for the specification of the deliverables and outcomes.

The Stakeholders

Like beauty, success is, to some extent, in the eye of the beholder. The stakeholders' assessment of a program is influenced both by their memories of how they felt as the program developed and, very strongly, by the way in which program completion was handled.

During the lifetime of the program, the set of stakeholders is likely to change; even if this is not the case, their expectations and even their memory of what has been agreed can evolve. For this reason, stakeholders are a key factor for program success and need to be considered at every step of the way.

Contractual Considerations

There are obviously many contractual clauses that are not directly covered by the considerations outlined above. These can help or hinder the delivery of the program and must be understood and complied with. Noncompliance with these clauses can have an effect on any of the other constraints outlined in the following subsections.

Interactions between Constraints

The management of the constraints—and therefore the work that needs to be carried out to achieve success—is made more difficult by the fact that the separate constraints are not independent of each other. In particular, the external constraints create dependencies between the internal constraints, as explained next.

Budget–Schedule

The cash flow considerations from Chapter 8 are one example of budget–schedule interactions. Other factors are shown in Figure 9.3.

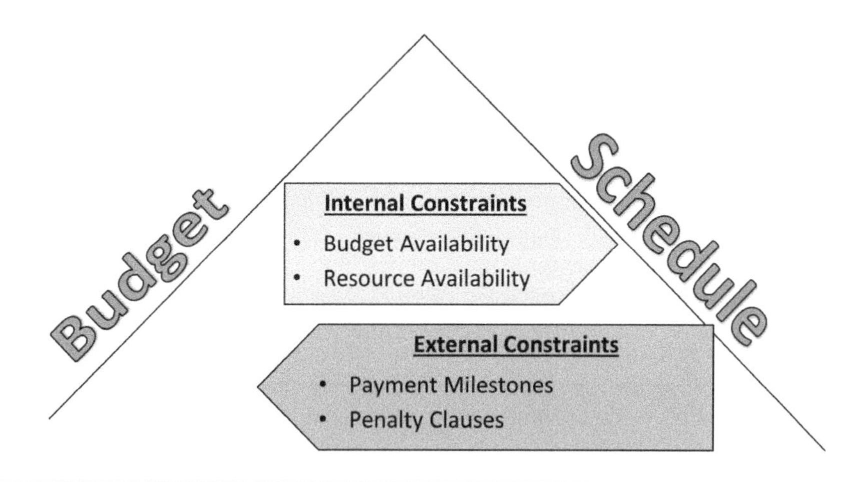

Figure 9.3 Constraint Effects on Budget and Schedule

One example of the possible effect of budget availability on the schedule is:

- Delay in procuring equipment until the budget is available
  - This delay can be due to approval requirements or because the funds are not available until the next budgeting cycle.

Resource availability may be limited because of decisions based on the relative priority of the current project or program with respect to others in the portfolio that need access to or use of the same resources. For example:

- Key experts are fully occupied elsewhere.
- Test equipment is currently reserved by another group.

These limitations will, of necessity, delay the work unless a workaround is found.

External, contractual constraints on the schedule as determined by the contract can affect the budget:

- The schedule must be developed and controlled in such a way as to capture the intermediate payments according to the business plan.
- All efforts must be made to avoid penalty clauses based on schedule overruns.
    - Similarly, any early termination incentives should be factored into the planning wherever feasible.

Schedule–Performance

The way in which some external constraints link schedule and performance success criteria is shown in Figure 9.4.

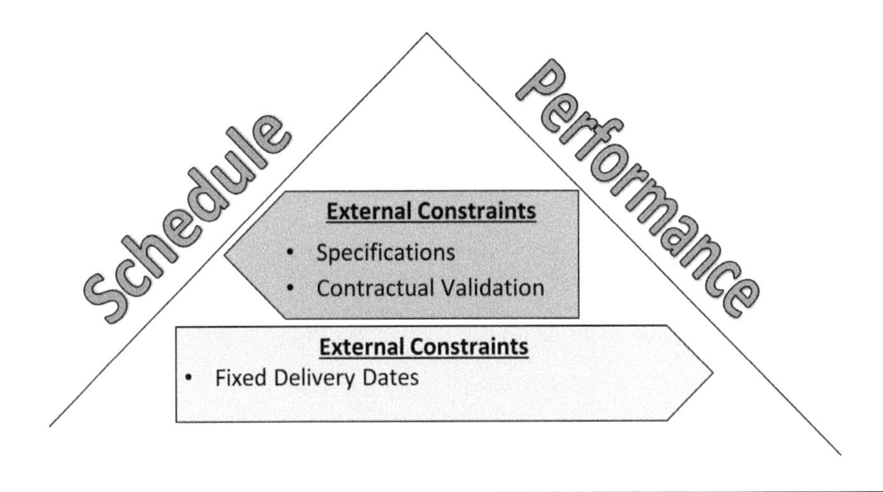

Figure 9.4 Constraint Effects on Schedule and Performance

The scope of the required deliverables or outcomes, as defined in the contractual specifications, is obviously the main contributing factor to the schedule. In addition, the work required to set up and administer the acceptance tests for the customer must be fully taken into account. These elements of the contract are the basis for developing the program's work breakdown structure, and the work breakdown structure is the basis for planning the schedule and budget for the program, as explained in Chapter 8.

In the opposite direction, if delivery is totally constrained by dates—if, for example, the results must be delivered in line with a legal commitment—and the development is behind schedule, then, for a given budget, a subset of the required capabilities is all that can be delivered.

Performance-Budget

The way in which some external constraints link performance and budget success criteria is shown in Figure 9.5.

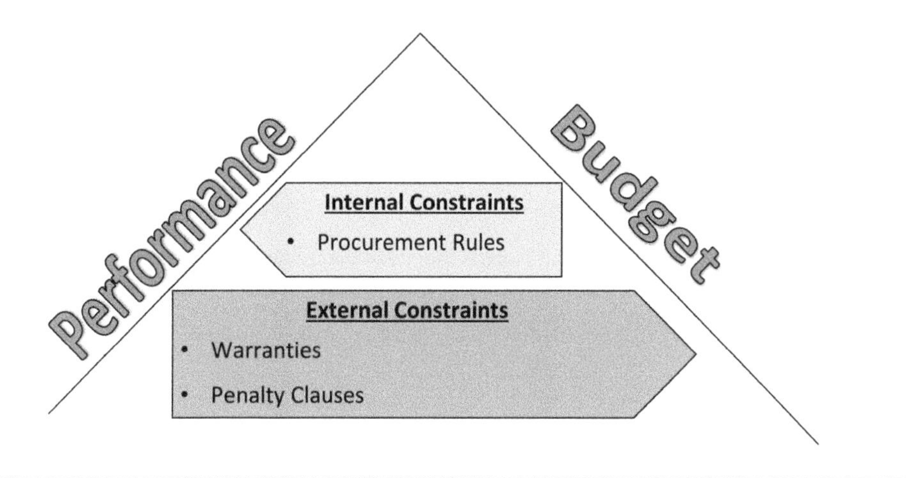

Figure 9.5 Constraint Effects on Performance and Budget

In some cases, the performing organization may have specific procurement rules that can limit the team's options for acquiring goods or services, thereby potentially limiting the capabilities of what can be delivered. Examples of such constraints include overall basic order agreements or partnerships set up at a higher level in the organization and imposed on all relevant work.

On the contractual front, equipment warranty agreements and performance-based penalty clauses have to be factored into the business case and the corresponding budget. For example, if equipment is purchased by the delivery team for later delivery to the client, the purchase agreement may need to include an extended warranty to ensure that the guarantee will be valid for the full contractual period starting from handover to the client, rather than simply from delivery to the performing organization.

9.3.3 The Dual Nature of Issues

Although the everyday understanding of an issue is that it creates obstacles, for the purpose of effective project and program management in general, and risk management in particular, the definition of issue needs to be extended to include positive as well as negative impacts. This definition of an issue will then mirror the way that the dual view of risk has now gained wide acceptance at a professional level.

The terminology also needs to be addressed: in the same way as risks are either *threats* (negative impact) or *opportunities* (positive impact), issues can be *problems* (negative impact) or *advantages* (positive impact).

The corresponding definition of an issue would therefore be: "A situation that is certain and will affect success in a positive or negative manner."

In the same way that the risk management process caters effectively for both threats and opportunities, as will be explained below, the issue management process can effectively encompass advantages (positive issues) as well as problems (negative issues).

An issue is therefore fully characterized by two independent features:

- The *situation*
- The *impact* on the objectives and corresponding benefits

In this way, with the addition of the uncertainty dimension, risk management becomes an integral part, or natural follow-on, of the issue management cycle, and a risk can therefore be seen as a *potential issue*.

9.4 Risks

The only difference between a risk and an issue is that risks are uncertain: compared to issues, a risk has uncertainty as an extra component. However, uncertainty has several features, which are all relevant for risk and issue management.

9.4.1 Uncertainty

One advantage of addressing uncertainty as a component of risk in its own right is that the different types of uncertainty described below affect risks—and issues—in different ways. In the literature on uncertainty, three main categories have been identified (Walker et al. 2003):

- Statistical uncertainty
- Scenario uncertainty
- Insufficient information

Each of these can range from total ignorance to full knowledge. Certainty would require that of these types of uncertainty to be eliminated together—which can normally only be approximated, rather than completely achieved, until the situation has occurred or been eliminated.

Several of these types of uncertainty can be present together, and each type affects each of the components of risks in different ways.

Statistical Uncertainty

Technically, statistical uncertainty is also known as *aleatoric uncertainty.* It refers to the inescapable variability of a specific situation, such as the result of throwing dice. It can be categorized by a probabilistic value or a probability distribution function and can apply to the occurrence of the situation and, independently, to the impact. The value can be based on historical information or, where this is not available, on a best-guess "prior" value that can be updated later, based on additional information. In a program, estimates of durations and costs need to take this type of uncertainty into account and specify the level of confidence that is associated with each estimate.

Scenario Uncertainty

Scenario uncertainty occurs when the context or the situation is not sufficiently understood. Put another way, scenario uncertainty arises when the program team's world view or model of the situation is incomplete or inexact. In the extreme, this type of uncertainty is the basis for the *black swan* type of risk (Taleb 2008) in which a situation is not predicted because it is inconceivable—that is to say, the idea of it never arises or is rejected as impossible. This type of uncertainly is sometimes called *ontological uncertainty*. For example, until it occurred, the idea of a crash in the housing market leading to a global recession was inconceivable to all but a very small number of people; this investment disbenefit risk was never factored into government plans.

Insufficient Information

When there is insufficient information, the uncertainty exists because there are available facts about the situation that have not been detected. The extremes go from total ignorance to complete knowledge. Whereas "total ignorance" can be the case in any endeavor, "complete knowledge" can never be fully achieved, and "sufficient knowledge" should be considered to be an adequate limit. The term for this type of uncertainty is *epistemological uncertainty*. Whereas aleatoric uncertainty is a fact of nature and associated with the situation itself, epistemological uncertainty is relative to a person or group: One person may have more information than other people and therefore have a lower level of epistemological uncertainty. A program manager needs to assess the level of reliable knowledge included in the program plan and ensure that it is sufficient to ensure a high level of confidence. Techniques such as interviews, surveys, lessons learned reviews, and benchmarking can be effective in reducing the level of epistemological uncertainty.

Example

An example of the difference between the types of uncertainty using the QERTS example is as follows.

An employment expert has told you that there is a 10 percent chance that the staff originally from the RTS company will call a strike due to the changes in working conditions. How should you interpret that information?

- Is the 10 percent chance epistemological—that is, the strike is actually going to happen, but the expert cannot find any confirmation?
- Is the situation aleatoric—that is, the decision is still in the balance, and in 10 percent of similar cases, the strike actually does occur?
- How large is the ontological risk—that is, how realistic is the assessment? The answer depends on the employment model, the accuracy of the information, and the historical examples that the employment expert has used in order to determine the risk. Making such a prediction, further information may change the ontological risk assessment even if the aleatoric degree of uncertainty of the actual outcome remains the same.

Although black swan risks are by definition virtually inconceivable, the following situation could fall into this category: The employment expert may have been bribed by a competitor to invent the strike, with the intention of causing conflict at QERTS management level, to get them to make the wrong decisions about integrating the workforce.

Uncertain Uncertainty

Ontological and epistemic types of uncertainty can even apply to statistical uncertainty. That is to say, there is either a lack of understanding or a case of incomplete information with respect to the probability distribution function that should be applied. As pointed out by Mandelbrot and Hudson (2008), faulty analysis of potential issues can be due to adopting, for example, a normal probability distribution when the physical reality corresponds to a significantly different distribution. In this case, the main differences usually affect lower probability situations and so are not easily detected initially, but become decisive when least expected, often when risks occur that had been discounted or accepted as totally unlikely. This is another source of black swan events. A number of recent disasters can be brought to mind to illustrate it. The design of the Fukushima nuclear plant in Japan is a warning to project and program managers (Center for Security Studies 2011); before ruling out a so-called "impossible" combination of extreme events, the potential impact of this impossible situation should be evaluated and factored into the total feasibility and viability assessment. If the possibility of flooding had not been rejected, backup generators would have been installed in secure locations.

9.4.2 Addressing the Three Components of Risk

As mentioned above, every risk needs to be characterized by three features:

- Uncertainty
- Situation
- Impact

In everyday use, risks are often referred to by only one of these characteristics:

- "There is a considerable risk that the airline will go on strike." In this case, the word risk refers to the degree of certainty.
- "The risk is that the airline could go on strike." In this case, the word risk refers to the situation.
- "The risk is that I will not be able to get to my meeting, and the project will be delayed." In this case, *risk* refers to the impact.

In the example above, context helps the reader understand which component of the risk is actually being described. The multiple uses of the term *risk* are not generally a cause of confusion in everyday life, but they can lead to faulty reasoning in a project environment if the concepts and context are not clearly defined.

9.4.3 Choose Any Two of Three

As mentioned above, a risk must comprise the three components of situation, impact, and likelihood. Taking only two of these components creates any of three specific entities.

No Uncertainty: Issues and Impossibilities

We know that, by definition, a risk that has occurred is an issue, and the probability of occurrence is 100 percent. The other way of removing the uncertainty in projects or programs is to make the occurrence impossible. This is the goal of "risk prevention" and is addressed in the section on Action Planning.

However, as will be explained later in the chapter, although the situation may have occurred (i.e., there is no uncertainty about its occurrence), the level of impact may not be known, giving risk to a degree of ambiguity. In this case, a link between analysis of the issue and additional risk identification is required. This linkage between issue and risk management is addressed in the TRIM process shown in Figure 9.11.

No Impact: Ambiguity

Given a situation and an associated degree of uncertainty but no defined impact, you are left with ambiguity: You cannot actually tell where you are or what is happening. Some stakeholders may characterize an ambiguous situation as an *issue*, and this can add to the confusion still further!

No Situation: Exposure

If you ignore the situation and focus only on the uncertainty and the impact, you can get a good idea of the "exposure" of the project or program to the undefined situation. This focused view can be a useful way of communicating status because it ensures that the people receiving the message concentrate on the information associated with the potential impact rather than discussing the current situation instead.

Another Way of Looking at It

All of the possible combinations are shown together in Figure 9.6 as different projections of the three-dimensional "risk" object onto each of the three planes, represented respectively by:

- Removing uncertainty (U = 0): issues
- Ignoring impact (I = 0): ambiguity
- Disregarding the situation (S = 0): exposure

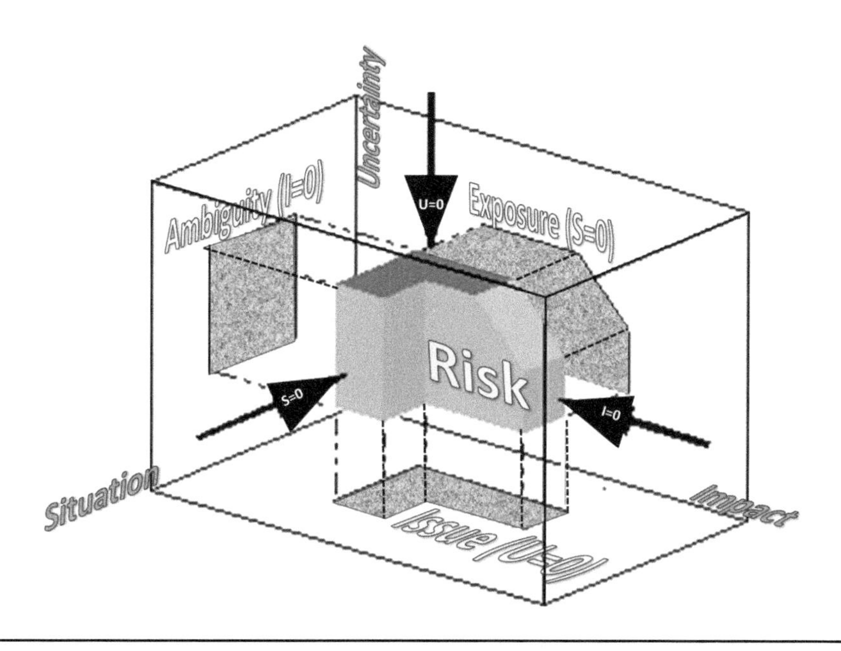

Figure 9.6 Three-Dimensional Risk and Its 2-D Projections

This means that the TRIM process automatically provides management not only of the two-dimensional issue entity but also of ambiguity and exposure. As will be shown next, control is achieved by acting on each and any of the dimensional components: The situation, the uncertainty, and the impact.

9.5 The TRIM Process

MSP® has a well-defined approach, based on a five-step process for addressing risks: Capture, Examine, Propose, Decide, and Implement. However, its issue management cycle is composed of four steps: Identify, Assess, Plan, and Implement. As mentioned previously, risk management and issue management should use a common approach and shared process. This section will therefore demonstrate that risk management is a natural consequence of proactive issue management and is part of a single TRIM process.

Since there is more published work available on managing risk than on managing issues, it seems sensible to base the TRIM process on the commonly accepted risk management process and then extend this process to incorporate issue management. The top-level view is shown in Figure 9.7.

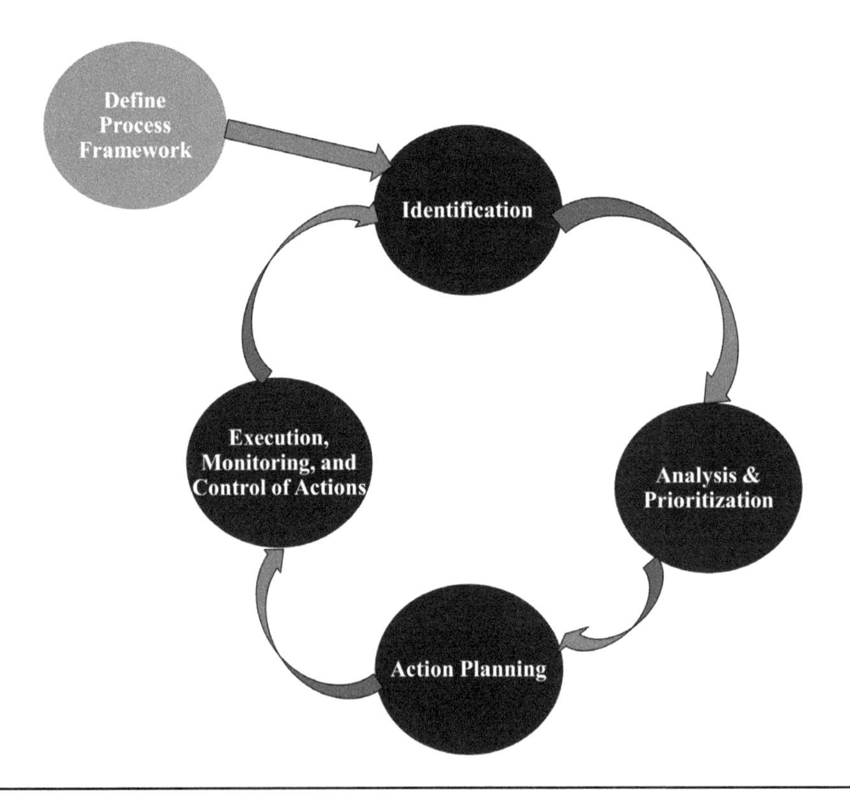

Figure 9.7 The Total Risk and Issue Management Process Stages

The main stages of a TRIM process are similar to those of the risk management cycle:

- Defining the approach: Tailoring and parameterization of the overall issue management process, as well as understanding the human, technical, and organizational environment
- Identification and documentation

- Assessment and prioritization
- Action planning
- Execution, monitoring, control—and issue resolution

It is important to understand that, although the diagram seems to indicate a sequential set of activities, there are considerably more interactions between the processes than shown in Figure 9.7. These interactions are analyzed in more detail at the end of this section. In addition, there can be subcycles running in parallel; for example, execution, monitoring, and control can remain active while extra action planning is going on.

Most of the steps are similar to the corresponding ones in the standard risk management process. However, there are important differences, which is why the processes have the generic names in Figure 9.7.

As shown in Figure 9.7, four processes are to be carried out repeatedly throughout the lifetime of the program—an action known collectively as TRIM assessment or TRIM reassessment. Although *define process framework* can and should be carried out for the whole TRIM environment of risks and issues, the other four processes have some features specific to either risk or to issue management because of the existence of the extra (uncertainty) dimension for risks. However, these differences can be incorporated into a single TRIM process, as will be explained by expanding each of the process stages.

9.5.1 Define TRIM Framework

For full consistency and to allow seamless treatment of risks as they become issues, a single, consistent framework needs to be established. This framework addresses people, processes, and tools as follows.

The Human Factors

It is important to analyze all of the people affected, either as direct participants or as external stakeholders. The roles of each of these stakeholders need to be identified or defined, and their attitudes toward risks and issues understood so that these can be taken into account when setting the rules and thresholds for applying the processes (see also Murray-Webster and Hillson 2008). Attitude can be affected by any or all of the components of risks and issues: the situation, the impact, and the probability.

The Influence of the Situation

There are many features of situations that can make people feel better or worse, independently of the level of impact:

- *Fairness/unfairness*—The person who gains or loses may be perceived as deserving or undeserving of the result.
- *Relevant area*—Some areas are required or expected to be under greater control than others, with a corresponding influence on the tolerance for error.
- *Controllability of the environment*—A negative result in an uncontrollable environment is often considered more acceptable than the same effect when it appears that something could have been done about it.

These and other situational biases need to be documented and factored into the analysis, prioritization, and action planning steps.

The Influence of the Impact

This area was extensively analyzed beginning in the 18th century (Bernouilli's work on Utility Theory, described in Bernoulli 1954) and was brought up to date by Kahneman and Tversky's Prospect Theory (1979). The basic findings are as follows:

- People's reactions to impacts are not mathematically rational; twice the impact does not always generate twice the emotional influence. As shown in Figure 9.8, the response curve is generally in four parts (Piney 2003):

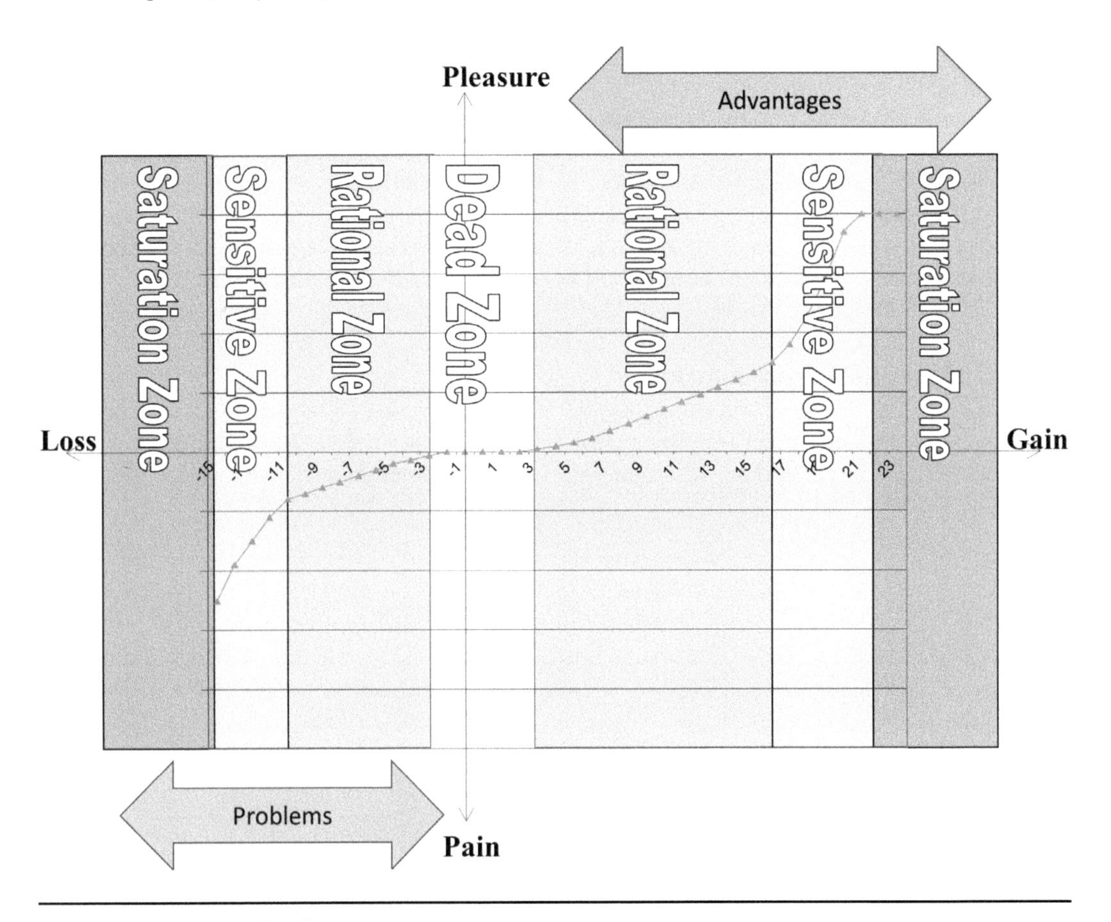

Figure 9.8 A Typical Utility Curve

  - The *dead zone*, where the impact is too small to matter
  - The *rational zone*, where the emotional influence follows the value of the impact
  - The *sensitive zone*, where the influence starts to climb faster than the impact
  - The *saturation zone*, in which the emotional influence is maximized, and a greater impact cannot increase it significantly

  This analysis is made more complex by the following facts:

  - The curve may be different for each stakeholder.
  - The curve depends on the current context: factors such as budget, sunk costs, etc., determine the relevant curve.

In fact, changing the "set point" with respect to which gains and losses are calculated results in a different curve; for example, the value at origin of the curve (given by the overall budget) can modify all of the parameters that determine the shape of the curve.

- Although it has the same four-zone shape, the curve relating to a profit is different from the curve for a loss, even when all of the factors above are identical. The emotional influence of a loss is generally greater than for a gain of the same amount. As an aside, this threat bias partly explains why risk management still focuses predominantly on the management of threats rather than of opportunities, and why issues almost exclusively address problems rather than advantages.

The Influence of the Uncertainty

The influence of uncertainty is of course only relevant to risks, where it adds to the difficulty of analysis and prioritization. Once again, for valid psychological reasons, people's reaction to uncertainty is non-linear with respect to the corresponding mathematical probability. This emotional aspect of uncertainty becomes obvious when you consider the two ends of the scale:

- The impossible has become possible: A 0 percent probability is now 3 percent, for example. This step from impossible to just possible leads to a major conceptual change; in one study quoted in Kahneman (2013), this was shown to have 10 percent of the influence compared with absolute certainty.
- Certainty is no longer absolute: A 100 percent probability is now 97 percent. The same study indicates a loss of confidence of over 15 percent for a shift of 3 percent in probability.

Correspondingly, the center of the range is relatively insensitive to changes in probability. This subjective importance is shown in Figure 9.9. Its shape roughly approximates a side-on view of a bathtub and shows a plot of the subjective importance (Y-axis) of each successive percentage of mathematical probability, for which I use the name *prospect weights*. In an unbiased, rational environment, the curve

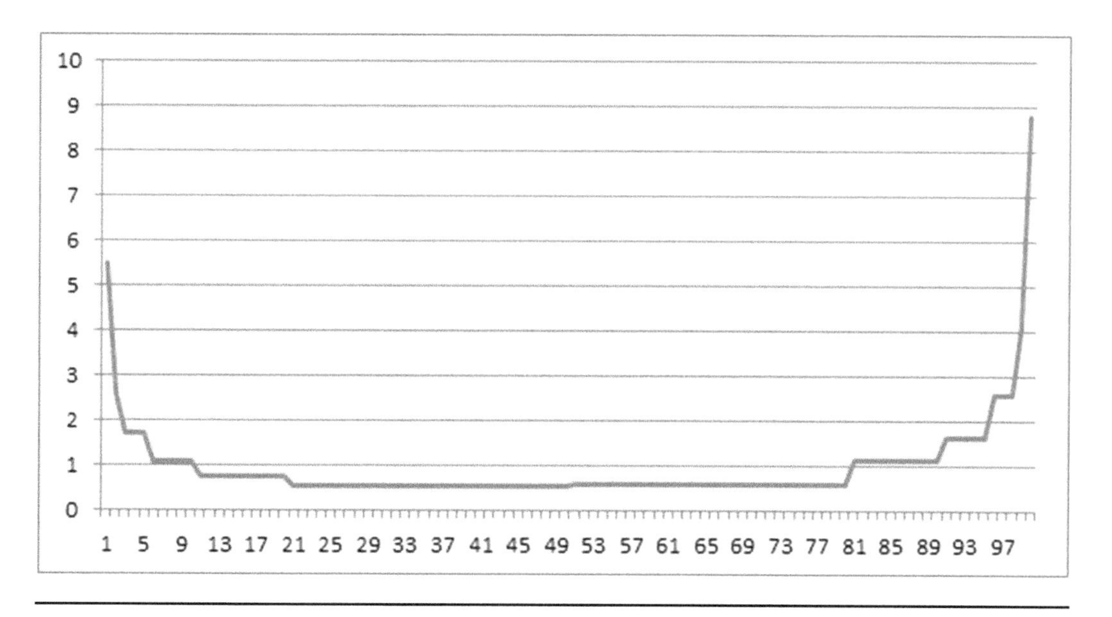

Figure 9.9 The Prospect Weights: Subjective Importance of Each Probability Percentage

should be a horizontal line at Y = 1 percent since each increase or decrease of probability would have an identical influence on the subjective importance.

Another way of presenting this is the prospect curve, which shows a plot of the subjective importance (Y-axis) against the mathematical probability (X-axis) in Figure 9.10. In an unbiased environment, this graph would be the diagonal line for x = y; however, studies show it to have the distinctive shape shown in Figure 9.10, which is concave for gains (implying risk aversion), convex for losses (risk seeking), and generally steeper for losses than for gains (loss aversion). Note that a good term for this concept of subjective uncertainty would be *confidence level* or *perceived likelihood,* as those terms incorporate the concept of subjectivity. The qualifier *perceived* is required because ISO3000 has already defined *likelihood* as "the chance of something happening," which does not carry the same subjective connotation.

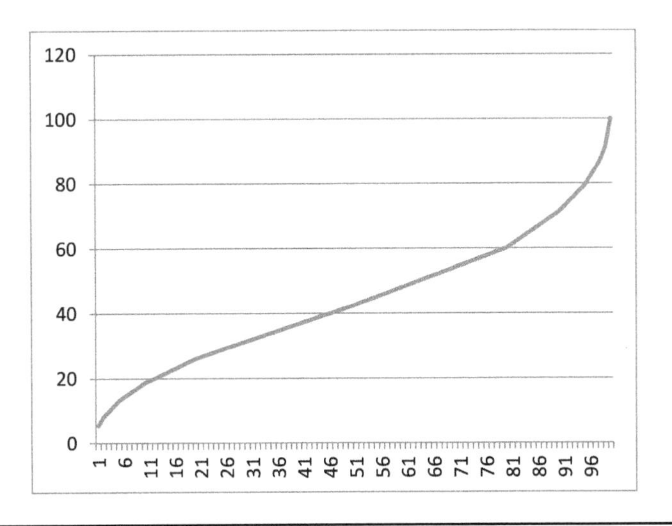

Figure 9.10 The Prospect Curve: Prospective Likelihood versus Mathematical Probability

It is interesting to note that the border between the advantages and problems areas in Figure 9.8 has a similar shape to the prospect curve in Figure 9.10. This shows the convergence between the utility approach and prospect theory.

In the QERTS example in Chapter 7, the disbenefit due to standardization was analyzed as follows: "the net financial benefit of the original plan is $7,210,000 – $5,600,000 = $1,610,000. The disbenefit has reduced this to $6,971,700 – $5,600,000 = $1,371,700—that is, a decrease of almost a quarter of a million dollars. The return on investment (ROI) has dropped from 129 percent to 124 percent, which does not seem so dramatic." The subjective impact of the potential loss of $238,000 from an estimated benefit of $1,610,000 is much greater than the purely mathematical change in ROI. This corresponds to the effect of loss aversion. Willingness to spend $100,000 to respond to this risk is enhanced by the differences between the gain and loss profiles of the prospect and utility curves: the attraction of "gaining" $100,000 by taking no action is dwarfed by the fear of suffering a loss of $238,000.

The role of the TRIM framework is to adapt the TRIM processes and tools to take these subjective feelings about risk into account.

TRIM Processes

Based on the information about the business environment and the human factors, it is important to define the way in which all of the TRIM processes will be applied. For example:

- The terminology to be used: The "metalanguage" for describing risks and issues
- The criteria for analysis: How the situation, impact and probability will be described and interpreted in a consistent manner by all relevant stakeholders
    - As part of this step, a review of the definition of the relevant objectives may be necessary to ensure that the corresponding metrics are relevant, realistic, and applicable.
- Who is responsible for what, and how the responsibilities will be applied
- Reporting formats and rules
- The amount of time and effort to be made available for the TRIM processes

TRIM Tools

The topic of tools includes:

- The documentation and storage of the TRIM information (i.e., the *TRIM Register*)
- The types of tools to be used; for example:
    - Probability-impact matrix
    - Simulation tools
- Parameterization for the tools
- Prioritization rules and thresholds

9.5.2 Identification

The identification step is composed of two main activities:

1. Discovering all knowable issues or risks
2. Documenting them formally into the TRIM register as specified in *define TRIM process*

This step needs to address issues separately from risks because the approach for each is different. Risk identification entails creating and analyzing scenarios of potential situations, whereas the identification of issues requires that analysis of the current situation.

Determining Issues

Issues arise in general from one of two reasons:

1. A previously identified risk occurs
2. An unforeseen situation arises that appears to require action

In both of these cases, the issue becomes obvious without the need for special investigation so long as the monitoring and communication channels are working effectively. The issue identification process therefore needs to evaluate the effectiveness of monitoring and communications. If any shortcomings are identified, these should be added to the list of problems and addressed accordingly.

Although issues are certain, they do not necessarily remain active throughout the program. Since the corresponding situation can be a condition or set of events, if the situation changes, which can happen, for example, as the program progresses, the issue can also change. For this reason, a time-window needs to be associated with each issue; such a window is often delimited by program milestones rather than precise dates. The definition of the window of each issue should be included in the TRIM register.

The Link from Risk to Issue Management

In the first case above, it is the risk owner who should first become aware that the risk has occurred (normally through the Execution or the Monitoring process, which is described later) and take the appropriate action to transition the risk and any relevant contingency plans into an issue for analysis, prioritization and action.

The Link from Issue to Risk Management

In some cases, it may not be clear whether the situation in question is actually as described. In these cases, a degree of uncertainty enters the analysis; the issue is better dealt with as a risk rather than as an issue—which leads into the process for identifying risks. In the QERTS example, the disbenefit due to overall standardization was identified as a negative issue and treated accordingly. However, this side-effect was not certain. It might therefore have been more prudent to consider it to be a threat and to treat it on the risk management side of the TRIM process.

Identifying Risks

As already mentioned, because of the need to attempt to imagine situations that are not currently visible, risk identification requires a degree of creativity and imagination that is not required when determining issues.

The three main reasons for the uncertainty about a situation are:

1. Lack of complete information (epistemic uncertainty)
2. Basic variability of some of the elements of the situation (aleatoric uncertainty)
3. Unreliable or unfounded assumptions (ontological uncertainty)

Current situations can also be potential sources of risks due to lack of information—for example, when the validity of a hypothesis or a constraint is in doubt. The addition of explicit hypothesis statements into the BRM can be used to identify areas of ontological uncertainty. Reviewing assumptions and constraints can identify potential ontological inaccuracies—leading, respectively, to the identification of threats and opportunities.

In common with issues, certain risks may exist during only part of the time—that is, if the effect can only have an impact on objectives when other, temporary conditions are present. As an example, bad weather is a greater threat to timescales in a building project while external work needs to be done.

The Link from Risk to Issue Management

There are situations in which the risk identification process will need to post a definite issue. In the QERTS case, for example, one area of focus for identifying risks is the objective of increased market share. If we find that the current market share figures are out of date or otherwise unusable, this will need to be addressed as an issue.

From Identification to Analysis and Prioritization

Once the identification step is (sufficiently) complete, the set of identified risks and issues must be recorded in the TRIM register. The register then needs to be analyzed and a decision made as to which of the risks and issues warrants deliberate action.

9.5.3 Analysis and Prioritization

Because risks and issues share two of their three dimensions, they also share some of the tools and techniques associated with the analysis of situations and of impacts. Only risk, however, takes the uncertainty dimension into account.

Analysis and Perception

The United Kingdom's Cabinet Office Strategy Unit (2002) states that risk refers to "uncertainty of outcome, whether positive opportunity or negative threat, of actions and events. It is the combination of likelihood [see ISO3000 definition provided earlier] and impact, including perceived importance" (p. 26). The significant addition in this definition is that risk perception is mentioned explicitly: The analysis of each risk or issue can never be fully objective and is affected by this perception and the stakeholder's attitude, as was discussed earlier with respect to the human factors in defining the TRIM framework.

Hillson (2010) provides an informal definition of risk as "uncertainty that matters" (p. 84). He also describes attitude as "a chosen response to a given situation" that is affected by perception of the situation (p. 155). He combines these two definitions to state that risk attitude is "a chosen response to uncertainty that matters, influenced by perception" (p. 155). Risk attitude is therefore central to the analysis and prioritization of the identified risks and issues.

By replacing the informal definition of risk in this definition with the definition of risk based on the three dimensions (situation, uncertainty, and impact), the definition of *risk attitude* becomes "a chosen response, based on perception, to an uncertain situation, the uncertainty of the situation, and the impacts (which can be positive or negative) that the occurrence of the situation would have on the success of the endeavor." The same approach can be taken in defining attitude to issues, simply by removing the reference to uncertainty, as follows: "a chosen response, based on perception, to a situation and the impacts (which can be positive or negative) that the situation would have on project success."

Since the chosen response is based on perception, issue and risk attitudes are influenced by the person's perception of each of the components of issues or of risks.

Perception of the Situation

Although it is difficult to separate the perception of a situation from its impact (Kahneman, Slovic, and Tversky 1982), this can be done by considering the affective factors associated with the situation: loss of a loved one, for example, is a situation that affects some people much more than others. In the QERTS example, lack of flexibility due to standardization could be seen as a minor disadvantage to QERTS executives but might make the working environment unbearable to some ex-QE employees.

Perception of the Effect

Envisaging the effect of a future situation often requires some form of storytelling or scenario analysis (Fahey and Randall 1997). The way that this is done can have a considerable influence on a person's perception of what the situation would actually imply. In the QERTS case, management will be motivated by a description of how easy it will be for them to track and control all of the work in a consistent manner; they will be interested in the effect of a situation on that goal. For the employees, however, seeing themselves as part of a reliable, considerate, and successful organization is a major factor in their analysis. A risk to the training program due to nonattendance might therefore be presented as two very different stories about the same situation, as follows:

- *For management:* The training has been carried out, and 80 percent of the customer-facing agents have participated. One agent, George, who missed the training, has completed a customer order but has forgotten to ensure that the inventory is updated accordingly. Another agent, Jose, at another location, has just selected the same article from the inventory, identified it as being available, and assigned the same article to his customer. The customer will wait in vain for the order, and a number of people will now have to be involved to rectify the situation; management will have to take steps to recover the situation with the customer.
- *For staff:* George has been with the company for many years and knows how to make things work for him. He missed the training but is confident that his skills are up to date. He is pleased to have put through a customer order in the same way he always did. He did not realize that he would now need to ensure that the inventory is updated. Jose, at another location, has followed the training and is confident that he can rely on the new tools and processes. He completes an order for a favorite customer. Unfortunately, this order will fail because it relies on the same item that George has included in his order. When this becomes clear, Jose loses confidence in the system and has an embarrassing session with his customer. George fears he has lost the respect of his colleagues and management.

Although storytelling is a powerful way of communicating a message, in a program environment, the BRM is also invaluable in order to provide an objective analysis.

Perception of the Impact

Perception of the impact falls into the realm of *utility theory*. That is to say, each person's perception of a given objective impact is strongly conditioned by his or her situation. For example, a loss of $10,000 on a $10 million program might be considered insignificant, whereas the loss of the same sum to an individual might be considered dramatic. The analysis of this area is covered by expected utility theory (Piney 2003) as well as the variant, prospect theory (Kahneman and Tversky 1979) and shown in Figure 9.8.

Perception of the Objectives and the Scope

Most programs have a number of stakeholders with different goals. Even when the stakeholders agree on the set of objectives and the specified scope, they are unlikely to hold identical views on the importance of each individual objective or component of the scope. For this reason, the perception of the importance of what is impacted is likely to differ considerably between stakeholders.

Perception of the Uncertainty

The effect of uncertainty obviously only applies in the case of risks. There are numerous studies that show how people's perception of the uncertainty of a situation can be influenced illogically by cognitive biases, most of which also apply to the other dimensions of issues or risks: lack of understanding, wishful thinking, anchoring, affect, framing, groupthink, confirmation bias, self-serving bias, and the like. In addition, the effect of this perception is also subject to the nonlinear influence of the prospect curve (Figure 9.10).

Common Analysis Tools and Techniques

The goal of analysis is not only to provide a basis for prioritization of use of resources for developing a response plan but also for assessing the viability of the action plan itself with respect to all of the other options, including the alternative of taking no action. The link between analysis and action is explained in more detail in the Plan Responses section.

Analyzing the Situation

The definition of the TRIM framework will have specified a model and the types of situation that can occur, along with the corresponding metrics and thresholds associated with them. This information for each identified situation needs to be evaluated and recorded in the TRIM Register.

Analyzing the Impact

As mentioned above, the TRIM framework plus the characterization of the situation determines the metrics, utility curve, and relevant tools to be used to translate the effect of the situation (e.g., a budget overrun) into an impact value in the corresponding category of objectives. Once the objective value of the impact has been estimated—or, more frequently, the possible range of values of the impact—this estimate can be translated into the subjective utility value or range of values of the corresponding gain or loss.

However, it can happen that there is not enough information about the impact to provide a valid basis for subsequent prioritization. In this case, the risk of misunderstanding the importance of the corresponding risk or issue has to be added to the list of risks.

Analyzing the Uncertainty

The feature that distinguishes risks from issues is uncertainty about the situation.

The TRIM framework should be used to determine the relevant prospect curve based on the situation and the impact that have already been evaluated. Once the mathematical value of the uncertainty has been estimated—or, more frequently, its possible range of values—this estimate can be translated into the subjective value of the corresponding gain or loss by applying the relevant prospect curve.

Prioritization

Because of the different metrics and dimensions, opportunities have to be prioritized separately from threats, and advantages from problems. It should be noted that the priorities assigned at this stage may be modified later once more information is available, such as the timing, cost, and potential effectiveness of response actions.

Prioritizing Issues

Although, as mentioned above, problems cannot be directly prioritized along with advantages, the two corresponding lists can be sorted separately using the same tools—as defined in the corresponding part of the TRIM framework.

Typical features to take into account are:

- Utility
  - Where the analysis provided a range of values, it will normally be the highest value in the impact range that is retained for problems, and the lowest value in the range for advantages.

- Urgency
  - If the effect will not occur for a considerable time, and action is not required in the near future, use of resources might be optimized by delaying any further work on this issue.

Prioritizing Risks

Prioritization is more difficult for risks than for issues because there are two (almost) independent variables to be taken into account: the impact and the uncertainty. However, by using both the utility value and the prospect curve, we can obtain a number that is much more meaningful than the conventional *expected value* (i.e., the product of impact by probability [Schuyler 2001]). That is because the expected value approach takes the *rational model* as hypothesis; that is, both impact and mathematical probability are a direct measure of the influence of these parameters on people's attitudes to the outcome. This is not how people think. On the other hand, utility and prospect take subjective influences directly into account.

For this reason, if a single prioritization criterion is required, individual risks can more reasonably be prioritized by using the product of the utility by the applicable value on the corresponding prospect curve, which will be known as the *prospective utility* of the risk. This approach entails more work than the *expected value* one, if only because the choice of utility curve and prospect curve to use depends on the overall situation, the category of impact, and of course the stakeholders. If the situation changes markedly, other versions of the curves may need to be used and complete reprioritization done. In a project, for example, once a risk has occurred, people's perception of the impact of a reoccurrence may be very different from the original perception associated with that risk. Unfortunately in most programs, this is not carried out and can lead to using outdated or totally inappropriate responses when major issues arise.

When evaluating the *overall risk* for the endeavor—that is to say, the integrated effect of the set of identified risks—a different approach needs to be taken. Modeling the overall risk requires the use of simulation tools. Normal probability and impact values should be used in the simulation. However, once the simulation is complete, it can be quite revealing to convert the purely mathematical curves of impact vs. probability into charts of utility vs. prospect value. The utility–prospect approach should then be used when carrying out sensitivity analysis to prioritize risks based on their effect on overall risk.

However, to ensure full consistency between the business case and the risk analysis, simulations such as Monte Carlo should be based on the BRM. In addition, as for issues, the urgency of taking action if required with respect to the risks also needs to be factored into prioritization.

Using the Prioritized Lists

At this stage, we have defined the framework, identified issues and risks, analyzed them, and decided which ones need to take priority for the treatment. Effective planning of the required response actions is now required and needs to address all of the data that have already been developed on each of the TRIM dimensions: the situation, the impact, and, for risks, the uncertainty.

9.5.4 Action Planning

It is important at this stage to remember the statement made at the beginning of this chapter: "Risks do not matter, issues do." This assertion means that the focus should be placed on the issue—whether it is certain or still uncertain (i.e., a risk)—when planning how to deal with it. The conclusion is that risk management can—and should—be considered quite simply as proactive issue management. Since there is considerably more literature available on risk response planning than on issues, these ideas will be used as a basis for the corresponding TRIM process. The six active response "strategies" currently used for risks are:

- For threats:
  - Mitigate
  - Avoid
  - Transfer
- For opportunities:
  - Enhance
  - Exploit
  - Share

Where no immediate action is considered necessary or viable, a seventh option is taken: *acceptance*. These strategies under various names are recommended in most standards (International Organization for Standardization 2009, Office of Government Commerce 2006, Standards Australia/Standards New Zealand 2004, Project Management Institute 2009, 2013a). They can be used for issues just as well as for risks and will be analyzed with respect to the relevant components (the situation, the impact, and, for risks, the uncertainty).

All of the approaches explained below for managing issues are therefore applicable to risks; for risks, however, the extra dimension (i.e., uncertainty) also needs to be taken into account. In this way, there is no disconnect or indecision on the corresponding responsibilities between risk owners and issue owners since they are one and the same role; the same remark holds true for action owners.

This action planning step is closely linked to the other processes in the TRIM cycle because any proposed response will:

- Have a set of potential side effects:
  - These side effects can be definite (issues) or potential (risks).
  - They will have to be taken into account (identified, analyzed, prioritized, and potentially provided with a response).
- Change the corresponding risk or issue being considered so that the "residual" risk or issue will then have to be addressed (i.e., *documented* [*identification*], *analyzed*, and *prioritized*). If necessary, it will then need its own action plan.
  - And so on until all residual risks and the resulting overall risk–issue structure complies with the tolerance threshold defined in the framework.

This structured way of thinking about potential responses encourages the development of multiple options and helps to find the optimal set of actions to be taken. The approach outlined below expands on the current set of risk response strategies of *avoid*, *mitigate*, *transfer*, (and *accept*) for threats and *exploit*, *enhance*, *share*, (and *accept*) for opportunities.

The novel idea in TRIM that is developed below is to consider the actions in each of these response categories that separately affect each of the dimensions. This divided approach has several advantages; it applies the ideas of utility and prospect that have already been explained and it also provides a consistent approach for both issues and risks. This commonality of approach is possible because, with respect to the dimensions of Situation and Impact, the options are identical for both issues and risks. Treatment options for risk apply these six categories in addition to the *uncertainty* dimension. In a similar way, this action planning process can also be used to address the exposure (impact + uncertainty) as well as the ambiguity (situation + uncertainty) presented earlier (see Figure 9.6).

Table 9.1 The Full Set of Active TRIM Response Categories and the Relationship with Current Risk Response Strategies

	Negative			Positive			
Current	Situation	Impact	Uncertainty	Situation	Impact	Uncertainty	Current
Mitigate	**Adapt:** work on the situation to make it less damaging	**Reduce:** lessen the impact	**Disable:** reduce the uncertainty	**Improve:** increase benefit of the situation	**Augment:** increase the impact	**Enable:** increase the uncertainty	Enhance
Avoid	**Refocus:** change the objectives so that the situation is less important	**Counteract:** make sure the impact does not affect any of the objectives	**Eliminate:** make the occurrence impossible	**Refocus:** change the objectives so that the situation is more important	**Capture:** ensure that the impact is optimal	**Ensure:** make the occurrence certain	Exploit
Transfer	**Reassign:** make somone else take reponsibility for dealing with the situation	**Shift:** find somene else to accept the impact in place of the project	**Outsource:** find someone else to work on reducing the uncertainty	**Reassign:** make someone else take reponsibility for improving the situation	**Partner:** work with someone who has more control over the impact	**Outsource:** find someone else to work on increasing the uncertainty	Share

The full set of response categories is shown in Table 9.1 and explained in more detail as follows. The columns of the table are described in turn, one row at a time. Negative and positive options are explained together.

Addressing the Situation

Mitigation/Enhancement: Adapting or Improving the Situation

The situation has an effect on people's perception and attitude; for example, if the outcome appears to be unfair, the utility value of a negative impact will tend to increase. By addressing the situation, the shape of the utility and prospect curves that apply can be modified to provide mitigation or enhancement by adapting the negative situation or improving the positive one.

Avoidance/Exploiting: Refocusing the Situation

The relevant part of the definition of an issue or a risk in this case is the effect of the situation on the relevant objectives. Refocusing entails repositioning the situation or the objectives in such a way that the situation no longer affects the relevant objectives in the case of a negative effect, and, for a positive effect, that the situation aligns completely with as many as possible of the relevant objectives.

Transfer/Sharing: Reassigning the Situation

Transfer and sharing are carried out in conjunction with one or both of the previous options: responsibility for working on the situation is assigned outside the current organization. The transfer of responsibility or ownership normally entails payment in one form or another—such as fees, profit sharing, etc.—to or from the person or company that takes the responsibility.

Addressing the Impact

Mitigation/Enhancement: Reducing or Augmenting the Impact

In the negative case, impact reduction at a physical level can be provided by protective devices (e.g., sprinklers and fire-retardant materials) or by direct response to the situation (e.g., negotiation with suppliers or with striking personnel).

Examples of mitigation of disbenefits are given in Chapters 6 and 7.

In the positive case, the impact of a situation can be enhanced if the organization is structured in such a way as to be able to react to the situation effectively.

Avoidance/Exploiting: Counteracting or Capturing the Impact

In the negative case, by taking an action that completely inhibits the impact, the issue or risk can be negated. For example, fruit growers use "smudge pots" to send up smoke to counteract the damaging effect of frost, or, depending on the expected conditions, use sprinklers to provide a protective coat of ice around the fruit.

In the positive case, taking immediate action to gain full advantage of a beneficial improvement can allow you to get ahead of the competition (e.g., purchasing a partner whose product is complementary to yours provides you with the full profitability and synergy from that product.)

Transfer/Sharing: Shifting or Partnering the Impact

There are many examples of transfer and sharing of risks and issues at a financial level. For example, insurance companies accept the insured impact (in exchange for a fee), and investment companies provide a return on your invested money (once again, in exchange for a fee). On the positive side again, joint partnerships are another valid and interesting approach for dealing with the case where neither company on its own can provide the full set of required services. A joint partnership can also be entered into in order to share the negative impact of a situation.

In another example of sharing, given in Chapter 11, the client rewards bidders for valuable ideas even when the bidder is not retained for the total contract.

Addressing the Uncertainty

As already explained, uncertainty is the component that distinguishes risks from issues, and risk management is simply proactive issue management. Therefore, the following approaches are undertaken to affect the uncertainty in such a way as to have the best possible effect on the corresponding result. Since threats are associated with problems (negative issues), and opportunities with advantages (positive issues), the responses designed to affect the uncertainty need to be chosen accordingly.

Mitigation/Enhancement: Disabling or Enabling the Potential Issue

For mitigation or enhancement, actions need to make problems less likely and advantages more likely. The "prospect weights" have to be taken into account (see Figure 9.9) because changes to the uncertainty values along the bottom of the "bathtub" do not have much influence on the perceived outcome and are frequently not worth the investment. This approach can therefore give rise to different strategies from the standard expected value technique.

Avoidance/Exploiting: Eliminating or Ensuring the Potential Issue

The goal of avoidance or exploiting is to remove the uncertainty completely. For threats, two main elimination approaches can be envisaged:

- Eliminating the cause of the risk
- Eliminating the link between the cause and the effect

Elimination is different from a) *risk mitigation*, in which the expected value of the risk is reduced but the risk still exists with some potential capability of affecting the project's objectives, and b) *risk avoidance*, in which the situation still exists but the scope or set of objectives are modified so that the situation is no longer relevant. Avoidance limits the range of options of the program by restricting the options to those that do not involve the stated risk.

As an example, assume that your program is dependent on the provision of a specific technical component that you will specify. You are concerned that any supplier might steal your design for this component, thereby damaging your competitive advantage. You can mitigate the threat by having binding agreements with a reliable supplier. You can avoid the threat by modifying your solution to remove the need for this component. You can eliminate the threat by developing the component in-house (i.e., removing the cause of the threat). Alternatively, you could eliminate the threat by acquiring the supplier, so that the transfer of your intellectual property to them would no longer damage your competitive objective (thus ensuring that the cause cannot create the threatened impact).

For opportunities, the opposite should be attempted:

- Making absolutely sure the cause does occur *and* ensuring that the cause–effect link is definitive
  - For example, obtaining a letter of intent from a client *and* preparing a credible and viable proposition in order to capture the business

Transfer/Sharing: Outsourcing the Work

Responsibility for working on the uncertainty is assigned outside the current organization. In the same way as for situation-related transfer and sharing actions, outsourcing of the work on the uncertainty dimension is carried out in conjunction with one or both of the previous options. This option normally entails payment in one form or another—such as fees, profit sharing, or the like—to the person or company that takes the responsibility.

Issue and Risk Acceptance

As outlined at the start of this section, in addition to all of the options described above, and once no additional action is considered necessary or viable at the moment the planning is carried out, the situation should be accepted. However, in this case, there are also two options: Prepare for the eventuality (active acceptance), or decide to ignore it (passive acceptance).

Active Acceptance

Active acceptance, being conditional, only applies in the case of risks. In this case, the plan for responding to the uncertain issue (the "contingency plan") needs to be prepared, along with the conditions under which it would be executed ("action triggers") defined.

Whereas risk management is *proactive* issue management, because actions are taken immediately, contingency planning can be considered to be *predictive* issue management, since the plan is developed for use in the future, if at all.

Passive Acceptance

Passive acceptance means that no specific action is planned in advance of the occurrence of the risk. For example, a certain number of defects and omissions are normally allowed for in a building project; these are dealt with in a final "snagging" phase. Once all of the viable steps to address a risk have been taken, then what is left, the residual risk, may be accepted passively. The costs in terms of time, resources, and money to address any passively accepted risks are normally covered by a general contingency buffer.

Passive acceptance of issues carries the risk that the impact has been miscalculated, but in general the issues need only to be documented in the TRIM register and no further action taken.

Consolidating the Action Plan

One of the key points about developing and maintaining an action plan that incorporates both risks and issues is that, in this way, the portfolio of actions can be optimized for maximum effectiveness. The frequently reactive approach that is adopted in many organizations normally leads to insufficiently planned, shortsighted actions that can generate greater threats, invalidate planned responses, interfere with each other, and use resources on low-importance risks or issues when they should be reserved for high-importance cases.

Secondary and Residual Risks

As outlined at the start of this section, TRIM planning must be carried out in an iterative way to take into account the potential side effects (secondary risks) and the resulting effect of the action on the corresponding risk (residual risk). Where a contingency plan has been developed, one secondary risk that should always be considered is the threat that the plan turns out to be unfeasible or ineffective. In this case, if this threat is important enough, it will justify the creation of a fallback plan to be used if the contingency plan fails to deliver the required result.

Modified Urgency

As was mentioned earlier, urgency can be used as one of the prioritization criteria. At the response planning stage, it may become clear that the corresponding action may take a long time either to put into practice or to have the desired effect. This time lag can therefore modify the urgency of the corresponding risk or issue and lead to reprioritization of the set of risks and issues.

Action Plan Review

An overall review is also needed because some of the proposed actions may be incompatible with each other, or on the contrary, some may be redundant because the secondary effects of other actions have the same effect. These interactions should become visible if the actions are integrated into the BRM. The existence of positive or negative interactions can then be cross-checked by use of simulation tools or, at least, a walk through with a new set of stakeholders, for sign-off by the sponsor.

9.5.5 Execution, Monitoring, and Control

Execution

Once the TRIM plan has been integrated into the overall program management plan and the risk/issue owners and action owners have been appointed and confirmed, the work should be controlled using management processes as described in Chapter 12. Although this ensures that execution is completely included with all of the other execution steps, some special consideration needs to be made for monitoring and control of risks and issues. These two subprocesses, monitoring and control, are explained separately.

Monitoring

A considerable number of TRIM-specific items need to be included in a monitoring checklist:

- The occurrence of previously identified risks
- The accuracy of the prior evaluation of the impact once the risk occurs
- The appearance of action triggers for contingency plans
- The start or end of risk windows
- The effectiveness of actions recently completed or currently under way
- The availability of additional information about the uncertainty
- The emergence of new risks
- The appearance of new issues (i.e., the occurrence of risks that were not identified beforehand)
- The stakeholders' attitudes

In addition, it is the responsibility of the monitoring process to ensure that the results of the monitoring and any related actions are fully documented and made available for any of the other processes that could require them—communications, lessons learned, and the like.

Each of the checklist items above can then give rise to one or more control actions, as described below.

Controlling and Resolution

The need for specific control actions is normally determined in the monitoring subphase and leads to executing planned (contingency) risk response actions or invoking other TRIM processes to carry out additional functions for managing features of risks or issues.

The Occurrence of Previously Identified Risks

When a risk occurs, by definition the risk becomes an issue. The change from risk to issue should be updated in the TRIM register and then prioritized along with any previously defined contingency plan, and a decision made on whether the planned action should be carried out or whether extra analysis and planning are required.

The Accuracy of the Prior Evaluation of the Impact Once the Risk Occurs

If the impact of the risk is different from what was expected, the issue needs to be reanalyzed and re-prioritized, and a new response plan may need to be developed.

The Appearance of Action Triggers for Contingency Plans

When an action trigger occurs, once the validity of the trigger and any associated assumptions have been checked, the corresponding contingency plan should be implemented.

The Start or End of Risk Windows

When a risk window opens, the corresponding plans should be reviewed by the risk owner with the action owners to ensure that everything necessary is in place to deal with the potential issue, should it arise.

When a window closes, the risk owner should verify that the risk is no longer relevant and, if that is the case, update the risk status and inform the program manager that the corresponding action owners can be released or reassigned.

The Effectiveness of Actions Recently Completed or Currently Under Way

If the recent or current set of actions is not having the expected effect, unless there is already a fallback plan to deal with this eventuality, this situation has to be treated as a new issue and managed accordingly: full identification, analysis, prioritization, and action planning.

The Availability of Additional Information

In general, additional information that becomes available needs to be evaluated to determine whether it affects any of the decisions so far made with respect to the TRIM environment.

In particular, as was mentioned earlier, the uncertainty of a risk my come from a lack of information about the situation or the impact. If this information later becomes available, it needs to be incorporated into the TRIM register, followed by re-analysis and reprioritization of the risk. As an example of additional information, the occurrence of a risk or a change to the reference environment can be used to update the value of the probability by use of Bayesian techniques (Piney 2014). The Bayesian approach can be compared to an iterative sampling technique, in which the result of each sample is used to update the expected frequency or likelihood of an event by taking into account the relative probability of not detecting the occurrence even if the situation or impact is as suggested.

The Emergence of New Risks

As the work progresses, the environment evolves and can generate new risks or make previously unknown risks visible. When this occurs, the full cycle—starting from identification—has to be repeated as required throughout the program lifetime. In addition, this cycle of TRIM reassessment should always be carried out at critical points in the life cycle, such as at the start of a phase, when a key stakeholder leaves, or with the identification of a major, previously unexpected risk.

The Appearance of New Issues

By definition, the appearance of an issue should normally correspond to the occurrence of a previously identified risk, as described earlier. However, it is impossible to forecast all risks, so it is likely that some issues will arrive unexpectedly. As just mentioned, this calls for a TRIM reassessment.

The Stakeholders' Attitudes

All of the previous control actions depended on applying the plan or reacting to deviations from the plan. However, because the TRIM framework is based on stakeholder attitudes and linked to the corresponding utility and prospect functions, changes in stakeholder attitudes can have an initially invisible effect on the validity of the assessment and prioritization of risks and issues.

Change in attitude can, for example, happen if market conditions improve and senior management becomes more "bullish" and willing to take greater risks in achieving success. On the other hand, a major challenge is the fact that, since the shape of utility curve depends on the set-point at the time the framework was established—that is, before the start—as the work progresses, and once the handover completes, a different curve may apply. In this case, a program in which previously "acceptable" threats occurred may be judged in retrospect as a failure. For a project in the pharmaceutical industry, for example, if a new drug fails in the final stage (stage three) of a set of clinical trials, it is not uncommon for management to feel that the risk of initiating this stage was unwise—although the same management team may earlier have recommended accepting the positive findings from stage 2.

If any changes in the program environment, the stakeholders, or the stakeholder attitudes are detected, therefore, the TRIM framework itself must be revalidated; if it has changed, TRIM reassessment needs to be carried out by executing each of the TRIM processes in turn.

9.5.6 Links between the TRIM Processes

The links and interactions between processes were mentioned in prior sections. To provide an integrated picture, they have been listed below with reference to the originating TRIM process and the earlier process descriptions. This information is also shown in Figure 9.11, where the links are tagged with the following subheadings.

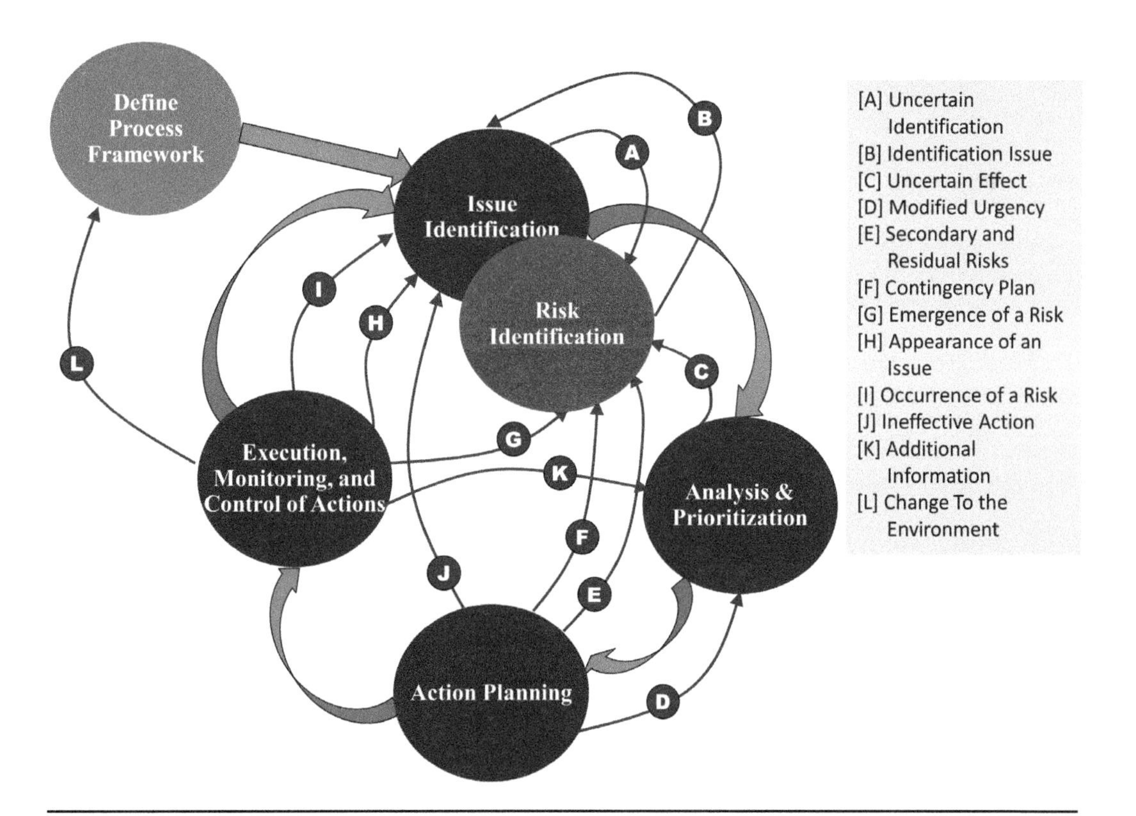

Figure 9.11 The Information Flows between the TRIM Processes

Identification

There are reciprocal links ([A] and [B]) between issue identification and risk identification.

[A] Uncertain Identification

As mentioned in Section 9.5.2, if there is not enough information to describe the issue sufficiently, the situation is uncertain and should be documented as a risk.

[B] Identification Issue

The converse effect is also mentioned in Section 9.5.2, where the work of identifying a risk reveals an issue, such as faulty data.

Analysis and Prioritization

Analysis and prioritization can lead back to risk identification ([C]).

[C] Uncertain Effect

As explained in Section 9.5.3, there is sometimes not enough information about the impact to provide a valid basis for subsequent prioritization. In this case, the threat of misunderstanding the importance of the corresponding risk or issue must be added to the list of risks.

Action Planning

Action planning can lead back to analysis and prioritization ([D]) and to risk identification ([E] and [F]) as well as to issue identification ([J]).

[D] Modified Urgency

A better understanding of the scheduling constraints associated with a response action can modify its urgency (see Section 9.5.4) as used in the earlier prioritization of this action.

[E] Secondary and Residual Risks

Any risk created by an action or left over once all of the actions have been determined must be documented in the TRIM register along with the other identified risks.

[F] Contingency Plan

The contingency plan with its trigger, as explained in Section 9.5.4, needs to be included with the identified risks because it is, by definition, a risk in its own right: an uncertain circumstance (the trigger) that would have an effect on one or more of the approved objectives (that is exactly why the plan is there!). In principle, therefore, the contingency plan should represent an opportunity to improve the corresponding risk situation.

Execution, Monitoring, and Control

Execution, monitoring, and control have links with risk identification [G], issue identification [H] (when an unforeseen issue occurs), and [I] (when a risk becomes an issue), with analysis and prioritization [K] (as more information becomes available) and to the definition of the TRIM framework [L].

[G] Emergence of a Risk

As explained in Section 9.5.5, as the program progresses, the program environment evolves. This can give rise to new risks or make previously unknown ones visible. These risks must then be fully identified and processed accordingly.

[H] Appearance of an Issue

Since it is impossible to forecast all risks, it is likely that some issues will arrive unexpectedly as the work progresses. These issues must then be fully identified and processed accordingly.

[I] Occurrence of a Risk

By definition, once it occurs, the risk becomes an issue. As explained in Section 9.5.5, the identification of this issue should be updated in the TRIM register and then prioritized along with any previously defined contingency plan (see Section 9.5.6), and a decision should be made as to whether the planned action should be carried out or whether extra analysis and planning are required.

[J] Ineffective Action

One of the tasks of monitoring is to verify whether or not the response action is having the expected effect (see Section 9.5.5). If this is not the case, the fact must be flagged as an issue and treated accordingly.

[K] Additional Information

As the program progresses, more information becomes available about the reference environment, so some of the earlier assumptions (see Section 9.5.2) may need to be reevaluated. For example, changes to the assumptions can then have an effect on the prioritization information as described in Section 9.5.5.

[L] Change to the Environment

If, for example, there are changes among the key stakeholders, or if the set point for the relevant utility curve changes for any reason (see Section 9.5.5), the TRIM framework will need to be reviewed; if this review leads to a change, the TRIM cycle starting from risk identification is carried out in case the change in environment has created new risks or modified the assessment criteria and prioritization of existing risks and issues.

9.6 Generic versus Specific Risks

To make risk management effective, it is important to separate the generic risks associated with all integrated project-type work from those particular to a specific endeavor. The generic risks should be addressed by use of organization-wide *controls* included in the operational procedures or the applied methodology. The more situations that such controls can deal with proactively, the more attention and effort that can be devoted to dealing with the specific risks. The BRM can be used to include information on generic aleatoric risks.

There is always a degree of uncertainty associated with any component of a model. Estimates of time and cost required for any initiative normally take estimation uncertainty into account. What is often overlooked, however, is that the links do not necessarily deliver the expected outcome:

- The outcome or its value is rarely certain in any case and will normally depend on the assumption made, which is documented in the description of the corresponding link. Since assumptions are, by definition, not certainties, each assumption can be used to identify one or more risks. Failing assumptions normally cause problems. In a similar way, being able to avoid a constraint may lead to an advantage.

The application of the uncertainty values can be used to generate the following information:

- Probability distributions
- Likelihood-based sensitivity analysis
    - This should be carried out with progressive left-to-right (bottom-up) focus. The approach in this case is to identify the key components and links for each category of component, starting from the initiatives, then their links, then the capabilities and their links, one step at a time across the entire map. This will provide additional information for prioritization.

9.6.1 Probability Distributions

There are a number of such distributions that can be of interest:

- Time-based distributions
- Cost-based distributions
- Value-based distributions
- Value–cost ratios

Each of the distributions can be presented in the same way: as a histogram of values that the variable (time, cost, etc.) can take, overlaid with the cumulative probability of not exceeding a value (see Figure 9.12). The actual values that were found during the simulation are given on the X-axis; the frequency is measured on the left-hand Y axis; the cumulative probability values for the heavy black curve is given on the vertical scale on the right.

Time-Based Distributions

The schedule information of interest normally relates to milestones (i.e., intermediate or final states of interest to specific stakeholders). Typically, these milestones are the achievement of intermediate or final (strategic) outcomes.

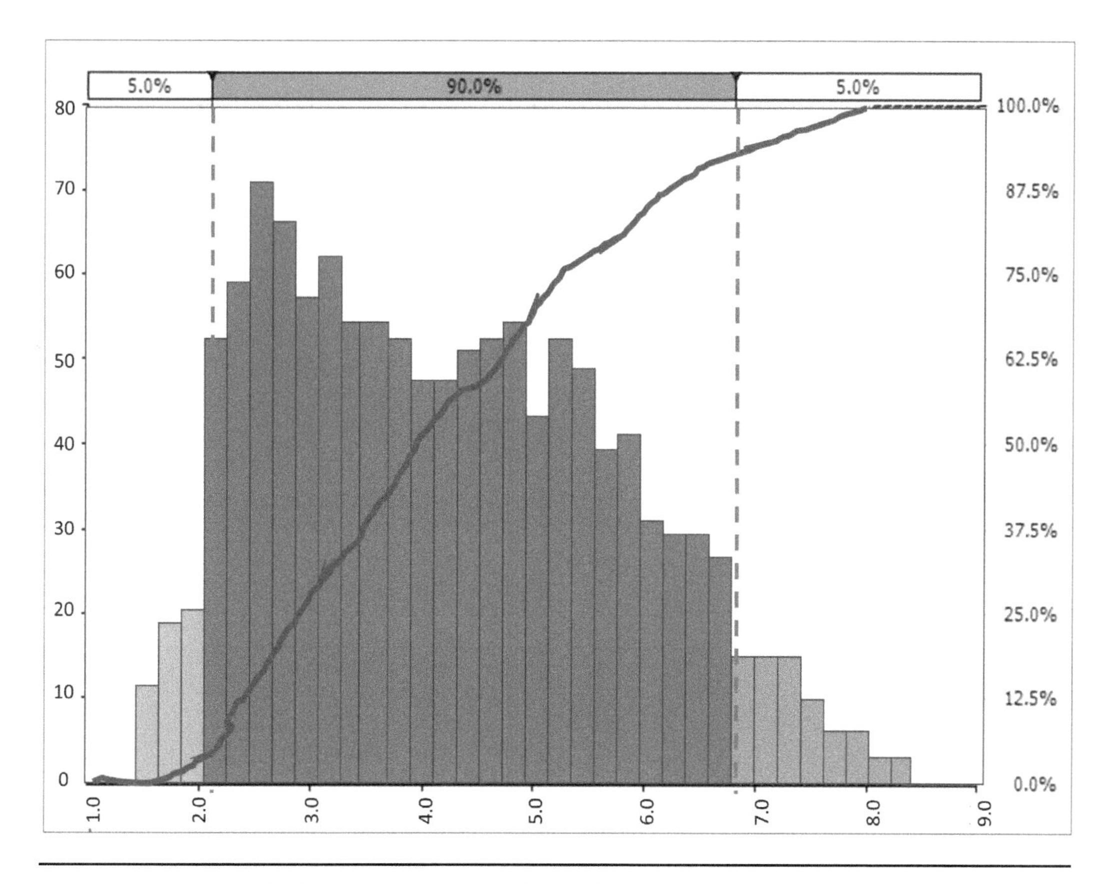

Figure 9.12 Typical Probability Distribution from a Simulation

Cost-Based Distributions

The budget information normally concerns the cost up to or between various milestones, such as those just defined, or the specific cost of various entities, based on the calculations using the costs and allocation percentages of the relevant entities.

Value-Based Distributions

The value-based information relates to the possible values of the various entities of interest, based on the value of the strategic outcomes and the contribution percentages.

Value–Cost Ratios

Value–cost ratios provide important information for validating the strategic viability of the overall plan.

9.6.2 Sensitivity Analysis

Sensitivity to changes in parameters can be evaluated progressively for each the level of the BRM: initiatives, capabilities, and business outcomes show the sensitivity of the strategic outcomes on these intermediate entities.

Sensitivity analysis can be carried out by eliminating entities one by one from the BRM, recalculating the result and comparing this with the result for the full BRM. This analysis will provide a means of ranking the entities based on their importance to the final results. Note that in this case, essential links (see Chapter 6) should be taken into account when the corresponding entity is removed.

In addition, the sensitivity of the forecasts to the values of the contribution fractions should be assessed in the case where participants in the corresponding workshop are uncertain or divided on the values to assign.

9.6.3 Hypothesis Analysis: The Confidence Matrix

Every link in the BRM carries one or more hypotheses. The hypotheses can be simple, such as whether or not a component project will succeed in creating its deliverable, or more hypothetical, such as whether a specific outcome will actually contribute as planned to the result (subsequent outcome or benefit). Although the initial formulation of the BRM is based on the assumption that the hypotheses are fully valid, alternative scenarios should be examined in which one or more of the hypotheses are not realized. This analysis can be carried out by overlaying a *confidence matrix* on top of the contribution matrix. This confidence matrix is initially created by taking a copy of the contribution matrix and then setting the non-zero cells to 100 percent—to show full confidence in the assumptions on the plausibility and value of each contribution. Alternatively, every cell in the confidence matrix can be set to 100 percent to indicate additionally that we are sure that there are no other links to be included in the BRM.

These confidence values can then be adapted either to investigate other scenarios or to carry out a set of hypothesis scenario analyses. Recalculation of the BRM taking into account the confidence matrix is carried out as follows.

As has been underlined several times, because of the break-even everywhere requirement (BEER), the BRM with all of the node contributions can be regenerated by setting the initiative values to their calculated contributions and then recalculating from left to right using the allocation fractions. The confidence matrix is taken into account by multiplying each allocation fraction by the percentage in the corresponding cell of the confidence matrix. When these are all set at 100 percent, this calculation obviously has no effect on the final result. However, confidence values lower than 100 percent will show how the final result depends on specific assumptions.

An example of ways in which to assess the credibility of assumptions is given in the discussion on key performance indicators (KPIs) in Chapter 13: By definition, KPIs are designed to provide a measure that can be used to assess the likelihood of achieving a predicted outcome.

9.6.4 Resource Loading

It is not only time, cost, and value that contribute to the uncertainty of a program or portfolio; depending on the organization, the availability of resources can vary from "fully committed" to "promised on a best-effort basis." This situation arises, for example, when some of the resources have day-to-day operational responsibilities in addition to their program and project work. Controlling this risk effectively is key to successful portfolio management in particular and, as such, also affects programs and projects. One approach for handling resource management in an uncertain environment is addressed in Chapter 10.

9.7 Summary

This chapter has developed an integrated approach incorporating risks and issues in a single model (TRIM). The definitions of risk and of issues were reworked. The various components of risks and of issues were described and explained, as well as the way in which these components work together. An integrated risk and issue management process, from identification to action planning and execution, was created and described. The way in which the TRIM can be used with the BRM as a basis for overall program risk and issue management was explained.

9.8 References

Bernoulli, D. "Exposition of a New Theory on the Measurement of Risk." *Econometrica*, 22, no. 1 (1954): 23–36.

Cabinet Office, Strategy Unit. "Improving Government's Capability To Handle Risk and Uncertainty," 2002. Retrieved from http://www.rdec.gov.tw/DO/DownloadControllerNDO.asp?CuAttachID=6279

Center for Security Studies (CSS), "Fukushima and the Limits of Risk Analysis," 2001. Retrieved from www.css.ethz.ch/publications/pdfs/CSS-Analysis-104-EN.pdf

CMMI Institute. "Capability Maturity Model Integration," 2010. Retrieved from http://cmmiinstitute.com/cmmi-models

Fahey, L., and Randall, R. M. *Learning from the Future: Competitive Foresight Scenarios*. Hoboken, NJ: John Wiley & Sons, Inc., 1997.

Hillson, D. A. *Exploiting Future Uncertainty*. Farnham, UK: Gower, 2010.

International Organization for Standardization (ISO®). *Risk Management—Principles and Guidelines* [ISO 31000:2009]. Geneva, CH: International Organization for Standardization, 2009.

Kahneman, D. *Thinking, Fast and Slow*. New York, NY: Farrar, Strauss and Giroux, 2013.

Kahneman, D., Slovic, P., and Tversky, A. *Judgement under Uncertainty: Heuristics and Biases*. Cambridge, UK: Cambridge University Press, 1982.

Kahneman, D., and Tversky, A. "Prospect Theory: An Analysis of Decision under Risk." *Econometrica*, XLVII (1979): 263–291.

Mandelbrot, B. B., and Hudson, R. L. *The (Mis)Behaviour of Markets: A Fractal View of Risk, Ruin and Reward*. London, UK: Profile Books, Random House, 2008.

Murray-Webster, R., and Hillson, D. A. *Managing Group Risk Attitude*. Aldershot, UK: Gower, 2008.

Office of Government Commerce (OGC). *An Introduction to PRINCE2: Managing and Directing Successful Projects*. Norwich, UK: The Stationery Office (TSO), 2009.

Office of Government Commerce (OGC). *Management of Risk: Guidance for Practitioners*. Norwich, UK: The Stationery Office (TSO), 2006.

Office of Government Commerce (OGC). *Managing Successful Programmes*. Norwich, UK: The Stationery Office (TSO), 2011.

Piney, C. "Applying Utility Theory to Risk Management." *Project Management Journal*, 34 (2003): 26–31.

Piney C. "A Matter of Life and Death—Applying Risk Management to Help Search and Rescue," 2014. Retrieved from http://www.pmi.org/learning/risk-management-help-search-rescue-1474

Project Management Institute. *A Guide to the Project Management Body of Knowledge* (*PMBOK® Guide—Fifth Edition*). Newtown Square, PA: Project Management Institute, 2013a.

Project Management Institute. *Practice Standard for Project Risk Management*. Newtown Square, PA: Project Management Institute, 2009.

Schuyler, J. R. *Risk and Decision Analysis in Projects*. Newtown Square, PA: Project Management Institute, 2001.

Standards Australia/Standards New Zealand. *Risk Management Guidelines: Companion to AS/NZS* 4360:2004. Sydney, Australia & Wellington, New Zealand: Standards Australia/Standards New Zealand, 2004.

Taleb, N. N. *The Black Swan: The Impact of the Highly Improbable*. New York: Random House, 2008.

Walker, W. E., Harremoes, P., Rotmans, J., van der Sluijs, J. P., van Asselt, M. B. A., Janssen, P., and Krayer von Krauss, M. P. "Defining Uncertainty: A Conceptual Basis for Uncertainty Management in Model-Based Decision Support." *Integrated Assessment*, 4, no. 1 (2003): 5–17.

Wideman, M. A. "Comparative Glossary of Project Management Terms." Retrieved from www.maxwideman.com/pmglossary/

Chapter 10

Resource Capacity Planning

When analyzing the reasons for failure of a number of internal change programs in different industries, we have found the need to address a common problem that the program managers are unable to resolve at their level. Although the program manager has a responsibility for ensuring the availability of adequate resources for the component projects, unplanned operational issues are frequently allocated a higher priority for the use of shared resources, to the detriment of project and program schedule management. This weakness in resource management is an organizational shortcoming that is rarely addressed effectively. This chapter proposes a data-driven approach for solving the issue and enhancing the capabilities of the entire organization.

10.1 The Challenge

In most organizations, internal change program teams need to include both dedicated and shared resources. Some of the key shared resources reside in the operations organization and, as such, have conflicting calls on their time. In such organizations, one challenge for the program manager is to use internal operational personnel as resources without jeopardizing the organization's day-to-day operations while maintaining the program's committed schedule.

The corresponding constraint to be addressed can be defined as follows: projects need to be able to rely on advance resource planning with the required resources available for a given period at a predefined time; operations need to have priority with minimum notice for applying the same resources to unforeseen situations. Program managers may have considerable control allocating from a pool of dedicated project resources, but have no direct authority over the allocation of the operational staff on whom they will have to rely for additional project work.

10.2 Simple Operations Resourcing Model

A practical staffing model is developed in this section, and an established theoretical model is adapted to address unpredictable staff assignments. The two models are then integrated to provide a consistent and quantifiable approach.

10.2.1 *Practical Operations Staffing Model*

The following model is suggested as an interface between the operational and the project resourcing worlds (Figure 10.1).

1. Ongoing operations (keep the business running, or KBRs): the day-to-day planned operational work
2. Scheduled operational activities (SOAs): project involvement of operational resources
3. Unplanned remedial actions (URAs): remedial operational work

Figure 10.1 The Three Categories of Operations Group Resourcing Needs

KBR and SOA resource requirements can be planned for or at least controlled in advance, in a deterministic manner using standard scheduling and resource planning tools (see, for example, Section 8.6 of *The Standard for Program Management*—Third Edition [PMI 2013c]). By definition, URA can only be forecast using statistical tools.

These three categories of resources are all staffed from operations' pool of human resources and normally used as follows:

URAs take priority over the other two components in operations if the required resources are not free at that moment. The relative priority of KBRs and SOAs to have their resources preempted is obviously a management issue for operations. This decision can only be addressed by answering a number of questions:

Q1. How is the level of staffing for KBRs and SOAs determined?
Q2. How is the level of staffing of URAs determined in a staffing prediction method?
Q3. How is staffing of specific URA events handled on an ad hoc basis?
Q4. Are some KBR or SOA activities "protected" from URA preemption?
An approach to addressing these questions is given below.

10.2.2 *Theoretical Resourcing Model for Unplanned Remedial Actions (URAs)*

The following theoretical model is based on existing resource planning models used in sizing resource requirements for handling random requests, as used for example in planning call centers (Erlang 1909). The following explanation adapts this model from the telecommunications environment to project resource planning. This Erlang-based approach will be used directly for analyzing resource requirements in the operations organization to make adequate provision for unplanned remedial requests (i.e., URAs) while also providing support for planned project and program activities (i.e., SOAs).

An Erlang is a unit of average resource loading measurement. Strictly speaking, an Erlang represents the continuous use of one resource. In practice, it is used to describe the total resource usage of one time unit (let us take one week, in the case of URAs). Simply put, if the available resources are less than the calculated number of Erlangs, then the queue of requests will increase uncontrollably. On the other hand, the greater the amount by which the number of resources exceeds the Erlang number, the shorter the expected time to wait before a resource (or set of resources) becomes free. Erlang traffic calculations are made to help telecommunications network designers cater effectively for traffic patterns within their voice networks. Calculations of this type are necessary for designing the capacity of the network correctly and determining the necessary trunk group sizes. The model allows calculation of parameters such as average waiting time and maximum waiting time. This can be applied to the operational staffing problem being addressed in this chapter.

For example, if statistically there are ten calls on a group or category of resources in one week, and each call has an average call duration of one day (i.e., one-fifth of a week) and occupies three people, then the number of Erlangs this represents is worked out as follows:

Days of resource usage in one week = number of calls * average call duration * resource loading
Total weekly load = 10 * 1/5 * 3 = 6 Erlangs

The number of Erlangs in our case will correspond to the minimum number of staff required to handle the average URA load: any fewer, and you will have to take resources from elsewhere (e.g., from SOA and KBR work)—and never catch up again. The typical question to be answered is: What level of staffing is required so that the average waiting time for the resources is one day without entailing preemption?

In our extension, Erlang traffic measurements or estimates can be used to work out how many resources are required to handle the asynchronous arrival of requests for support, with an acceptable level of service and without impacting committed ongoing operational or project-based work.

There are several Erlang models; the one that is most suitable is Erlang C. This model assumes that all blocked requests stay in the system until they can be handled. Erlang C can be applied to the design of call center staffing arrangements where, if calls cannot be immediately answered, they enter a queue. When applied to operational staffing, it corresponds to the fact that the work that needs to be carried out does not generally disappear if it cannot be addressed immediately.

For the example above, as currently available tools are designed for call center telecoms planning, we need to map the Erlang model from telecommunications environment to our program staffing environment.

- The number of resourcing requests corresponds to the number of calls.
- Duration in the standard models and the corresponding tools is normally in seconds, so we need to convert duration and allowed waiting time accordingly (I have taken 10 minutes in the telecoms model = one day in the project world).

For example, applying a calculation tool available from the Web (KoolToolz), you get the following: For a telecom environment with 12 calls per hour (i.e., 10 calls in 50 minutes), 30 minutes per call (three people for 10 minutes), and accepting an average call delay of 10 minutes, you would need eight agents. Translating this into the program environment, to deal with the random arrivals of 10 URA events a week (five days), each event requiring three resource-days of effort, and accepting an average service delay of one day, because of the potential clustering effect of random arrivals, you would need eight URA operations staff in addition to the KBR and SOA staff—although the Erlangs as calculated earlier are six (10 * 1/5 * 3). Note that the delay (one day) is the *average*; the tool also shows that 22 percent of all calls will have a greater delay due to lack of available resources. This potential overrun underlines the danger of relying only on an *average*, because an average gives no information about the total range of values (see also Savage 2009). In reality, as explained later, you also need to have a contingency plan for dealing with the eventuality—which is bound to occur in the long run—that the actual delay will be unacceptably long.

The other point to understand is that, since we have to have more than the absolute minimum number of agents to service the calls (in this case eight instead of six), some agents will be unoccupied for a part of their time, although, of course, there will be times when all agents are busy and a queue of outstanding requests is waiting to be serviced. The tool calculates this "free time" to be 25 percent on average in the present case. It is often difficult to get line management to accept any approach that leads to "idle time," as managers often consider idle time to be a symptom of overstaffing. Modeling and simulation examples need to be used and explained to help management accept the concepts and the value of using such an approach to reduce staffing crises. Training and lower priority operational work can be scheduled on an ad hoc basis to make use of this potentially idle time.

Once these concepts for setting resource levels have been accepted, they can be applied to the organization as follows.

10.2.3 Calculating Staffing Levels

We should assume that the KBR levels are predictable and have been determined in line with an agreed-upon operations plan.

The URA level should now be established using the Erlang-C model, based on historical figures of support requests and an agreed-upon service level for responding to these requests.

The sum of these two numbers will determine the base level of staff required by operations to perform its operational tasks in line with a specified set of service levels.

For the operations group also to be able to take an effective role within projects, the level of investment—and therefore involvement—in that type of work (i.e., the SOA staffing level) needs to be determined and the corresponding number of extra resources added to the operations group.

10.3 Applying the New Model

We have made considerable progress in answering the first two of the four questions at the start of this chapter (see Figure 10.2), and we can now answer question 3 ("How is staffing of specific URA events handled on an ad hoc basis?"). The answer to question 4 ("Are some KBR or SOA activities 'protected'

1. How is the level of staffing for KBR and SOA determined?
2. How is the staffing of URA managed in a staffing prediction method?
3. How is staffing of specific URA events handled on an ad hoc basis?
4. Are some KBR or SOA activities "protected" from URA pre-emption?

Figure 10.2 The Questions To Be Answered for Using Operations Resources in Projects

from URA preemption?") will depend on the financial model and the relative value of the actions in contention for URA resources.

10.3.1 Servicing URA Requests

The model assumes that you can delay servicing asynchronous service requests (i.e., the URA requests) on average for the specified time. However, it is not always possible to allow any delay, since some requests may be too urgent to wait. In this case, a formal preemption scheme needs to be agreed upon in advance. Consider one example:

1. Take highest priority URA request.
2. Identify the operations (SOA) activities to be delayed.
3. Preempt the corresponding, previously committed project resources.
4. If this is still not enough (this situation will be rare if the Erlang model has been correctly applied and regularly updated), then consider preempting KBR resources.

The impact to the program—and therefore to the organization—of the SOA preemption needs to be considered explicitly before making a final decision.

10.3.2 Staffing for Operational Staff in Projects

A financial model needs to be established, which provides a viable approach for supporting program and project work with additional resources from within the operations group. This financial model should be based on a forecast of the number of forthcoming initiatives, their value to the company, and the involvement required from within the operations group. In addition, of course, portfolio management must take into account the constraints imposed on SOA resources by this model.

10.3.3 Planning SOA Resources

To make this staffing model work, the total load of SOA resources promised in any category to initiatives should not reduce the number of operations' resources available for URA activities to a level such that the average waiting time would exceed a prespecified limit, unless there is a joint agreement between the Project Sponsor and the operations manager to accept this risk, based, if necessary, on a specific contingency plan.

Once all of these concepts are accepted, steps can be taken to adapt the organization accordingly.

10.4 Implementing the Erlang-Based Approach

10.4.1 Advantages of the Erlang-Based Approach

The advantages of the approach of including the skilled project resources within the operations group include the following:

- A broader exchange of experience within the operational team

- The availability of additional resources for operations when those resources are not required by projects
  - . . . and vice versa!
- A larger set of operational resources to call on in emergencies
- A larger number of resources will smooth out statistical exceptions
- More effective handover from project work into operations

The changes need to be planned and should be based on accurate and relevant historical data.

10.4.2 Planning the Required Organizational Changes

Planning for the Erlang-based approach needs to be able to model the environment, using reliable, quantitative data per resource skill category:

- The number of staff required to handle the KBR activities
- The number on SOA activities per week
  - The average duration of each such activity
    - or, better, a three-point estimate
      - shortest, longest, most likely values
- The number of URA calls per week
  - The average duration of each such call
    - or, better, a three-point estimate
      - shortest, longest, most likely values

It would also be beneficial to refer to a cumulative histogram based on weekly figures of these three parameters plus the number of times activities are delayed for each type of activity and by how much. This information would help to validate the accuracy of the model.

Once these parameters are known and the model is validated, it can be extended to take into account forecast and committed project resources in each category. This forecast information will provide the inputs required to evaluate the number of additional resources required in each skill category to cater for project needs, without negatively affecting operational service levels.

10.5 Organizational Change Program

The Erlang-based capacity modeling ideas should be discussed and, if agreed, integrated into an organizational change initiative in order to ensure a consistent and smooth company-wide adoption of the concepts and capabilities. This organizational change initiative should, naturally, be organized along the lines of a program, as recommended in this book!

10.6 Summary

This chapter has explained the challenge of using shared resources within projects and programs. It adapted a probabilistic approach (the Erlang formula) to a simple model of staff assignment. An example showed how the Erlang-based approach allows an organization to adapt their staffing levels to their needs for rapid access to key resources. It proposed that organizations wishing to adopt this approach should do so by means of a formal organizational change program.

10.7 References

Erlang, A. K. *The Theory of Probabilities and Telephone Conversations* [*Nyt Tidsskrift for Matematik B*], Vol. 20, 1909.
KoolToolz, Call Center Calculator. *KoolTools* website: www.kooltoolz.com/ccm.htm
Project Management Institute. *The Standard for Program Management—Third Edition.* Newtown Square, PA, USA: Project Management Institute, 2013c.
Savage, S. L. *The Flaw of Averages: Why We Underestimate Risk in the Face of Uncertainty.* Chichester, UK: John Wiley & Sons, 2009.

Chapter 11

Procurement

This chapter explains how the benefits realization map (BRM) can be used to support value-added procurement management. It is based on an award-winning program for the London Underground that was managed successfully by Simon Addyman. I would like to thank him for allowing me to use his formal report and to reprint of some key tables, as well as for providing comments to improve this chapter. Any errors of fact or concept are, however, entirely my responsibility.

11.1 Background: Innovative Contractor Engagement and the London Underground Bank Station Capacity Upgrade

This chapter is inspired by the pioneering approach known as innovative contractor engagement (ICE), which was adopted in the London Underground program to upgrade the capacity of the major London Underground interchange stations (Mayor of London and the London Underground 2014). The project manager, Simon Addyman, was awarded the Association for Project Management (APM) Project Professional of the Year award for successfully developing and implementing the methodology, after the overall concept had been developed from lessons learned within the Stations Capacity Programme. When I heard about this approach, I was immediately struck by the way it could fit into the integrated results-chain-led program management methodology that is the subject of this book.

To quote from the introduction to a report on the project (Mayor of London and the London Underground 2014):

> With the objective of procuring better value whilst delivering projects, Innovative Contractor Engagement (ICE) has been conceived to ensure that the good ideas the market has in response to project requirements can be bought forward and developed with the client as soon as possible for maximum benefit.
>
> ICE has been pioneered on a major upgrade project at Bank Station and the results demonstrate the spectacular increase in value that the industry can achieve when we—client, designer, Tier 1 contractors and their supply chain—get it right. The winning bid is a clear demonstration that good ideas from the market will deliver better value and win bids.
>
> The successful execution of ICE on Bank has provided London Underground (LU) with a platform for future development which addresses our historic challenge of how to control costs,

speed up the works and reduce the impact on the travelling public. It also reinforces our aspirations to be an intelligent, innovative and efficient client that can build strong relationships with the supply chain whilst delivering value to the public purse, for the travelling public, and for London. A sure and significant step to a better future.

11.2 Tendering Lifecycle

11.2.1 Process Overview

Once the project requirements had been clearly defined, the project team developed their *base case* and shared it in detail with the bidders. In fact, all project documentation that had been developed prior to tender was shared with the bidders. During a negotiated dialogue, the bidders were encouraged to provide innovative changes to the clients' base case. The project team then reviewed these suggestions and assessed their value against the client's original business requirements with the aim of enhancing the value of the initial proposed solution. The confidentiality of individual bidders' work was preserved throughout the tender process, and the opportunity to incorporate their innovative ideas into their final bid proposals was facilitated through careful wording of the tender documentation. Where appropriate and of use to the client, unsuccessful bidders were rewarded financially if any of their innovative ideas were adopted for mitigating threats or enhancing benefits in the project.

All of the selection criteria were made clear to the bidders in advance of the negotiated dialogue and then formalised before the formal tender documentation was issued. "The tender scoring targeted a more 'Effective Product' (70%), the long term viability of the investment, and (30%) on an 'Efficient Method,' the short term service provision" (Mayor of London and the London Underground 2014). On contract award, the design and build contractor worked with the client to revise the concept design, seek statutory planning approvals, and complete detailed planning before gaining final financial approval for the construction stage.

The steps outlined above were managed within a formal five-phase life cycle: (1) pre-dialog, (2) dialog, (3) interim, (4) invitation to tender (ITT), and, finally, (5) evaluation and award.

11.2.2 Pre-Dialog Phase

The pre-dialog phase focused on setting up the overall governance for the program. It ensured acceptance of the ICE approach and the associated documents and templates. Once the acceptance was granted, the project could be officially launched with the potential bidders.

11.2.3 Dialog Phase

The dialog phase allowed the bidders to understand the base scheme and provided them with all information they might need, to discover value-added innovations. It provided the bidders with details of the evaluation model by means of a calibrated scale of priorities.

The tender evaluation model comprises three parts as shown in Table 11.1.

It is important to underline that, as shown in Table 11.1, the potential effects, threats, and opportunities are included in the tender evaluation model. This ensures that risk management is built into the overall solution, rather than being added as an afterthought. The value of threat- and opportunity-related items in the proposals is reflected in the concept of *quality points* in the evaluation model. The same approach is used to encourage value-added innovation.

Table 11.1 The Three Sections of the Tender Evaluation Model

1. Mandatory Questions (financial, organisational, HSQE, Value for Money (VFM)),
2. Core Requirements as below:
   - CR1 – Capacity Enhancement;
   - CR2 – Reduction in Journey Times;
   - CR3 – Disruption during construction;
   - CR4 – Step Free Access;
   - CR5 – Fire and Evacuation Plan;
   - CR6 – Time; and
   - CR7 – Value for Money.
3. Management of Risks and Opportunities:
   - RO1 – Risk Management and Employer's Risks;
   - RO2 – Transport and Works Act Order;
   - RO3 – Design and Construction; and
   - RO4 – Opportunities.

Note: Reprinted with permission from Mayor of London and the London Underground, 2014.

The value of a quality point is determined based on comparing the estimated costs for the base case against the monetary benefits they provide in the business case. In terms of the benefits realization map (BRM), the estimated costs are termed *allocations* and the monetary benefits are called *contributions*.

As a convention, to allow bidders to provide solutions that deviate (for better or worse) in a number of features from the base case, each such feature is given a baseline quality point score out of ten.

The evaluation of the value of the quality scores generally involves the project team applying quantitative weightings to the set of requirements. Some of the monetary benefits or disbenefits in the business case can be directly monetized to elements of the base case. For example, the disbenefit costs of shutting the railway to do work ("disruption during construction" in Table 11.2) have a definite financial impact that is directly proportional to time. In this case, if the base case is scored as 5/10 and closes the railway for 22 weeks at a disbenefit of –£54 million, the quality point for closure of the railway would be £54 million / 5 = £10.8 million. A score of 10/10 would be assigned to no closure of the railway, with an added value of £54 million.

Once you have the assessment method, you can run a sensitivity analysis on the other criteria, either based on a similar monetary relationship or on value management workshop techniques, including representatives from several functions from the client organization. In all, across the full range of criteria, there are 100 possible points.

Once bidders understand the value of a quality point, they become less focused on controlling cost come what may. They may propose an innovative solution that costs £5 million extra but brings a benefit of a single quality point worth £10.8 million. In this case, all other things being equal, that bidder will still be considered to be less expensive by £5.8 million (£10.8 million – £5 million) than everyone else. Without this kind of evaluation, why would the bidder knowingly bid £5 million higher, even though their proposal provides a better solution to the requirements?

The quality point approach provides a total monetary value for a proposal (the *benefit score*) that takes into account the degree of conformity with the baseline set of requirements plus the value of innovative improvements. The *value for money* of each bid can then be calculated as VFM = (benefit score)/(whole life cost). The VFM will then be used to rank the bidders' submissions.

Once the details of this process have been agreed upon, the actual content and values of the evaluation for the specific project must be defined and approved.

11.2.4 Interim Phase

The interim phase links the ICE weighting and scoring concepts to the standard ITT process in the following phase. In the interim phase, the bids are submitted.

The qualitative and quantitative bids are submitted in separate envelopes. The qualitative criteria are evaluated against the scoring and evaluation model. The quantitative information is reviewed and assessed for credibility and accuracy. The results of the *independent* evaluation of the two envelopes are then combined in the evaluation model to determine the MEAT (most economically advantageous tender).

Once the results of this phase are approved, the standard ITT processes can be invoked.

11.2.5 The Invitation to Tender (ITT) Phase

The ITT phase has two stages:

Stage 1: Compliance, evaluation, due diligence, and identification of unique outputs proposed by the bidder

Stage 2: Negotiation and agreements for any unique outputs, development of specific key performance indicators (KPIs), development of draft contracts, and preparation of recommendations for awarding contracts

11.2.6 The Evaluation and Award Phase

This phase provides the link from the ICE to the program implementation itself.

The recommendations from the previous phase are reviewed. Bidders are asked to confirm elements in their proposals. Contracts for unique value-added options that have been retained are negotiated. Compensation to losing bidders for their proposed innovations is also considered.

A formal submission is prepared and submitted to the strategic board. At the same time, the program board is formally notified and engaged for the forthcoming work.

11.3 ICE and the Benefits Realization Map (BRM)

Although the whole of this procurement approach is interesting in its own right, this chapter will now focus on comparing the BRM with the ICE supplier assessment and evaluation approach.

ICE and BRM have a number of similar goals:

- The determination of cost and value independently to ensure an honest and unbiased model as a means of discouraging "strategic bids" (i.e., low initial cost compensated by expensive change orders), as the ICE report puts it.
- The evaluation of overall benefits based on value-added intermediate outcomes.
- The trade-off between benefits, disbenefits, threats, opportunities, value-added capabilities, and total cost.

- The management of complex stakeholder relationships; as the ICE report states: "The key challenge . . . was in managing multiple internal stakeholders and maintaining objectivity."

The specific compatibility between the ICE approach and the BRM technique is analyzed in more detail below.

11.4 Modeling the Value of the Options

The ICE approach (Mayor of London and the London Underground 2014) stated:

> Approval of the ICE Strategy at the TfL Board in March 2012 set the project a target of 15% additional value to be achieved through a combination of cost savings, improved benefits and reduction of dis-benefits (blockade). A scoring and evaluation model was required that would reflect the purchasers' core requirements and priorities in all circumstances. As far as reasonably practicable, the model also had to discourage any opportunity for the bidders to "strategically bid" the project.

The concepts of benefits, disbenefits, and an honest model align completely with the goals and capabilities demonstrated earlier for the results chain approach. In the Bank Station Capacity Upgrade Project (BSCU), the bidders were measured on a well-defined set of contributions to overall success, comprising specific project requirements as well as further project parameters deemed to be of value by the client.

The ratings are shown in Table 11.2.

Table 11.2 BCSU Tender Evaluation Ratings

Product	
Capacity Enhancement	17.0%
Reduction of Journey Times	17.0%
Design & Construction Layout & Approach	15.0%
Step Free Access	10.0%
Fire and Evacuation Plan	10.0%
Subtotal Product	69.0%
Method	
LU Project Business Case risk reduction	2.5%
Transport and Works Act Order	5.0%
Disruption during construction	12.5%
Time DfT Milestone	2.5%
Design to Cost	2.5%
Opportunities	6.0%
Subtotal Method	31.0%
Total	**100%**

Note: Reprinted with permission from Mayor of London and the London Underground, 2014.

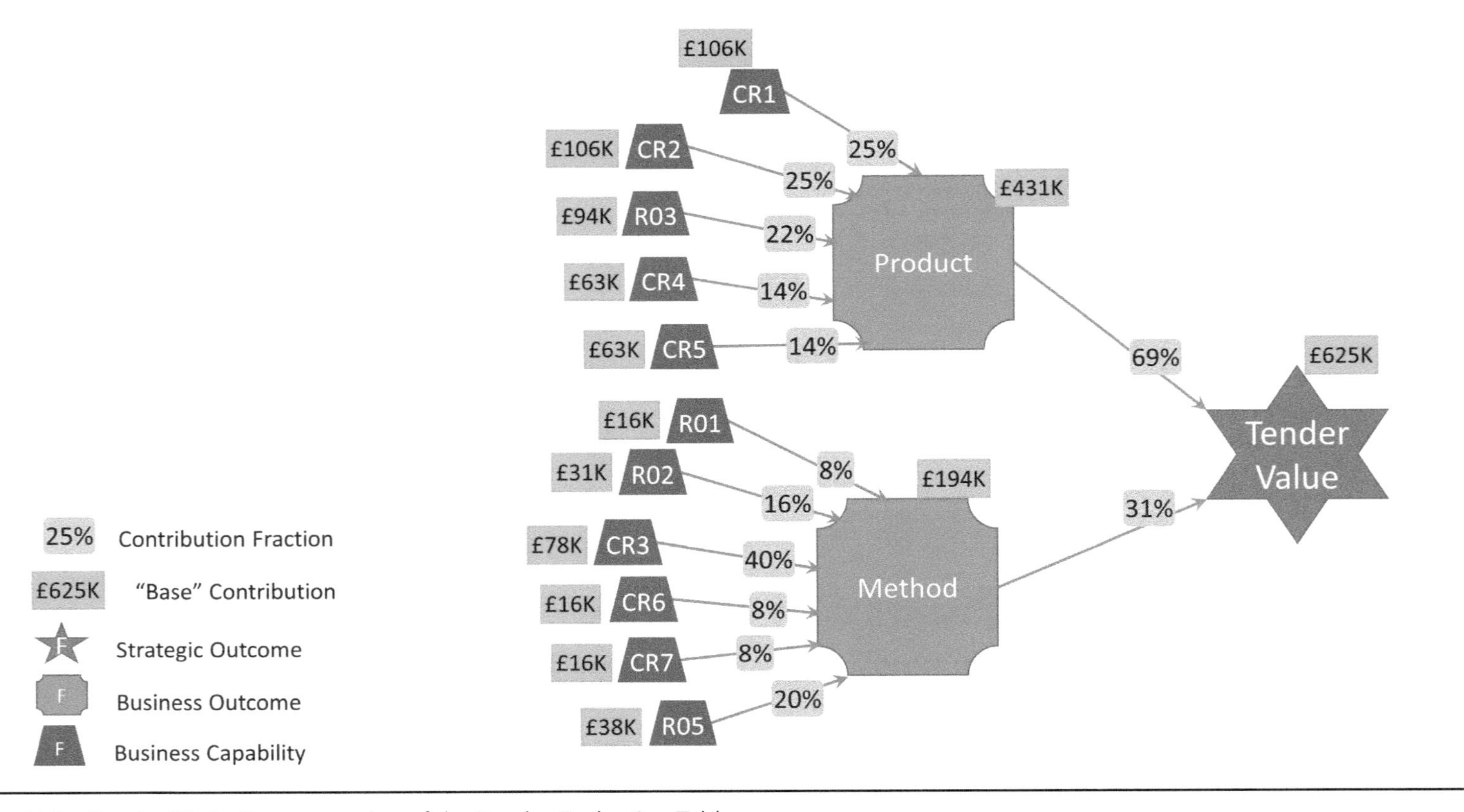

Figure 11.1 Results Chain Representation of the Tender Evaluation Table

The criteria in Table 11.2 can be associated with strategic outcomes in a benefits map. The technical solution for achieving these outcomes would then be detailed and costed in the technical proposals from the bidders and correspond to the initiatives in the results chain approach. A simplified BRM corresponding to these ICE criteria is shown in Figure 11.1.

There is another important similarity with the approach recommended in earlier chapters for developing the BRM. In the ICE ITT phase, "Bidders were required to submit qualitative and quantitative responses in separate envelopes." The qualitative responses correspond to the contributions in the results chain. The quantitative numbers provide the allocations. "The outputs for the evaluation of the two envelopes are then combined in the evaluation model to deliver a MEAT result."

Once the ratings have been established, they can be applied to the various submissions. An example of such an assessment (not actual bid figures) is given in Table 11.3. In this example, each proposal is

Table 11.3 BCSU Tender Evaluation Example

		BASE	1	2	3	4
Product		£625,000	£625,000	£625,000	£625,000	£625,000
CR1: Capacity Enhancement	17.0%	8.50	10.00	8.50	12.00	10.00
CR2: Reduction in Journey Times	17.0%	8.50	8.50	8.50	12.00	8.50
RO3: Design & Construction	15.0%	7.50	8.50	7.50	7.50	7.50
CR4: Step Free Access	10.0%	5.00	5.00	10.00	5.00	7.00
CR5: Fire and Evacuation Plan	10.0%	5.00	5.00	5.00	5.00	5.00
Subtotal Product	69.0%	34.50	37.00	39.50	41.50	38.00
Method						
RO1: LUL Project Business Case	2.5%	1.25	1.25	1.25	1.25	1.25
RO2: Transport and Works Act Order	5.0%	2.50	2.50	2.50	2.50	2.50
CR3: Disruption during construction	12.5%	6.25	6.25	6.25	8.00	6.25
CR6: Time	2.5%	1.25	1.25	1.25	1.25	1.25
CR7: Design to Cost	2.5%	1.25	1.25	1.25	1.25	1.25
RO5: Opportunities	6.0%	3.00	3.00	3.00	3.00	3.0
Subtotal Method	31.0%	15.50	15.50	15.50	17.25	15.50
Total	100%	**50.00**	**52.50**	**55.00**	**58.75**	**53.50**
Total Rank		**5**	**4**	**2**	**1**	**3**
Value Rating – £100m = 8.00 Quality pts if P = £625m		8.00	8.40	8.80	9.40	8.56
VR Rank		**5**	**4**	**2**	**1**	**3**
VFM		85.11%	89.36%	93.62%	100.00%	91.06%
P equiv		£531,915	£558,511	£585,106	£625,000	£569,149
Lost Value		£93,085	£66,489	£39,894	£0	£55,851
Value of a quality point (rank 1 .v. rank 4) £k		£10,638				

Note: Reprinted with permission from Mayor of London and the London Underground, 2014.

evaluated against the previously defined criteria (Table 11.2) and the base case. This allows the value of each proposal to the client to be calculated based on the previously determined value of the base case (being scored as a baseline of 50/100). Each proposal, therefore, has its own specific *notional value* (P). At this point, the ROI can be calculated based on the submitted price.

Innovative features from the other proposals can then be "bought" from the proposing company for incorporation into the final solution, if they are shown to increase the overall ROI.

Although this was not the case for the BCSU project, the BRM technique could be used to develop a tender evaluation table. A subset of the corresponding results chain is shown in Figure 11.1. The contribution fractions into node "Tender Value" are taken directly from the tender evaluation table (Table 11.3). The values of the contribution fractions into the nodes "Product" and "Method" are taken from the same table but normalized to sum to 100 percent for each node. The technical details required to deliver the business capabilities (CR1, etc.) will then be specified by the selected suppliers.

11.5 The Next Step: Tracking the Work

Once all of the planning has been completed and all of the key stakeholders agree that the plan has an acceptable chance of success, the contracts with the suppliers can be signed. It is then time to execute the plan. One of the key challenges in program management is to track this execution, though not merely with respect to the work carried out and the deliverables produced; the sponsor and strategic teams are most interested in how the program is progressing in relation to achieving the benefits specified in the foundation documents and quantified in the business case. This is the domain of *earned benefit*, as explained in the next chapter.

11.6 Summary

This chapter integrated the ICE approach with the BRM as a means of obtaining the optimal result of program tendering that is fully compatible with all of the other stages of earned benefit program management. An example from the London Underground Bank extension was described. A part of the BRM for this example was then developed.

11.7 References

Mayor of London and the London Underground. *Innovative Contractor Engagement*. SECBE, 2014. Provided to the author by Simon Addyman, and also available from: https://www.secbe.org.uk/content/panels/Report%20-%20Innovative%20Contractor%20Engagement%20Procurement%20Model%20-%20Bank%20Station%20Capacity%20Upgrade-6d5f2a.pdf. Downloaded May 12, 2015.

Chapter 12

Implementation Tracking—Earned Benefit

For total program management, implementation tracking must be fully compatible with the tools and concepts used for strategic alignment, business justification, benefits mapping, and action planning. This integrated and compatible progress analysis and forecasting approach is provided by the application of the earned benefit techniques that will be developed, explained, and applied in this chapter.

Even as a project manager, I had been concerned about the meaning of the term "value" in "earned value management." The basic question I asked myself was: "Whose value is it?" Once I started working on programs I had a sudden, fundamental insight: Since project managers are expected to act and behave as suppliers, the term "value" represents the value of progress at any point in time *to the supplier*—that is, the estimated *budget* for the corresponding work. For a program manager, "value" needs to represent the value from the point of view *of the receiving organization*—that is, the potential *benefit* of the work accomplished to date. For this reason, earned benefit provides a more realistic measure of progress than can be obtained from earned value management (Project Management Institute 2011a) for a program or portfolio because it is based on the forecast value of the planned outcome, rather than the cost of the planned work. The initial earned benefit concept, based on a *component–benefit matrix*, was described in Piney (2013) and will be used to introduce the approach. However, a more integrated mode of calculation that uses some of the additional concepts such as essential links that were developed earlier in the book will be presented once the initial approach has been explained.

12.1 The Earned Benefit Approach

The concept of earned benefit for a program is a natural extension of the project management *earned value* management to program management.

The approach is to recognize that projects aim at deliverables, whereas programs aim at benefits. For a program, therefore, the goal is to translate from progress on delivering the scope of the program components into progress on realizing the benefits. This innovative progress monitoring approach is described below and then explained by use of a worked example which helps to demonstrate how earned value on its own can give a misleading view of actual program progress. The corresponding curves are also shown along with a glossary of earned benefit terms and abbreviations, plus the numerical value of each formula with respect to the example.

Whereas the component project managers report progress with respect to work accomplished (i.e., percentage complete), program stakeholders are interested in progress toward achieving the strategic objectives based on the benefits. For this reason, the earned benefit approach translates component tracking information into potential benefits. This mapping is represented by a *component–benefit matrix*.

12.2 The Component–Benefit Matrix

As explained earlier and shown in the diagrams of the results chain, each component can contribute to multiple benefits. These relationships are reflected in a component–benefit matrix.

Each cell of the matrix indicates the relative contribution of a component (row) to a benefit (column); a value of zero indicates that the component does not contribute to that benefit. The values are then normalized so that each column sums to 100 percent.

The values in the matrix can be represented as V_{ij}, as shown in Table 12.1.

Table 12.1 Generalized Component–Benefit Matrix

Component \ Benefit	B1	B2	B3
C1	V_{11}	V_{12}	V_{13}
C2	V_{21}	V_{22}	V_{23}
C3	V_{31}	V_{32}	V_{33}
C4	V_{41}	V_{42}	V_{43}
C5	V_{51}	V_{52}	V_{53}
C6	V_{61}	V_{62}	V_{63}

The component–benefit matrix is the core of this approach to earned benefit and will be explained along with the numerical example shown in Figure 12.1.

12.2.1 Calculating the Component–Benefit Matrix

To calculate the components of the matrix, using the benefits realization map (BRM), each strategic outcome (benefit) is considered separately. The value of each of the other strategic outcomes is set to zero, and the value (contribution) of each of the initiatives is then calculated from right to left using the contribution fractions. The ratio between to full strategic outcome (benefit) and the value calculated for each initiative provides the value of the corresponding component–benefit matrix cell.

In Figure 12.2, the focus is on outcome F (with a value of $40,000), so outcome G has been set to zero. The values are therefore:

C = 40% F = 40% * $40,000 = $16,000
D = 60% F = 60% * $40,000 = $24,000

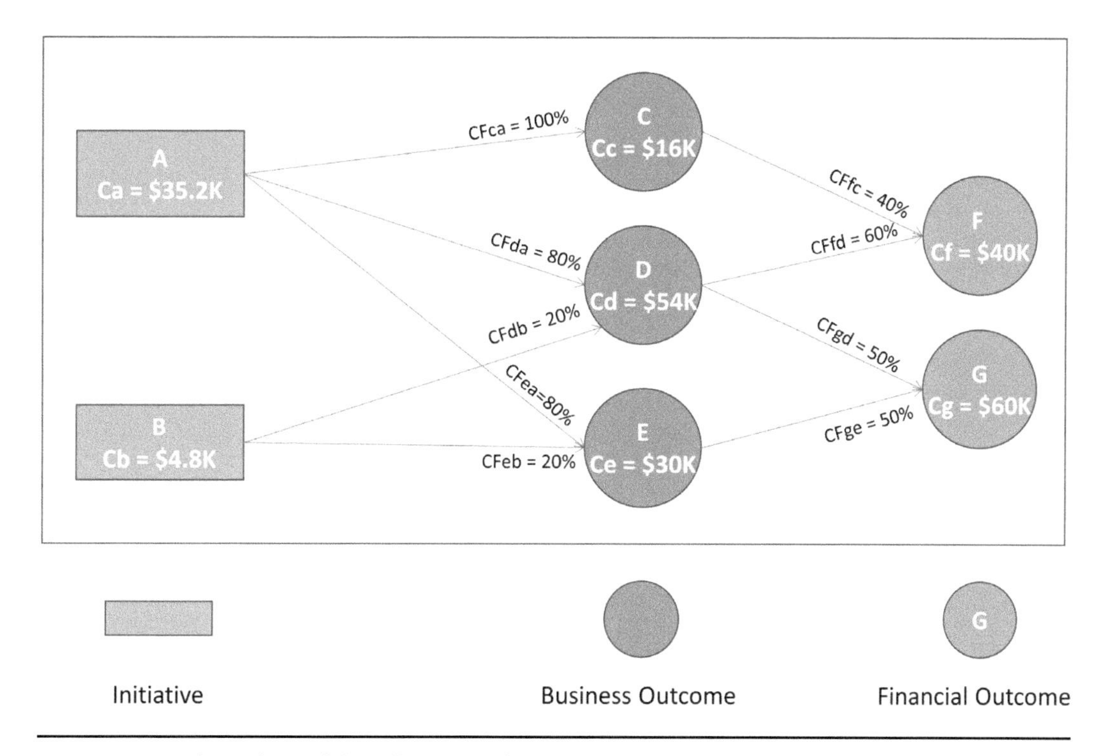

Figure 12.1 The Values of the Allocations for the Initial Earned Benefit Example

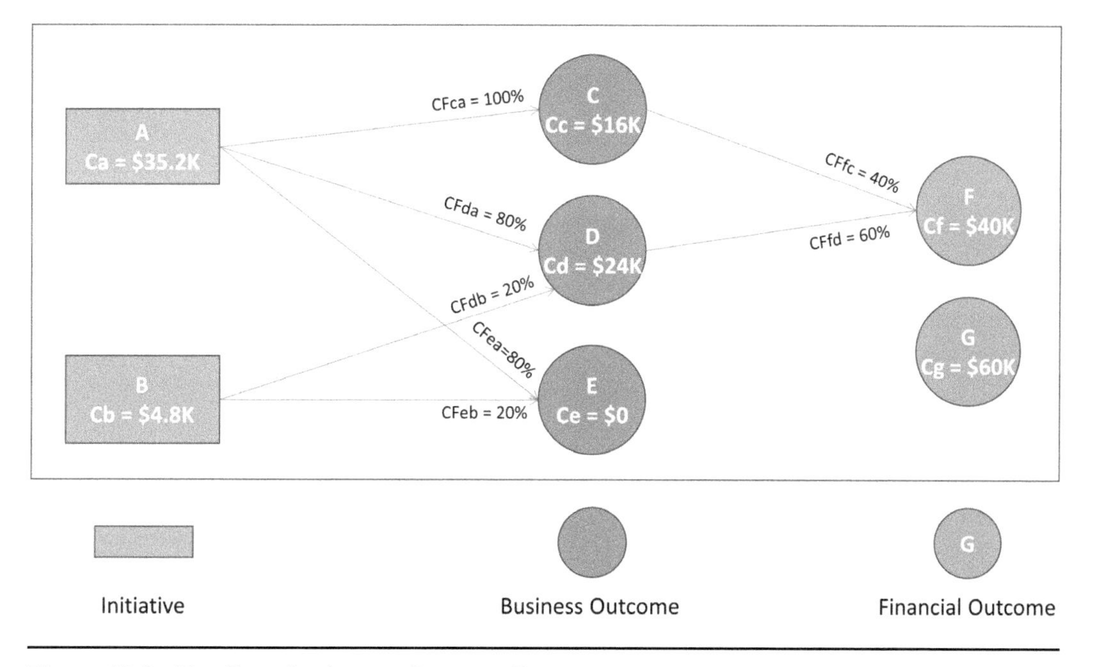

Figure 12.2 The Contributions to Outcome F

E = depends only on G so by hypothesis, E = $0
A = 100% C + 80% D + 80% E = $16,000 + $19,200 = $35,200
B = 20% D + 20% E = $4,800

Similarly, in Figure 12.3, focusing on G (with a value of $60,000) by setting F to zero gives the following values:

C = 40% F = $0
D = 60% F + 50% G = 50% * $60,000 = $30,000
E = 50% G = 50% * $60,000 = $30,000
A = 100% C + 80% D + 80% E = $0 + 80% * $30,000 + 80% * $30,000 = $48,000
B = 20% D + 20% E = 20% * $30,000 + 20% * $30,000 = $12,000

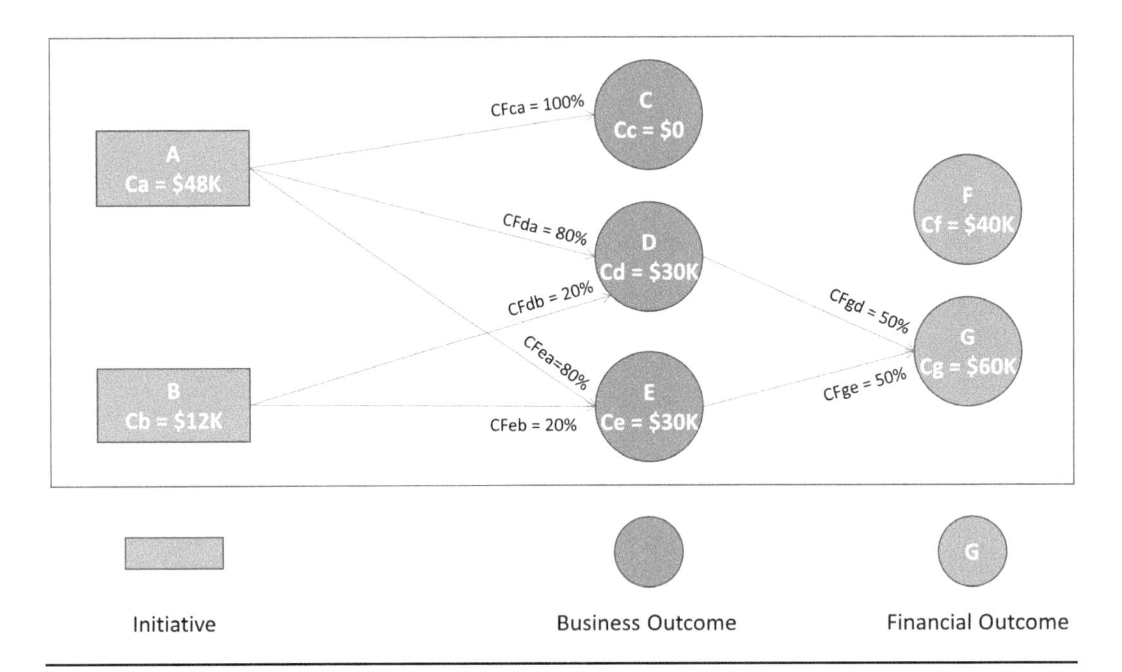

Figure 12.3 The Contributions to Outcome G

Each of the components (V_{ij}) of the component–benefit matrix can be evaluated from Figure 12.2 and Figure 12.3.

- Figure 12.2 (focus on F):
  - V_{AF} = 35.2 / 40 = 88%
  - V_{BF} = 4.8 / 40 = 12%
- Figure 12.3 (focus on G):
  - V_{AG} = 48 / 60 = 80%
  - V_{BG} = 12 / 60 = 20%

These numbers give the following component–benefit matrix (Table 12.2) for the simple example:

Table 12.2 The Component–Benefit Matrix

	F	G
A	88%	80%
B	12%	20%

This algorithm enables the various earned benefit formulae to be evaluated at each stage of the implementation once the roadmap—showing planned dates—and the cumulative cost curve are available. Similar calculations can also be carried out with respect to intermediate outcomes, using a more general *component–outcome matrix* evaluated in a similar manner to the component–benefit matrix, as shown below (Table 12.3).

The component–benefit matrix and component–outcome matrix are, as explained, based on the contribution of the initiatives to the specified outcomes. In contrast to the earned benefit approach, which is based on contributions, earned value works with the allocations (i.e., the costs) and focuses solely on the initiatives.

Table 12.3 The Component–Outcome Matrix

	C	D	E	F	G
A	100%	50%	80%	88%	80%
B	0%	50%	20%	12%	20%

12.2.2 Calculating the Earned Benefit

By definition, completed component (i) contributes to each of the benefits B_j according to the elements in row (i) of the component–benefit matrix, so, when partially complete, the component contributes a correspondingly reduced amount (the component earned benefit, or CEB_i). Using the component–benefit matrix as shown for example in Table 12.1:

$$CEB_i = \%complete[component(i)] * \sum_{cols\ j} (V_{ij} * B_j)$$

To calculate the total earned benefit, sum all of the CEB_i contributions, as follows:

$$EB = \sum_{rows\ i} \left(\%complete[component(i)] * \sum_{cols\ j} (V_{ij} * B_j) \right)$$

12.3 Earned Value

Whereas earned benefit focuses on the strategic outcomes (F and G in Figure 12.1), earned value is based on the initiatives (A and B in the same example). To compare the two approaches, we can calculate an *allocation–component matrix* showing how the cost (allocation) of each node (i) depends on each initiative (j). The cells of this matrix are given as A_{ij}. The result entails calculating the allocations from each initiative separately from left to right across the model. For the current example, Figure 12.4 gives the allocation values from A on its own.

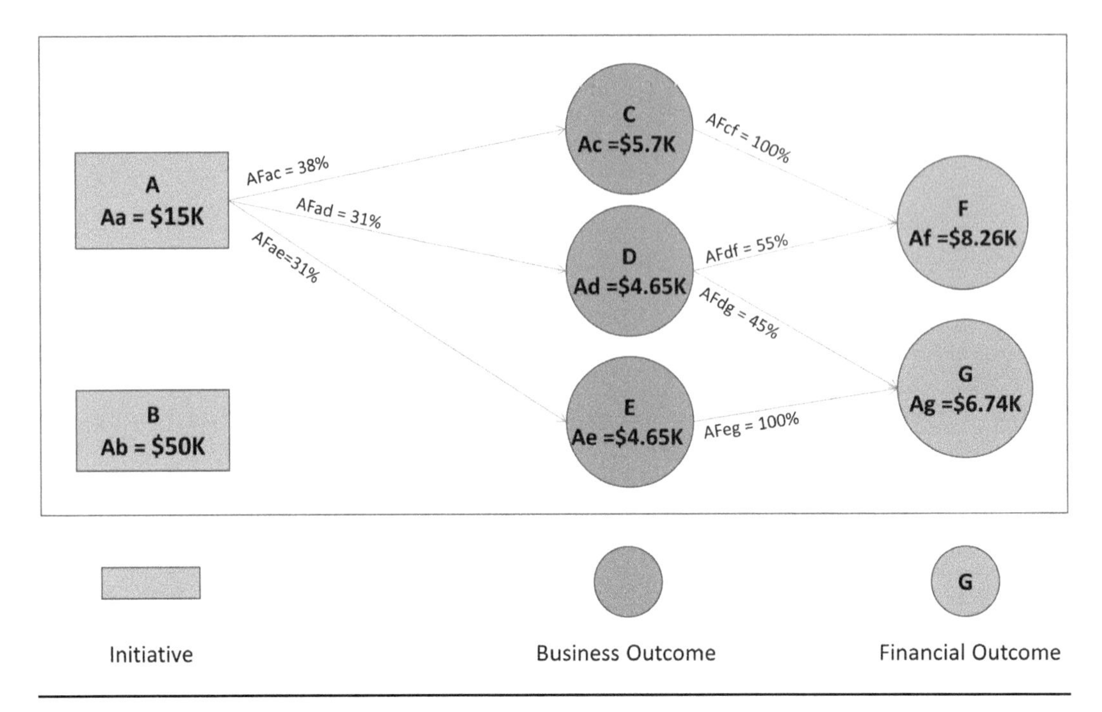

Figure 12.4 Allocation Distribution from Initiative A Alone

Comparing those values with the total allocations in Figure 12.1 provides the values in the allocation–component matrix (Table 12.4) as follows:

- A_{AC} = 5.7 / 5.7 = 100%
    - A_{BC} = 1 – 100% = 0%
- A_{AD} = 4.65 / 29.6 = 16%
    - A_{BD} = 1 – 16% = 84%
- A_{AE} = 4.65 / 29.6 = 16%
    - A_{BE} = 1 – 16% = 12%
- A_{AF} = 8.26 / 22 = 38%
    - A_{BF} = 1 – 38% = 62%
- A_{AG} = 6.74 / 43 = 16%
    - A_{BG} = 1 – 16% = 84%

Table 12.4 The Allocation–Component Matrix

	C	D	E	F	G
A	100%	16%	16%	38%	16%
B	0%	84%	84%	62%	84%

This analysis indicates, for example, that only about one-seventh of the expenditure on G comes from the investment in A (A_{AG} = 16%), whereas Table 12.2 indicates that four-fifths of the benefit comes from this same investment (V_{AG} = 80%). This result underlines once again the synergy and interworking of components in a program.

12.4 Further Development of the Earned Benefit Calculations

The component–benefit matrix approach explained above for calculating earned benefit was the basis for the development of a number of the ideas in this book. The challenge that the concept raised was to find an effective means of generating the component–benefit matrix. However, as these ideas were being developed—the BRM with contributions, allocations, and additional concepts such as disbenefits and essential links—some additional concepts were discovered that showed that the initial earned benefit method based on the component–benefit matrix was incomplete. These ideas were the break-even everywhere requirement (BEER) algorithms, and essential links.

The development of these additional ideas will be explained alongside the following case study.

12.5 Earned Value and Earned Benefit Case Study

To make the earned benefit features more obvious, a new case study is provided. The analysis of the business opportunity for this case study is deliberately less sophisticated than for the QERTS case; the initiatives are linked directly to the strategic outcomes, as mirrored directly by the component–benefit matrix in Table 12.5. In this simple case, the component–benefit matrix cells provide the contribution fractions of the components.

Table 12.5 The Component–Benefit Matrix for the Example

Component \ Benefit	B1 €100,000	B2 €200,000	B3 €300,000	Percent Complete	Actual Cost
C1 €200,000	0.25	0.25		100%	€200,000
C2 €40,000	0.5			100%	€30,000
C3 €60,000		0.083	0.625	75%	€50,000
C4 €50,000		0.333		75%	€50,000
C5 €20,000	0.25	0.083		100%	€20,000
C6 €100,000		0.25	0.375	0%	€0

The case study will be used to contrast the ways in which earned value and earned benefit describe program status. Some questions raised by the initial results from this case study will be used to explain why and when earned value and earned benefit can provide significantly different messages and, potentially, lead to different decisions.

12.5.1 The Garden Services Business Plan

In this case study, you are planning to set up a gardening company. Your benefits will come from three different groups of clients:

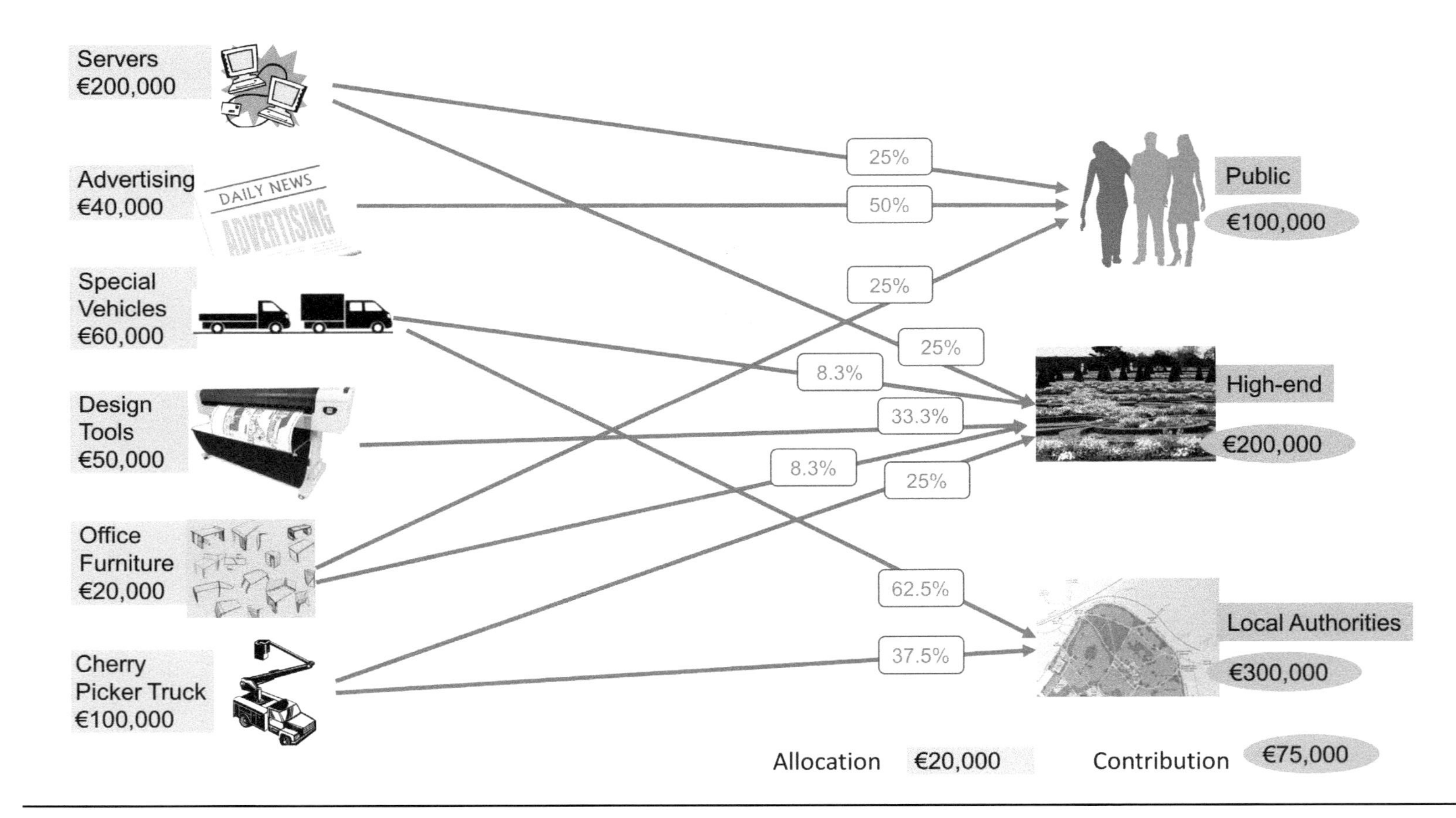

Figure 12.5 Garden Services Business Plan

B1—The Public: €100,000 per annum
B2—High-end clients and developers: €200,000 per annum
B3—Local authorities: €300,000 per annum

To achieve these benefits, you have identified that you need six areas of investment:

C1—Information technology (computers, server, plus a 3-D printer; more about that later): €200,000
C2—Advertising: €40,000
C3—Tennis court development and repair equipment: €60,000
C4—Architectural design tools and plotter: €50,000
C5—Office furniture: €20,000
C6—Cherry-picker truck: €100,000

This business analysis is depicted in Figure 12.5, showing the contributions from the public, high-end, and local authority markets. It also shows the allocations to servers, advertising, special vehicles, design tools, office furniture, and the cherry-picker truck. The contribution percentages are also indicated.

The schedule is specified as follows (Figure 12.6).

The program starts on January 1st and is composed of three stages, aligned with the realization of benefits:

Stage 1 (Benefit B1): (The Public) completion of components Servers, Advertising, and Office Furniture, planned for April 1st
Stage 2 (Benefit B3): (Local Authorities) adding components Vehicles and Cherry-Picker, complete by September 1st
Stage 3 (Benefit B2): (High-End Clients) adding component Architectural Design Tools, on-board for September 1st

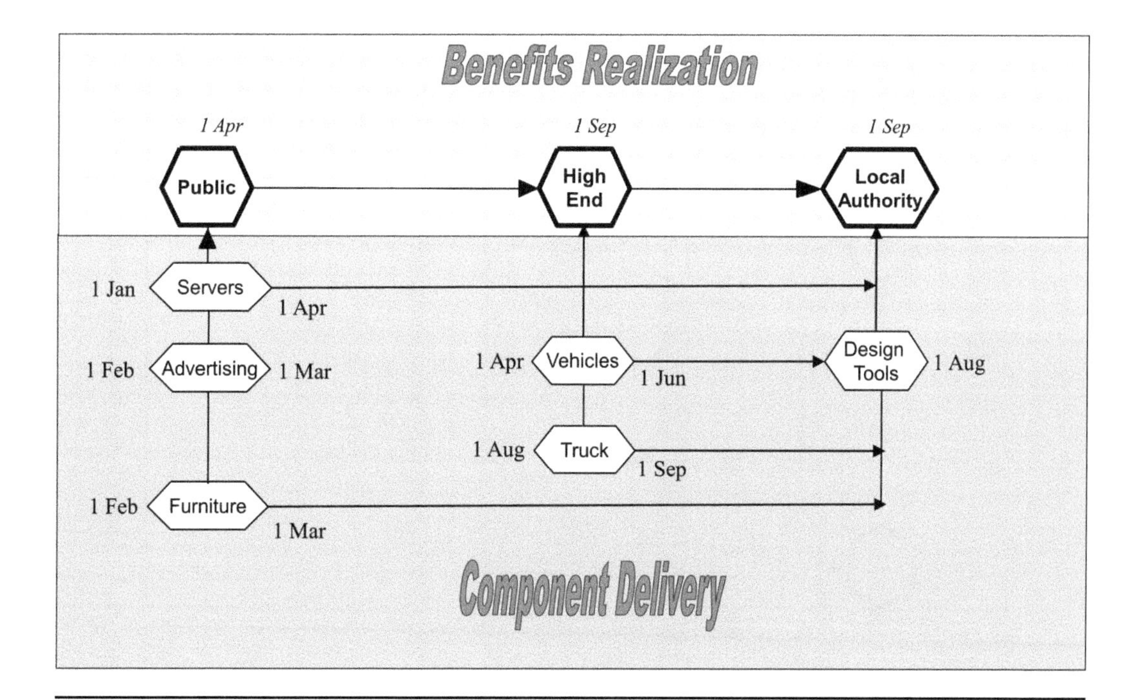

Figure 12.6 Schedule for the Garden Services Program

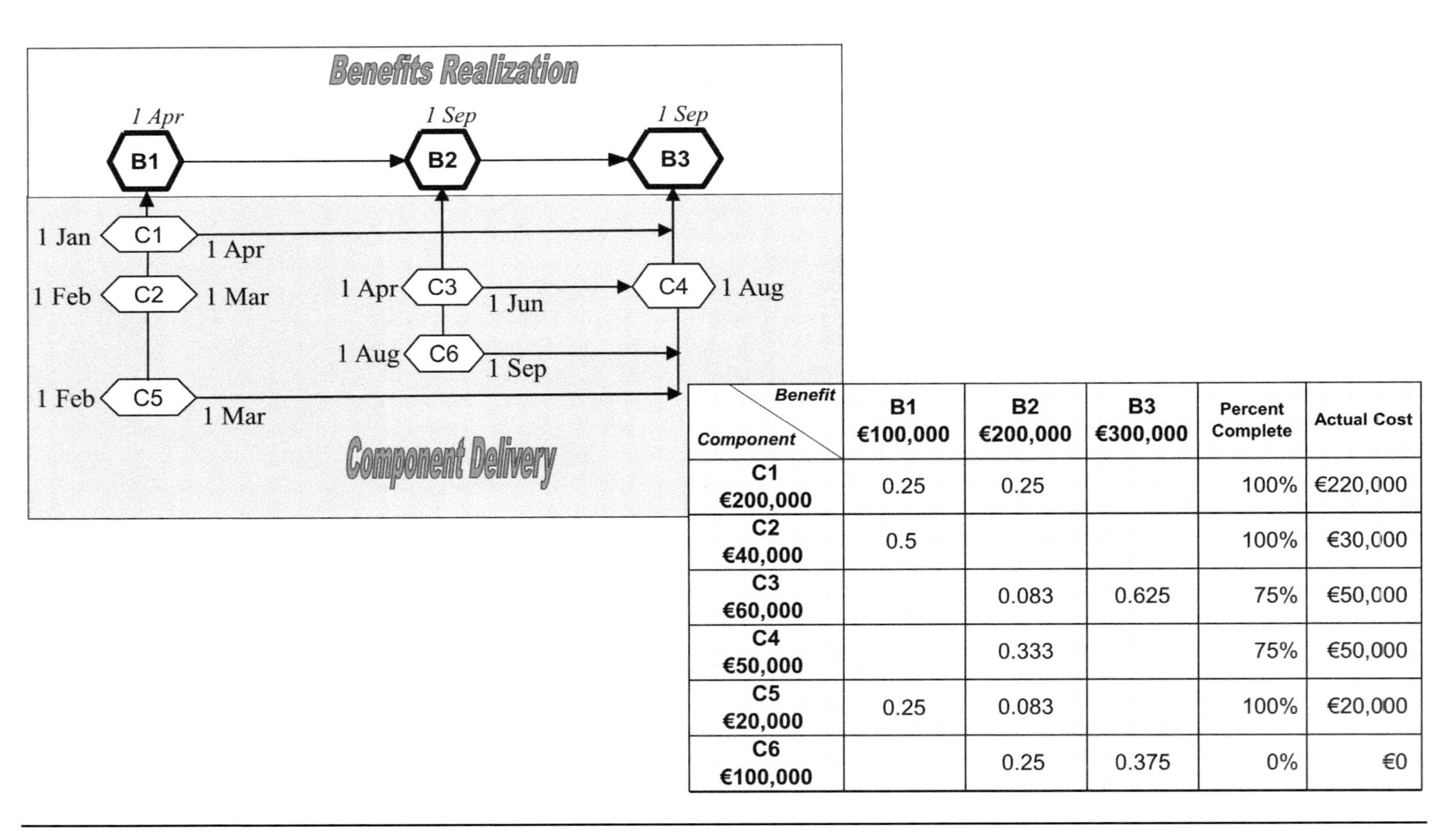

Component \ Benefit	B1 €100,000	B2 €200,000	B3 €300,000	Percent Complete	Actual Cost
C1 €200,000	0.25	0.25		100%	€220,000
C2 €40,000	0.5			100%	€30,000
C3 €60,000		0.083	0.625	75%	€50,000
C4 €50,000		0.333		75%	€50,000
C5 €20,000	0.25	0.083		100%	€20,000
C6 €100,000		0.25	0.375	0%	€0

Figure 12.7 Example Program Roadmap and Component–Benefit Matrix

12.5.2 The Initial Parameters for the Calculations

Because of the direct connections between initiatives and benefits, the component–benefit matrix can be read directly off the contribution fractions in the BRM of the business plan shown in Figure 12.5. The corresponding component–benefit matrix is given in Table 12.5.

The monetary value of the benefits (based on a one-year return on investment from the start of the program) is forecast to be:

B1—€100,000
B2—€200,000
B3—€300,000

Therefore, the total benefit is €600,000.

The estimated cost of the components (applied pro rata over the relevant implementation time) is:

C1—€200,000 from January 1st to April 1st
C2—€40,000 from February 1st to March 1st
C3—€60,000 from April 1st to June 1st
C4—€50,000 from June 1st to August 1st
C5—€20,000 from February 1st to March 1st
C6—€100,000 from August 1st to September 1st

Therefore, the total cost is €470,000.

The program starts on January 1st and is composed of three stages, aligned with the realizations of benefits:

Stage 1: (B1) completion of components C1, C2, and C5 planned for April 1st
Stage 2: (B3) adding components C3 and C6. Component B3 to be realized by September 1st
Stage 3: (B2) adding component C4. B2 to be realized by September 1st

This plan is shown schematically in Figure 12.7.

12.5.3 Current Status

It is now July 1st, and the following has been reported:

- C1, C2, and C5 are complete.
- C3 and C4 are 75 percent complete.
- C6 is not yet started.
- Spend to date (actual cost) is €370,000.

This case study will now be used to illustrate the total program management earned benefit system.

12.6 The Earned Benefit System

Ideally, to avoid confusion around the term *value*, the term *earned scope* should be preferred to *earned value*, since it does in fact represent the meaning of the concept more accurately than earned value. However, I will try to avoid confusion by careful use of terms.

In the following explanations, the term *indicator* is used for the result of calculations that can be used to monitor progress with respect to the plan; *parameter* is used for variables that are needed in the calculation of indicators.

For completeness of the analysis, the earned value terms are described first, followed by those for earned benefit. It should be noted that a number of the concepts included in the earned benefit section could be applied to extend conventional earned value or earned schedule ideas.

12.7 Earned Value Abbreviations, Parameters, and Indicators

All of these parameters and indicators with their case study values are also listed in a table at the end of this chapter (Section 12.15).

12.7.1 AC: Actual Cost

The actual cost is the amount of money (contribution) spent to date.

In the case study, AC is €370,000.

12.7.2 AD: Actual Date

The actual date is the date to which all of the parameters and indicators are evaluated.

In the case study, the AD is July 1st.

12.7.3 EV: Earned Value

The earned value is the basic indicator in earned value management (EVM) and is the sum of the planned budgets for the competed tasks or components. For incomplete tasks, the *task percent complete* (TPC) is taken into account and a convention agreed on how to calculate the value earned up to that point. This amount is frequently calculated as pro rata TPC.

In the current case study:

EV = C1 + C2 + 75% * C3 + 75% * C4 + C5 =
€200,000 + €40,000 + €45,000 + €37,500 + €20,000 =
€342,500

12.7.4 PV: Planned Value

The planned value is a standard EVM indicator and is the sum of the earned value of the tasks or components that should be achieved by the given date. Because the planned value is based directly on the budgeted cost of the tasks or components, it is numerically identical to the planned expenditure.

In the current case study:

PV = C1 + C2 + C3 + 50% * C4 + C5 =
€200,000 + €40,000 + €60,000 + €25,000 + €20,000 =
€345,000

The Planned Value Curve

This planned value curve (Figure 12.8) is composed of the summed planned value of the components (i.e., the cumulative expenditure); here, it is cumulated on a monthly basis. To provide a connection back to the program roadmap, the duration of each program component (C1 to C6) is indicated on the same diagram.

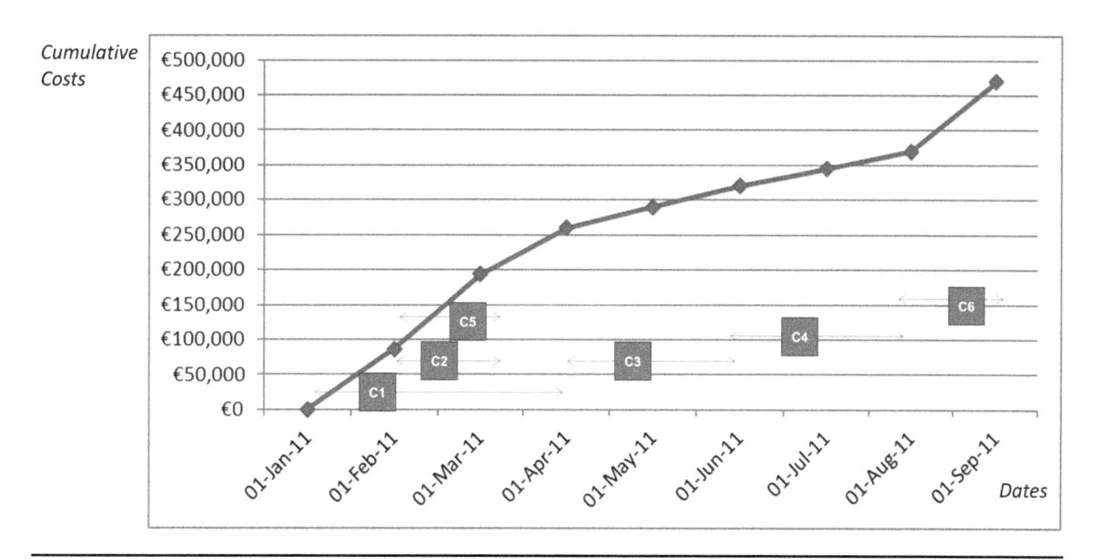

Figure 12.8 The Planned Value (Cumulative Cost) Curve

12.7.5 CPI: Cost–Performance Index

The cost–performance index is the standard EVM parameter: CPI = EV / AC

In the current case study, CPI = €342,500 / €370,000 = 93%

12.7.6 SPI: Schedule–Performance Index

The schedule–performance index is standard EVM indicator. It is given as SPI = EV / PV.

In the current case study, SPI = €342,500 / €345,000 = 99%.

12.7.7 BAC: Budget at Completion

The budget at completion is the budgeted cost for total work—that is, how much was originally planned to be spent.

The original plan in the case study was for a total program spend of BAC = €470,000.

12.8 Earned Benefit Abbreviations and Values

12.8.1 ADD: Actual Date Duration

The actual date duration is the number of time periods from the start until the actual date.

By July 1st, the program in the case study has been running for an ADD = 181 days.

12.8.2 EB: Earned Benefit

Earned benefit has been defined and explained in detail in the preceding section and is calculated as follows:

The completed component(i) contributes CCB_i to the total benefit, so when partially complete, it contributes a correspondingly reduced amount (the component earned benefit, or CEB_i):

$$CEB_i = \%\text{complete}[component(i)] * \sum_{cols\ j} (V_{ij} * B_j)$$

For the case study, the earned benefit for C3 at the review date (75 percent complete) is:

$$CEB_3 = 75\% \ \{(.083 * €200{,}000) + (.625 * €300{,}000)\} = €153{,}125$$

To calculate the total earned benefit, sum all of the CEB_i contributions, as follows:

$$EB = \sum_{rows\ i} \left(\%\text{complete}\left[component(i)\right] * \sum_{cols\ j} (V_{ij} * B_j) \right)$$

Giving, as total earned benefit on July 1st, based on the values in the case study:

$$EB\ (\text{July}) = €369{,}792$$

12.8.3 EBAC: Earned Benefit at Completion

The earned benefit at completion value of a stage is the value of the planned benefit at the end of that stage. The earned benefit at completion of the program is the sum of the benefits:

$$EBAC = €600{,}000$$

12.8.4 PB: Planned Benefit

The planned benefit is the value of the earned benefit planned to be achieved by a given date.

The planned benefit curve (Figure 12.9) is composed of the summed planned benefits; here, it is cumulated on a monthly basis based on the earned benefit formula (see 12.7, Earned Value Abbreviations, Parameters, and Indicators). In the same way as for the cumulative cost curve (Figure 12.8), key information from the program roadmap could be provided on the same diagram.

A similar diagram was calculated in Chapter 8 with a focus on cash-flow.

The next diagram (Figure 12.10, the benefit–value curve) gives a more informative contrast between planned benefit and planned value.

The planned benefit at any point in time is the forecast value of earned benefit up to that point. The planned benefit can be plotted against time as a cumulative benefit curve (benefit values on the Y-axis, dates on the X-axis).

For the case study, the planned progress of each component is given in Table 12.6.

The planned benefit as of the data date of July 1st would therefore be:

$$PB(\text{July}) = EB_1 + EB_2 + EB_3 + 50\%\ EB_4 + EB_5 = €404{,}167$$

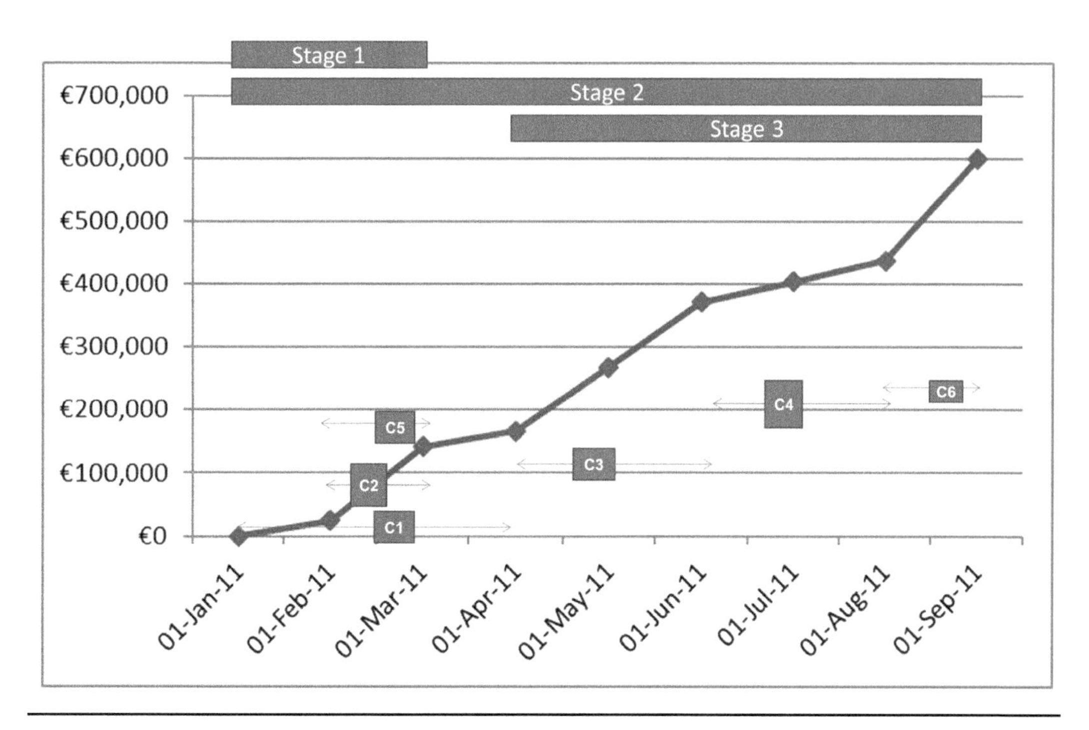

Figure 12.9 The Planned Benefit Curve

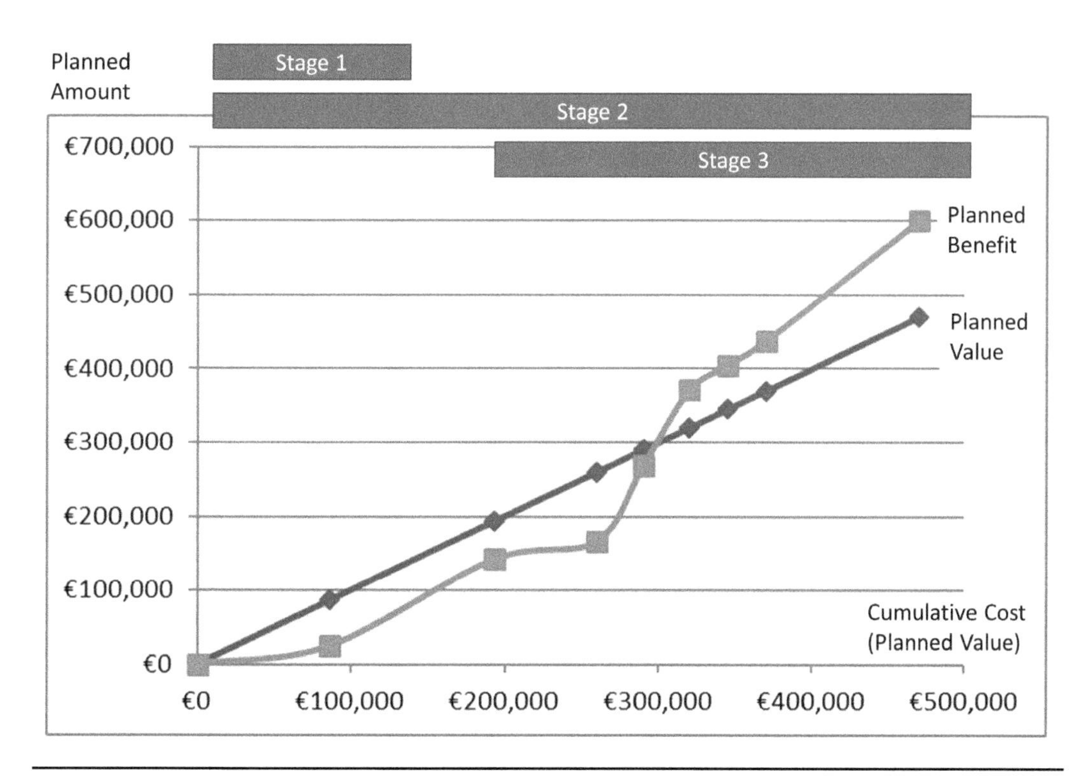

Figure 12.10 A Benefit–Value Curve Showing Planned Benefit and Planned Value (or Cumulative Cost)

Table 12.6 Planned Progress

	January 1st	February 1st	March 1st	April 1st	May 1st	June 1st	July 1st	August 1st	September 1st
C1	0%	33%	67%	100%	100%	100%	100%	100%	100%
C2	0%	0%	100%	100%	100%	100%	100%	100%	100%
C3	0%	0%	0%	0%	50%	100%	100%	100%	100%
C4	0%	0%	0%	0%	0%	0%	50%	100%	100%
C5	0%	0%	100%	100%	100%	100%	100%	100%	100%
C6	0%	0%	0%	0%	0%	0%	0%	0%	100%

12.8.5 EBEV: Earned Benefit Equivalent Value

The earned benefit equivalent value is used to link earned benefit and earned value. At each point in time, as shown in Figure 12.10, you can evaluate the amount of the earned value that corresponds to a given earned benefit (and vice versa).

To track performance with respect to costs, this benefit–value curve is developed as follows:

- Start with the cumulative benefit curve.
- Take the program cumulative cost curve, a deliverable of the program planning process.
- Replace each date on the X-axis of the cumulative benefit curve with the corresponding planned value at that date.

Note: The planned value curve on this graph is a diagonal line represented by (X-value) = (Y-value), since the planned value at any point is defined to be equal to the cumulative cost. The case study uses the same (financial) units for earned benefit and earned value. (Note: this curve can only be included if earned benefit and earned value use the same units.)

The reason for showing the two curves together is to underline the fact that the cumulative costs at certain times during the program can exceed the potential financial benefits that have been realized up to that date, even though the program is on schedule, on budget, and has a valid, profitable business case. It is important to make sure that the sponsor and other decision makers are aware of this possibility so that they do not erroneously take this situation as a justification for canceling the program.

The converse can also be the case. For the case study, reading off from the curves in Figure 12.11, for an earned benefit of €370,000, the value of EBEV is €320,000. That is to say that, given the current calculated earned benefit of €370,000, the corresponding earned value would be expected to be about 15 percent lower, at €320,000.

These calculations are used to measure the cost performance as shown in the next section.

12.8.6 BCPI: Benefit–Cost Performance Index

The benefit–cost performance index shows whether what has been delivered so far is in line with what was expected to be spent on achieving it. It is defined as the cost performance index (CPI) of the earned value that corresponds to the current earned benefit. Tracking this value can be of more interest to strategic managers than directly tracking the CPI itself because it measures the link to benefits achievement.

In the case study, we have achieved an earned benefit of €370,000, which was planned to correspond to an EV of €320,000. Given that the actual cost (AC) was reported as (by coincidence) €370,000, the BCPI (EBEV / AC) is 86 percent. The fact that EB = AC indicates that the net benefit at this point in time is precisely nothing, and that the rate of overspend ({1 / BCPI} – 1) is 16 percent. This situation could still lead to an overall net benefit because the ratio of the total final benefit (EBAC) to the planned budget (BAC) given by €600,000 / €470,000 is slightly better than 127 percent, so an increase of 16 percent to the BAC will still leave a net benefit. In fact, the detailed calculations do predict a final net benefit of €54,800, or about 10 percent on the total spend.

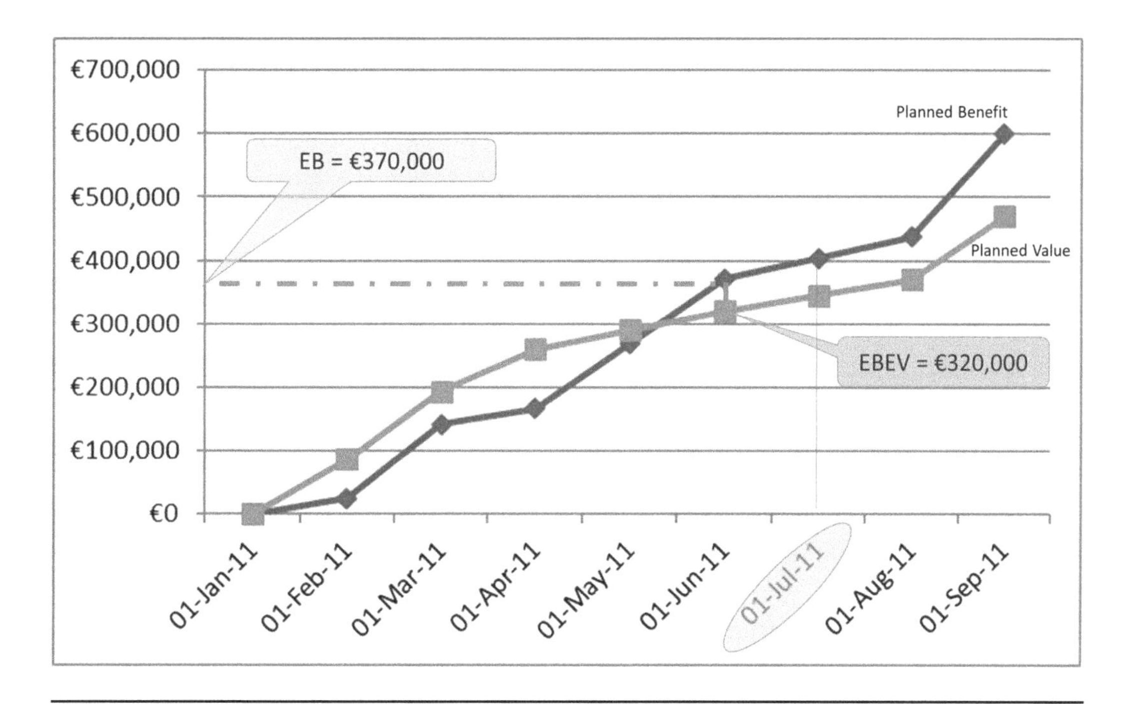

Figure 12.11 Mapping between Earned Benefit and Earned Value

12.8.7 BPI: Benefit Performance Index

The benefit performance index is used to track performance with respect to the benefit realization plan, in a similar manner to the EVM schedule performance index. It is calculated as the ratio of earned benefit to planned benefit.

For the case study, BPI = €370,000 / €404,000 = 92%, so we are behind schedule with respect to the benefits we planned to earn. A measure of how late we are is given by the benefit value schedule variance (BVSV, see Section 12.8.15) of 61 days.

12.8.8 EBED: Earned Benefit Equivalent Date

The earned benefit equivalent date is the date at which the current earned benefit was planned to be achieved. EBED can be translated into earned benefit equivalent date duration (EBEDD, see Section 12.8.9)—that is, the elapsed time by which that benefit should have been earned. The evaluation can be carried out using the cumulative benefit curve by mapping the earned benefit across to the point at which the curve reaches that value. The X-axis date gives the EBED. The process is shown in Figure 12.12: the EBED for the benefit earned by July 1st (€369,762) is just before June 1st.

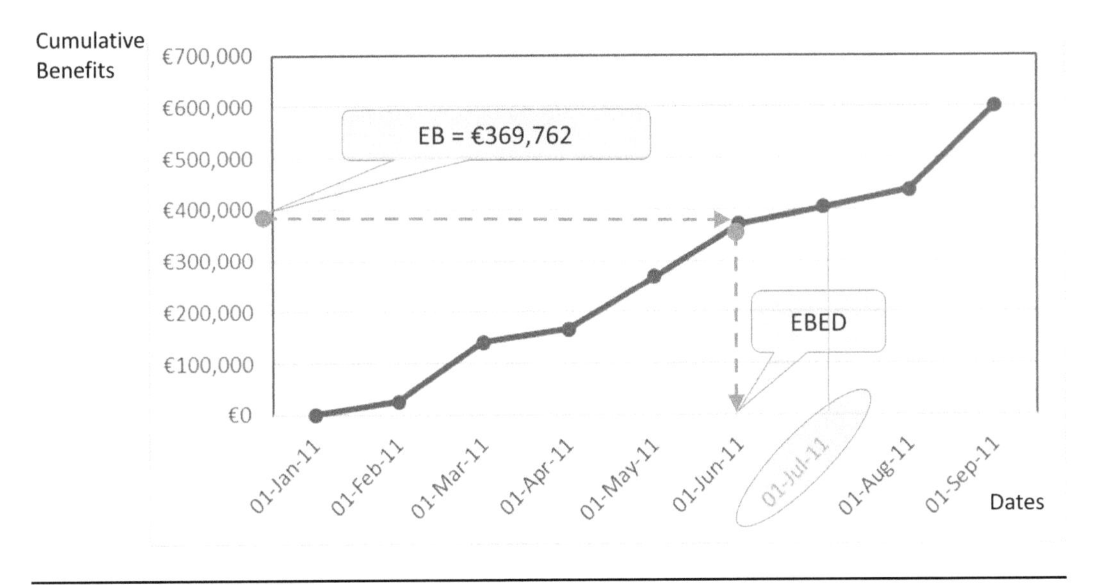

Figure 12.12 Calculating Earned Benefit Equivalent Date

EBED is the date at which the specified earned benefit was planned to be reached and can be used to assess the schedule variance with respect the benefits realization plan.

12.8.9 EBEDD: Earned Benefit Equivalent Date Duration

The earned benefit equivalent date duration is a parameter used by indicators relating to elapsed time. It measures the number of days after which the current earned benefit was planned to have been reached (Figure 12.12).

In the case study, the EBED is June 1st, so the EBEDD (number of program days to this date) is 151 days.

12.8.10 BV: Benefit Variance

The benefit variance indicator measures the difference between planned and actual values from the point of view of earned benefit: EB – PB.

For the case study, we have a shortfall of –(€370,000 – €404,000); that is, €34,000 to date.

12.8.11 BSV: Benefit Schedule Variance

The benefit schedule variance indicator is similar to BSPI except that it measures the variance between actual and planned dates, rather than their ratio: BSV = EBEDD – ADD.

The benefit schedule variance is the difference between the data date and the EBED: the date at which the current earned benefit was planned to be achieved. This value can be read from the planned benefit curve (see Figure 12.13), and shows we are approximately one month late.

Alternatively, the situation can be analyzed in terms of the benefit schedule performance index.

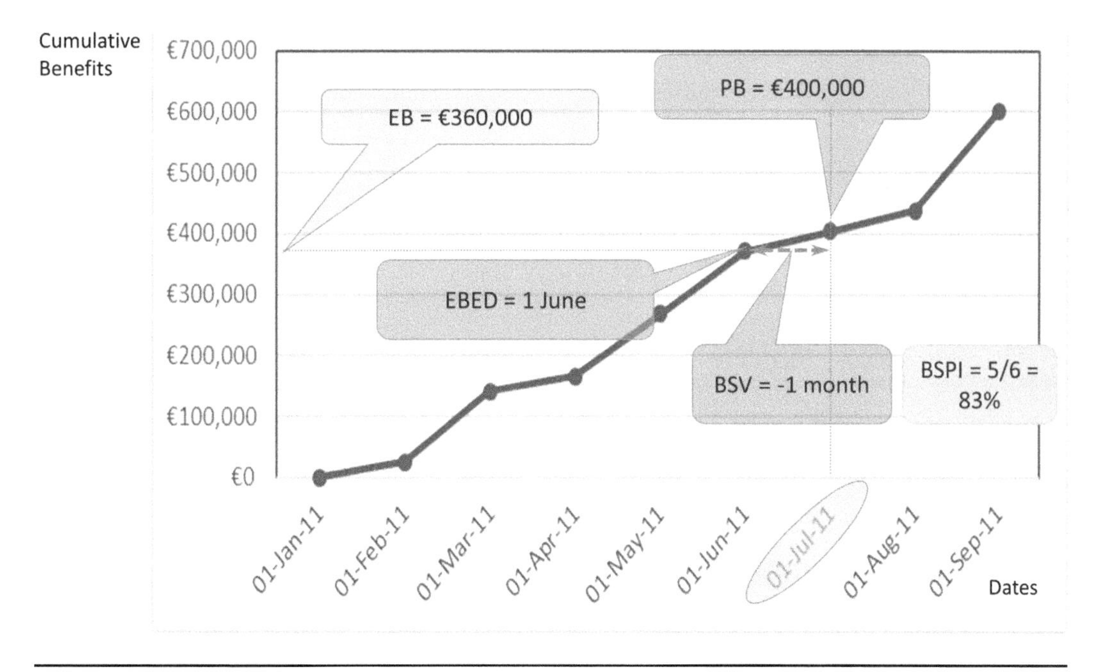

Figure 12.13 Benefit, Time, and Cost Analysis

12.8.12 BSPI: Benefit Schedule Performance Index

The benefit schedule performance index measures actual progress with respect to planned progress from the point of view of time. It is the ratio: BSPI = EBEDD / ADD.

For the case study (see Figure 12.13), BSPI is therefore 151 days / 181 days = 83%, showing that we are progressing at 83 percent of the rate that we had planned for achieving benefits.

Looking at this result another way, the program has been running for six months and has slipped by one month, so the benefit schedule performance index (BSPI) is 5 months / 6 months or 83 percent.

12.8.13 VED: Value Equivalent Date

The value equivalent date is the date at which a given actual cost was planned to be spent (similar in concept to VEB).

In the case study, with an AC of €370,000, the VED can be read from the PV curve (Figure 12.14) and gives August 1st.

12.8.14 VEDD: Value Equivalent Date Duration

The value equivalent date duration indicator measures the number of days from the start at which the current earned value was planned to have been reached. It is similar in concept to EBEDD.

For the VED of August 1st in the case study, VEDD is therefore 212 days.

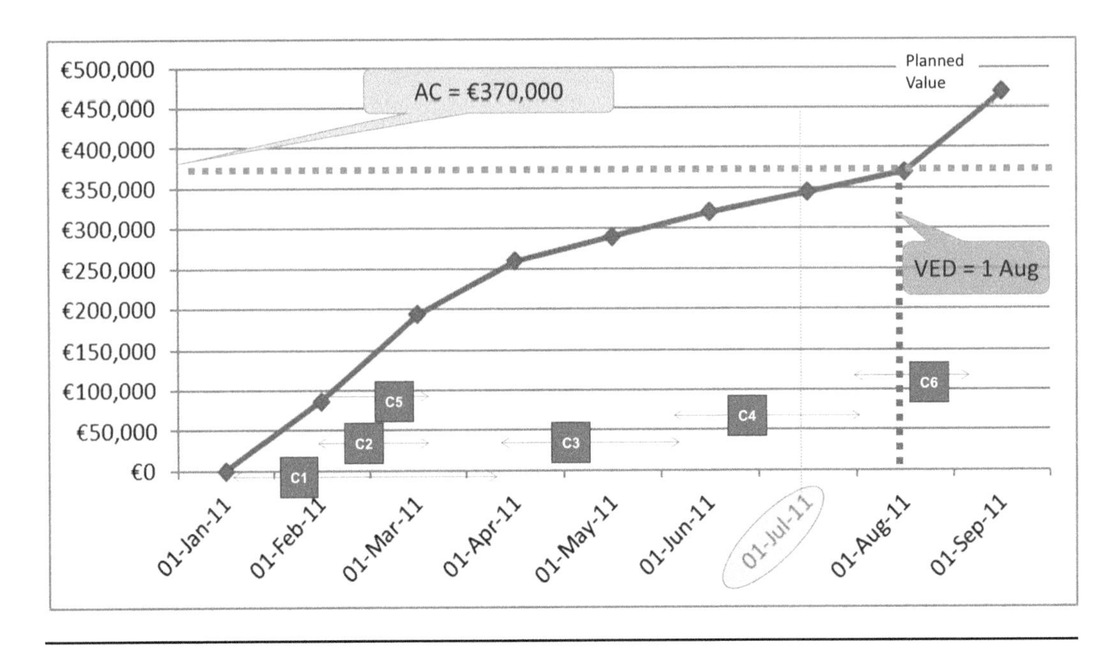

Figure 12.14 Value Equivalent Date—Corresponding to the Actual Cost

12.8.15 BVSV: Benefit Value Schedule Variance

The benefit value schedule variance indicator measures the how early (positive values) or late (negative values) the program is in achieving its earned benefit with respect to the planned schedule. BVSV measures the time gap (which should ideally be > 0) between when the current earned benefit should have been achieved and the point at which the actual cost should have been spent. This indicator adds together the amount that the program is ahead of schedule from the point of view of the benefit and the amount that the program is under budget from the point of view of spend. Negative values are an indication that either schedule or budget or both are problems. So, BVSV = EBEDD – VEDD.

The various indicators are shown in Figure 12.15, giving BVSV = 151 days – 212 days = 61 days late.

12.8.16 BVPI: Benefit Value Performance Index

The benefit value performance index shows the rate of progress toward achieving the benefits with respect to the planned rate. It is represented by {(time spent so far) + (how late we are)} / (time spent so far).

If EDD represents the number of time intervals since the start of the program, the formula is:

$$\text{BVPI} = (\text{EDD} + \text{BVSV}) / \text{EDD} = (181 - 61) / 181 = 66\%$$

That is to say: Taking into account the schedule and cost overruns, the program is actually only achieving two-thirds of the planned progress.

For our case study, a superficial analysis might conclude that the actual cost is not greater than the earned benefit, so we may have slipped a bit. In addition, earned value and planned value are virtually identical so, we are not doing too badly, are we? The following, more realistic (benefit-related) analysis puts a less optimistic view on the matter.

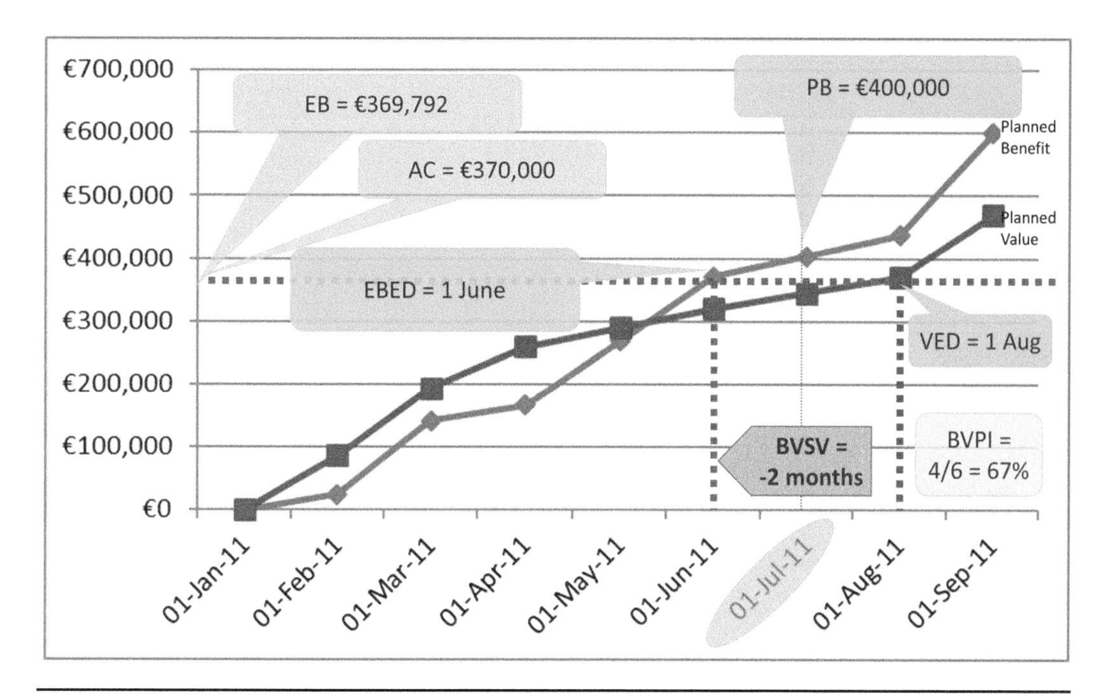

Figure 12.15 Benefit, Time, and Value Analysis

An examination of Figure 12.15 shows that the actual cost (AC = €370,000) corresponds to the planned value for one month later (the value effective date, or VED); whereas the earned benefit corresponds to the planned benefit for one month earlier (the earned benefit effective date, or EBED), so we are, in essence, two months adrift after six months, which equates to 4/6 or a 67 percent benefit value performance index. For comparison, the benefit schedule performance index (BSPI) is (6 – 1) / 6 or 83 percent at this point, since it ignores the budget overrun.

These indicators are shown in Figure 12.15.

12.8.17 CBM: Component–Benefit Matrix

The component–benefit matrix quantifies how each component contributes to each benefit.

A component–benefit matrix was shown in Table 12.5.

12.8.18 CCB_i: Component Contribution to Benefit

The component contribution to benefit parameter measures the total contribution of component(i) to the overall benefit.

12.8.19 CEB_i: Component Earned Benefit

The component earned benefit is the amount contributed to date by component(i) to the overall earned benefit.

12.8.20 EBEC: Earned Benefit Equivalent Cost

The earned benefit equivalent cost indicator measures the planned cost for a given value of earned benefit. it can be used as the basis for calculating the planned return on investment for that specific EB value. it is the planned value corresponding to the earned benefit equivalent date.

For the case study, and taking the graph showing planned benefit and planned value, as shown in Figure 12.16, the EBEC is €320,000.

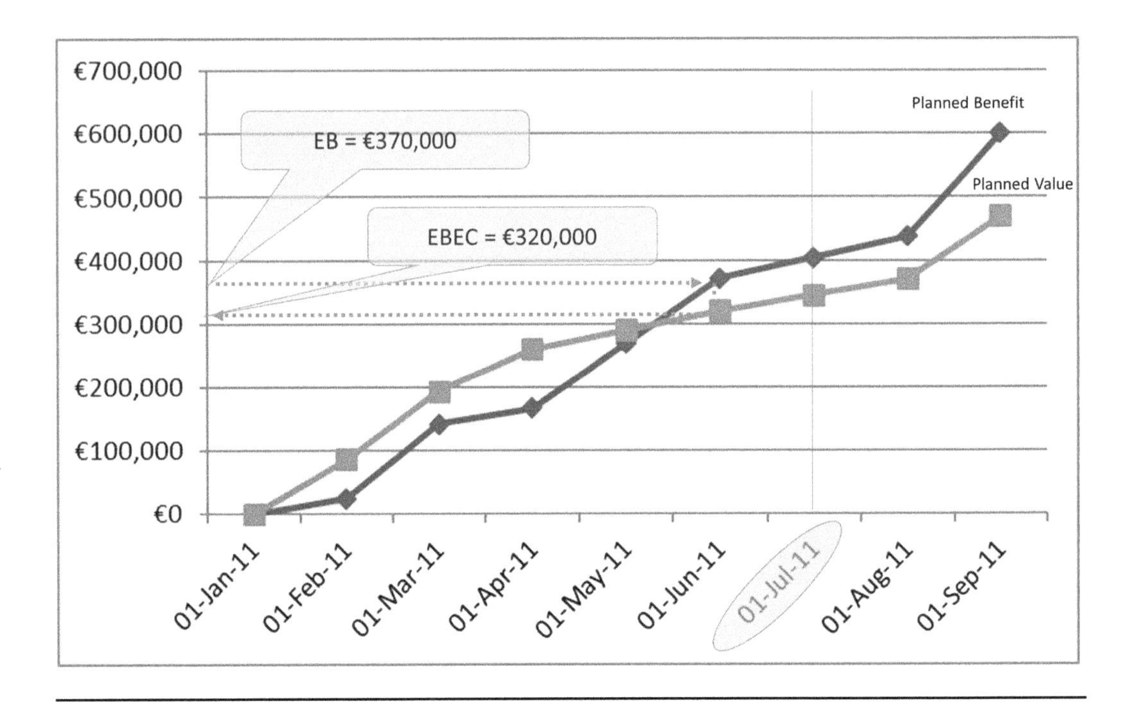

Figure 12.16 Evaluating the Earned Benefit Equivalent Cost

The EBEC can be used as the basis for calculating the planned return on investment for that specific EB value: ROI = EB / EBEC. These values can be plotted against time, as shown in Figure 12.17.

12.8.21 VEB: Value Equivalent Benefit

The value equivalent benefit is the value of the earned benefit that is planned to be achieved for a given actual cost. The VEB is the converse of EBEC (see Section 12.8.20).

12.8.22 PPC: Program Percent Complete

Since programs aim to achieve benefits, the program percent complete indicator is the ratio between the earned benefit and the earned benefit at completion: EB / EBAC.

For the case study: PPC = €404,000 / €600,000 = 62%.

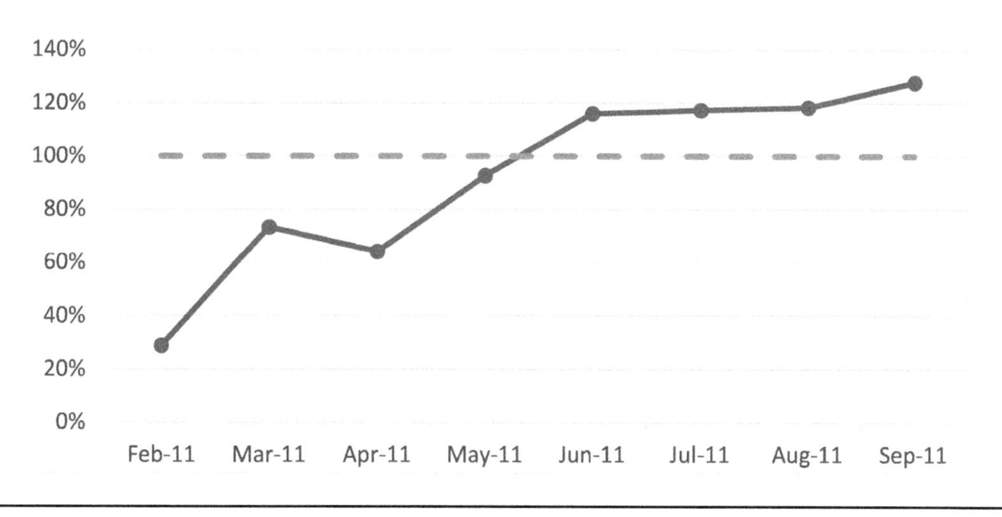

Figure 12.17 The ROI Curve for the Planned Progress of the Program

12.8.23 PED: Planned End Date

The planned end date is the date at which the program is scheduled to complete.

For the case study, PED is September 1st.

12.8.24 PEDD: Planned End Date Duration

The planned end date duration is the number of time units from the planned start to the planned end of the stage or the program.

For the case study, PEDD is 243 days.

12.8.25 PSD: Program Start Date

The program start date is used as the origin for all duration calculations.

For the case study, PSD is January 1st.

12.8.26 $TCPI_X$: To-Complete Performance Index

There are separate to-complete performance indexes for time ($TCPI_T$), cost ($TCPI_C$), and benefit ($TCPI_B$). They measure the performance index required to achieve the planned value of time, cost, or earned benefit respectively by the end of the program. They measure the ratio of the "amount" left to achieve by the "amount" remaining for achieving it:

$$TCPI_T = (PEDD - EBEDD) / (PEDD - ADD)$$
$$TCPI_C = (BAC - EBEC) / (BAC - AC)$$
$$TCPI_B = (EBAC - EB) / (EBAC - PB)$$

For the case study, the values of these indices are:

$TCPI_T$ = 148%
$TCPI_C$ = 150%
$TCPI_B$ = 117%

These calculations show that the shortfall in achieving time and cost targets gives a more pessimistic view than is justified when considering the real objective of the program—that is, in achieving the benefits.

12.8.27 $TCPF_X$: To-Complete Performance Factor

Although the TCPI numbers are useful, they only indicate the relative performance required with respect to the planned performance. A more informative value is the to-complete performance factor: the amount by which the current performance has to change with respect to the current performance to achieve the planned values by the end of the program:

$TCPF_T = TCPI_T / BSPI$
$TCPF_C = TCPI_C / BCPI$
$TCPF_B = TCPI_B / BPI$

The values for the case study are:

$TCPF_T$ = 178%
$TCPF_C$ = 173%
$TCPF_B$ = 128%

As for TCPI, the status from the point of view of benefits does not seem to be as much of a concern when considering the schedule or the cost dimensions.

12.9 Further Analysis of the Garden Services Case

Based on the business plan, the algorithms described earlier for evaluating the allocation fractions and the allocations can be applied to this model. This approach results in the diagram shown in Figure 12.18, and the figures in Table 12.7 serve as the basis for the earned benefit calculations.

The comparison between the costs (allocations) and the benefits (contributions) is shown in Table 12.7. This analysis is thought-provoking, as the mismatch between allocations and contributions is striking:

- For the strategic outcomes, both the Public and the High-End markets are shown to be taking a loss.
- For the initiatives, the Servers with the 3-D Printer are a major drain on the resources, costing well over twice what they are worth to the business.

However, the overall cost of €470,000 is considerably less than the envisaged return of €600,000.

This analysis provides an insight into the causes of the major difference between the earned value and the earned benefit indicators in the Garden Services case study.

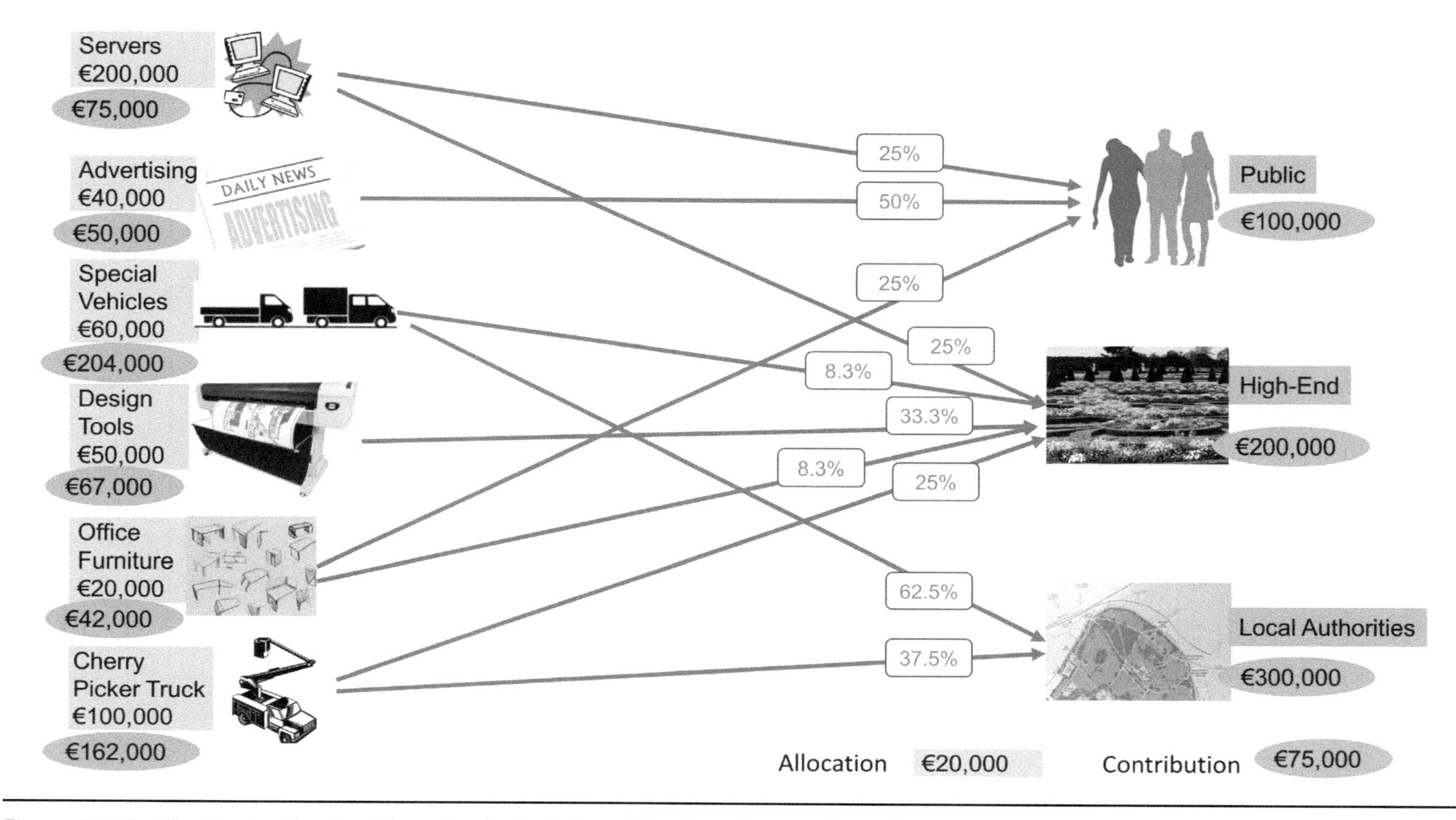

Figure 12.18 The Garden Services Plan with the Inclusion of the Contributions from the Components

Table 12.7 The Business Plan Analysis for the Garden Services Program

	Allocation	Contribution
Servers	€200,000	€75,000
Advertising	€40,000	€50,000
Special Vehicles	€60,000	€204,100
Design Tools	€50,000	€66,666
Office Furniture	€20,000	€41,667
Cherry Picker Van	€100,000	€162,500
Public	€118,667	€100,000
High-End	€226,983	€200,000
Local Authorities	€124,351	€300,000

12.9.1 Understanding the Earned Benefit Results

The reason why the planned benefit curve initially drops below the planned value line is now clear: One of the components provides a benefit that is lower than the budgeted cost, so it earns less than it costs from the benefits point of view. Since all of the other components do have a benefit at completion that is greater than the corresponding cost, once they are online, they make up for the shortfall from the first component.

12.9.2 Revising the Business Case for the Garden Services Program

Given this financial analysis, the initial reaction might well be to drop the plans for the Sophisticated Servers and 3-D Printer. This change would remove the corresponding allocation of €200,000 for the loss of the contribution of €75,000. This modification to the plan would lead to the analysis shown in Table 12.8.

Table 12.8 Analysis of the Revised Business Proposal

	Allocation	Contribution
Servers	€0	€0
Advertising	€40,000	€50,000
Special Vehicles	€60,000	€204,100
Design Tools	€50,000	€66,666
Office Furniture	€20,000	€41,667
Cherry Picker Van	€100,000	€162,500
Public	€52,000	€75,000
High-End	€93,649	€149,933
Local Authorities	€124,351	€300,000

This revised approach for the Garden Services would seem ideal, as all of the strategic objectives are achieved profitably.

However, if you are informed that you will never capture any of the High-End market without a 3-D Printer, you will need to set the corresponding link as *essential*, as shown in Figure 12.19.

The effect of applying the algorithm described in Chapter 6 for catering for essential links to this revised model is shown in the amended business proposal in Table 12.9.

This essential link analysis shows that losing the High-End market would mean that without the Servers, there is no need for the Sophisticated Design Tools, thereby "saving" an extra €50,000. Although we will now have a business plan that takes into account the essential link and ensures that each initiative has a positive net return, the total size of the business has dropped from the original €600,000 to €375,000. However, the profit has risen from €600,000 – €470,000 = €130,000 to €375,000 – €220,000 = €155,000.

In contrast, the revised business plan in Table 12.8 that overlooked the essential link promised a revenue of €525,000 for an expenditure of €270,000. The expenditure would certainly happen, but as shown in Table 12.9, the return would only be €375,000. This is a typical example of why so many programs are deemed to "fail" based on false premises, especially when changes are made to the original business model.

Table 12.9, however, sets a strategic question to the management team: Is business size (i.e., total revenue) more important than profitability? The wise approach may be to launch the business aiming only at the Public and Local Authorities while building up reserves through the improved profitability. The expansion into catering for High-End customers and capturing more of the market can be put off to a later date, when their cost may be more compatible with the structure of this business.

12.10 Revisiting Earned Benefit

There are two points to consider based on the analysis of the Garden Services case study.

The Garden Services case was simplified so that the initiatives were directly connected to the strategic outcomes. This leaves two questions to be answered:

1. What should be done in the general case of initiatives leading to deliverables, deliverables leading to capabilities that lead to outcomes?
2. How should one cater for essential links?

One answer to the first question is that you can create the component–benefit matrix from the detailed results chain. The way in which to do this is explained in the next section.

However, the creation of the component–benefit matrix does not address the question on essential links, and the approach for this provides an alternative answer to both of the questions. The solution will come from linking earned benefit to the automated approach for evaluating allocations and contributions that was developed and applied in Chapters 5, 6, and 7.

12.10.1 Evaluating the Component–Benefit Matrix from the Results Chain

The approach for evaluating the component–benefit matrix is based on the initial contribution table from the results chain developed using the BEER algorithm—that is, the matrix of the contribution percentages between the nodes. To generate the component–benefit matrix or the component–outcome matrix, each initiative is considered separately from the others as follows:

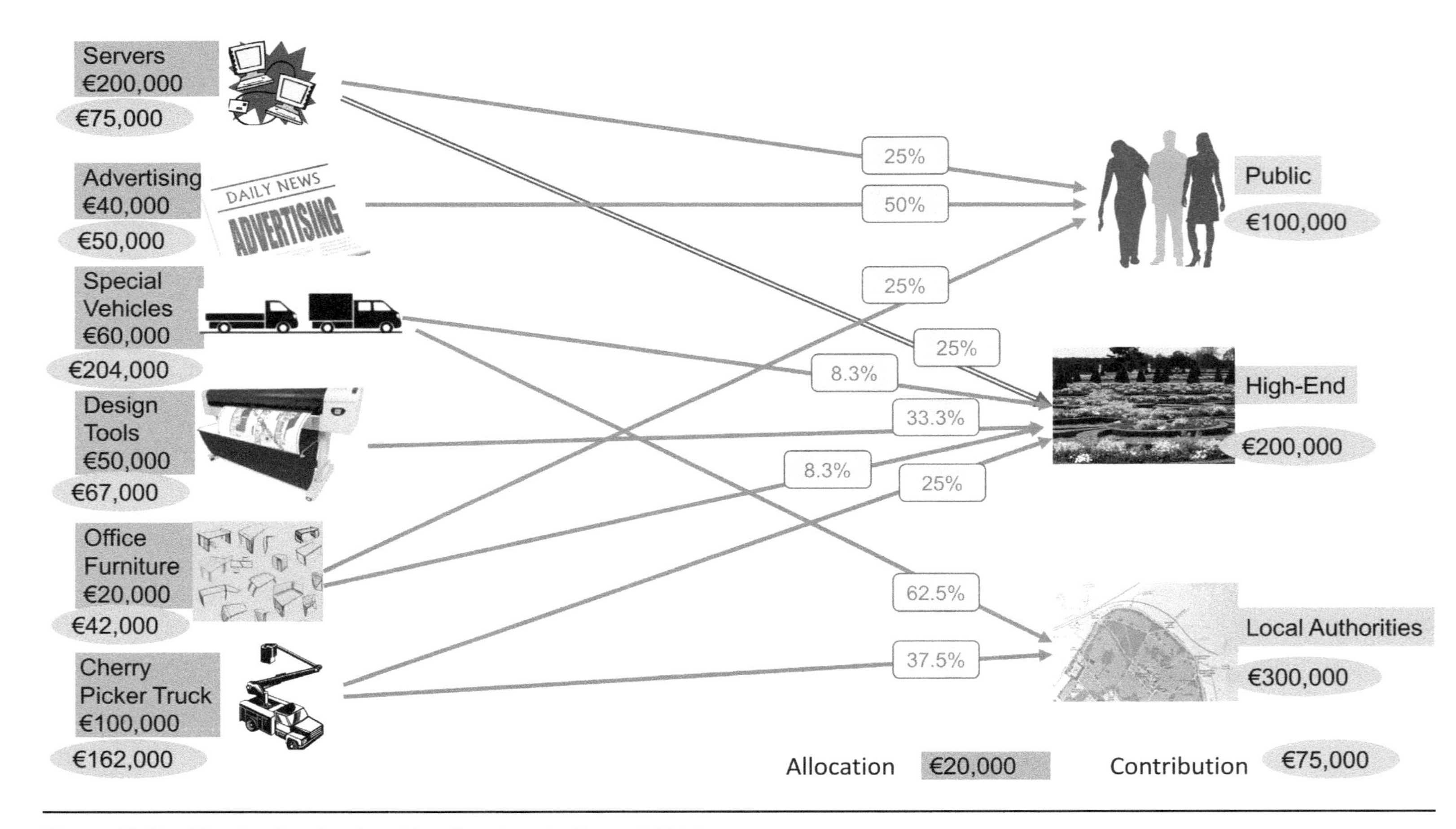

Figure 12.19 The Garden Services Plan Showing the Essential Link

Table 12.9 Revised Business Proposal Catering for the Essential Link

	Allocation	Contribution
Servers	€0	€0
Advertising	€40,000	€50,000
Special Vehicles	€60,000	€187,500
Design Tools	€0	€0
Office Furniture	€20,000	€25,000
Cherry Picker Van	€100,000	€112,500
Public	€60,000	€75,000
High-End	€0	€0
Local Authorities	€160,000	€300,000

- Create one row of the component–benefit matrix, by setting the contribution of the selected initiative to €100 and the contribution values of the other initiatives to zero.
- Then evaluate all of the other nodes of the results chain based on this set of contributions using the BEER algorithm from left to right.
    - The values this generates for the strategic outcomes will be the percentage values that this initiative provides to each of these strategic outcomes.
    - Each set of values will automatically sum to 100 percent because the input on the left of the results chain (i.e., €100 for the selected initiative and €0 for the others) is distributed to the strategic outcomes according to the BRM.
- Carry out this process for each initiative in the BRM.

For the QERTS case, the resulting component–outcome matrix is given in Table 12.10.

Table 12.10 The Component–Outcome Matrix for the QERTS Case Study

	Higher Market Share	Higher Revenue	Operational Savings	Less Turnover	More Profit
Standardize Processes	15%	15%	23%	5%	16%
Integrate Store Records	15%	15%	23%	5%	16%
Adapt Tracking System	26%	26%	46%	10%	30%
Develop Joint Marketing	12%	12%	0%	0%	10%
Align Compensation	7%	7%	8%	80%	7%
Train Staff	26%	26%	0%	0%	20%

This result can be useful in providing additional information about the business strategy or to help validate the logic in the BRM. In this particular case, Table 12.10 allows us to check certain characteristics of the resulting model, such as:

- Marketing and training have no effect on staff turnover; a targeted compensation scheme seems to be key staff retention.
  - This may indicate a need to involve the marketing group in addressing the turnover problem in their initiative and in training sessions.
- An integrated tracking system provides the greatest set of benefits overall, although staff training is just as important for increasing market share and revenue.
  - This underlines the general lesson that in any change program you should concentrate equally on people and on tools.

If all of the characteristics shown in the component–outcome matrix are deemed to be plausible, the model can be accepted.

Although the new cost–benefit matrix could be used for calculating earned benefit in the QERTS program, as mentioned earlier, this approach cannot take the concept of essential links into account. The way in which the full power of the BRM concepts can be used, including essential links, is addressed in the following sections.

12.10.2 Earned Benefit Based on the Results Chain Calculations

The BEER algorithm for analyzing the BRM can provide an alternative method for evaluating the earned benefit in all cases. In the case where there are no essential links, this approach is based directly on the BEER algorithm. In the case of essential links, more thought and computation are needed; this is addressed in Section 12.10.3.

Based on the algorithms that apply the BEER, the allocation for each program component is set to the amount that this component adds to the strategic benefit; the sum of these amounts is identical to the sum of the strategic objectives.

This feature implies that an adaptation of the normal earned value approach can be applied in order to calculate the earned benefit, as follows. Whereas the earned value for each component is the BAC (budget at completion = allocation) times the percent complete of the component, for earned benefit, you multiply the earned benefit at completion (EBAC = contribution) by the percent complete of the component. The sum of the component amounts (i.e., the corresponding allocation or contribution) measures the total amount earned (earned value or earned benefit).

This BEER-based approach can be applied to the Garden Services using the figures provided in Table 12.11.

Whereas earned value based on percent complete of the program initiatives (Servers, Advertising, . . . Cherry Picker) would take the values from the Allocation column, for earned benefit, the amounts in the Contribution column should be used.

For example, for April 1st, the following components are planned to be complete: Servers, Advertising, and Office Furniture. The others will not have started, so:

Planned value: PV = €200,000 + €40,000 + €20,000 = €260,000, whereas
Planned benefit: PB = €75,000 + €50,000 + €41,667 = €161,667

These figures agree with the calculations based on the component–benefit matrix shown in the graphs in Figure 12.16, reproduced in Figure 12.20.

Table 12.11 Copy of the Initial Garden Services Business Analysis

	Allocation	Contribution
Servers	€200,000	€75,000
Advertising	€40,000	€50,000
Special Vehicles	€60,000	€204,100
Design Tools	€50,000	€66,666
Office Furniture	€20,000	€41,667
Cherry Picker Van	€100,000	€162,500
Public	€118,667	€100,000
High-End	€226,983	€200,000
Local Authorities	€124,351	€300,000

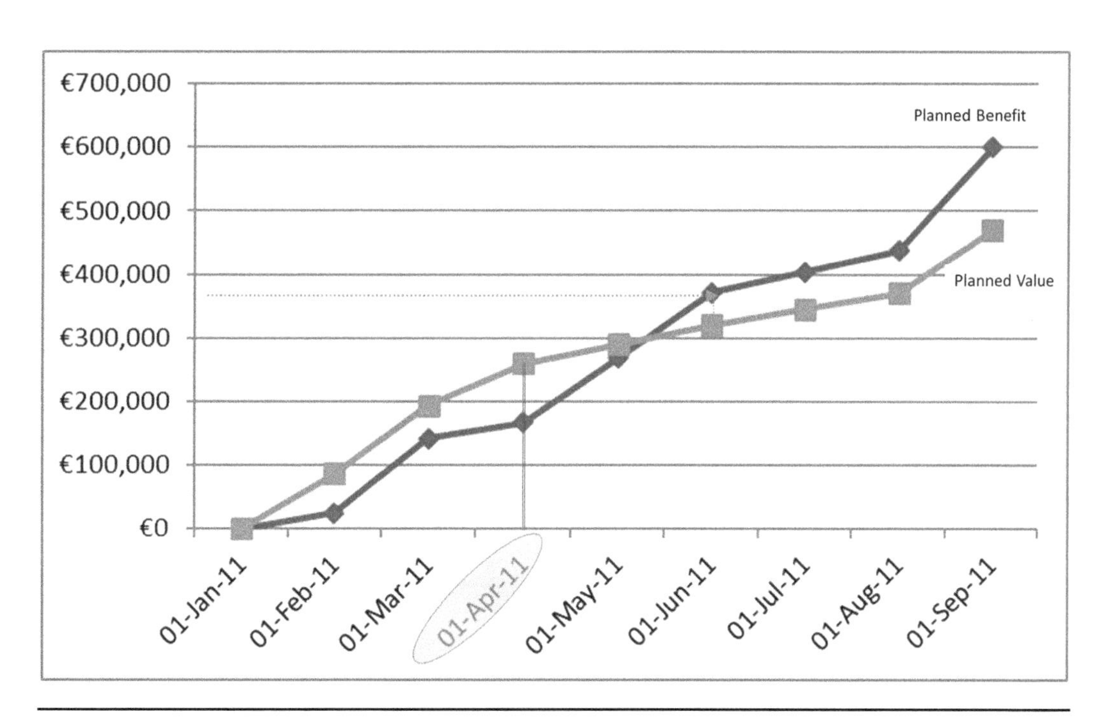

Figure 12.20 Planned Value and Planned Benefit Curves

However, the situation becomes much less straightforward once essential links are taken into account. Essential links have no equivalent in earned value management. However, they can make a considerable difference to the earned benefit, as will be explained based on the Garden Services case study.

12.10.3 Earned Benefit Including Essential Links

To deal with essential links, one additional convention needs to be defined: how to address the situation where an essential node is partially complete. This question was addressed—and answered—in

Chapter 8, in connection with cash-flow calculations by the definition of the algorithm for link evaluation (ALE) addition to the BEER method.

The approach was as follows. In the simplest case, the decision could be taken to set the progress of its essential successor to zero until the essential node is complete. This would be the most radical view, but it misrepresents the true situation, since as the work on the essential node progresses, our confidence in its forthcoming achievement increases. This progressive build up is in line with the overall earned value approach, in which tasks "earn" value even before they are in a completed state—that is, before they can actually be used in a valuable way.

The approach that will be taken for the successors of essential links is therefore the following, as was applied also in the calculations for the cash flow in Chapter 8:

- The percent complete of the successor node is not applied directly to the earned benefit calculation. Instead, an *effective percentage* value is defined.
- The effective percentage is set to 100 percent if there are no essential links into the node.
- In the case of one or more essential links going into the node, the effective percentage is calculated as the minimum of the earned benefit percent complete values of the essential source nodes.
- The earned benefit of a node is then calculated as its contribution based on the standard BEER algorithm times its effective percentage.

This method serves to smooth out the potential discontinuous application of the essential link effect, while ensuring that, as long as the essential node is missing, all of its essential successors will also be missing, since their contribution is multiplied by the percent complete of the essential node—that is, by 0 percent.

The effect of this algorithm on the calculations of earned benefit can be considerable, especially in the case where the scheduled progress of an essential node starts to slip. It is important for the impact analysis to take these factors into account in status and progress reviews.

In the absence of essential links, this approach generates the same results as the approaches described earlier.

As an example, this effective percentage approach is applied to the Garden Services case and then to QERTS.

Garden Services Case Study

Take the Garden Services example, in which the availability of Servers is essential for the high-end services (Figure 12.19); the effect will only be seen once the Servers are available. The business plan is to develop the Servers starting in January for completion in April, and to add the Advertising and the Office Furniture during March. The following scenario shows some of the effects of essential links:

- Assume that the program manager, in order to give the impression of "quick wins," decides to complete the Advertising and obtain the Furniture during the first month. The situation at the end of the first month would be as follows:
    - Assume that work goes to schedule, that the cost for the Server work is €30,000, that Advertising is under budget at €35,000, and that the Furniture is slightly over-budget at €25,000.
        - The percent spent is **PS = €90,000 / €470,000 = 19.15%**.
    - From the simple earned benefit point of view this would provide: 1/3 contribution of Servers + contribution of Advertising + contribution of Furniture.
        - The calculations give **EB = €75,000 / 3 + €50,000 + €41,667 = €116,667**.
        - The program percent complete is **PC = €116,667 / €600,000 = 19.4%**.

- Taking the essential links into account, the calculation is more complicated and gives the following result:
  - Earned benefit: **EB = €105,555** (i.e., about 10 percent less).
  - Percent complete: **PC = 17.6%**.

This financial analysis indicates that, when ignoring the effect of the essential link, the Program Manager could argue that anticipating the spend is valid since, even before any service can be provided, the percent spent (PS) is still less than the percent complete (PC) of the program:

PS / PC = 19.15 / 19.4 = 98%

However, if the correct earned benefit approach is adopted, and the essential link is considered, the result is not as comforting:

- The situation is now: **% Spent / % Complete = 19.4 / 17.6 = 111%.**
  - We can see that we are really spending over 10 percent more money than we are earning in benefit. To quote Mr. Micawber in Charles Dickens's *David Copperfield*: "Annual income twenty pounds, annual expenditure twenty pounds ought and six, result misery." In the current situation, the full earned benefit approach taking essential links into account indicates a steeper slope towards a miserable outcome than even Mr. Micawber (or the simplistic program manager) could have anticipated.

The QERTS Case Study

The BRM for the QERTS case study is shown again in Figure 12.21, with the essential links indicated by double arrows.

This case study is more complete than the Garden Services example, since it shows deliverables, outcomes, and the like; however, it is also more complicated because it contains essential nodes connected to essential nodes.

The numerical results of applying the full algorithms is shown in detail in the cash-flow calculations in Chapter 8.

However, the same algorithms can be applied, as they will provide the program manager and the key stakeholders with indicators that take the whole model into account while also giving an overview of progress and trends. The use of this information is explained in more detail in Chapter 13 on key performance indicators.

The earned benefit indicators are critical to the success of program status reviews.

12.11 Applying Earned Benefit for Status Reviews

At each status and progress review, calculate the following:

- cost performance index = earned value / actual cost (i.e., the standard earned value cost tracking approach)
- benefit performance index = earned benefit / planned benefit
- benefit schedule performance index = benefit equivalent duration / current program duration

For completeness, the benefit value performance index, combining the time and cost information as defined previously (see Section 12.8.16), should also be calculated.

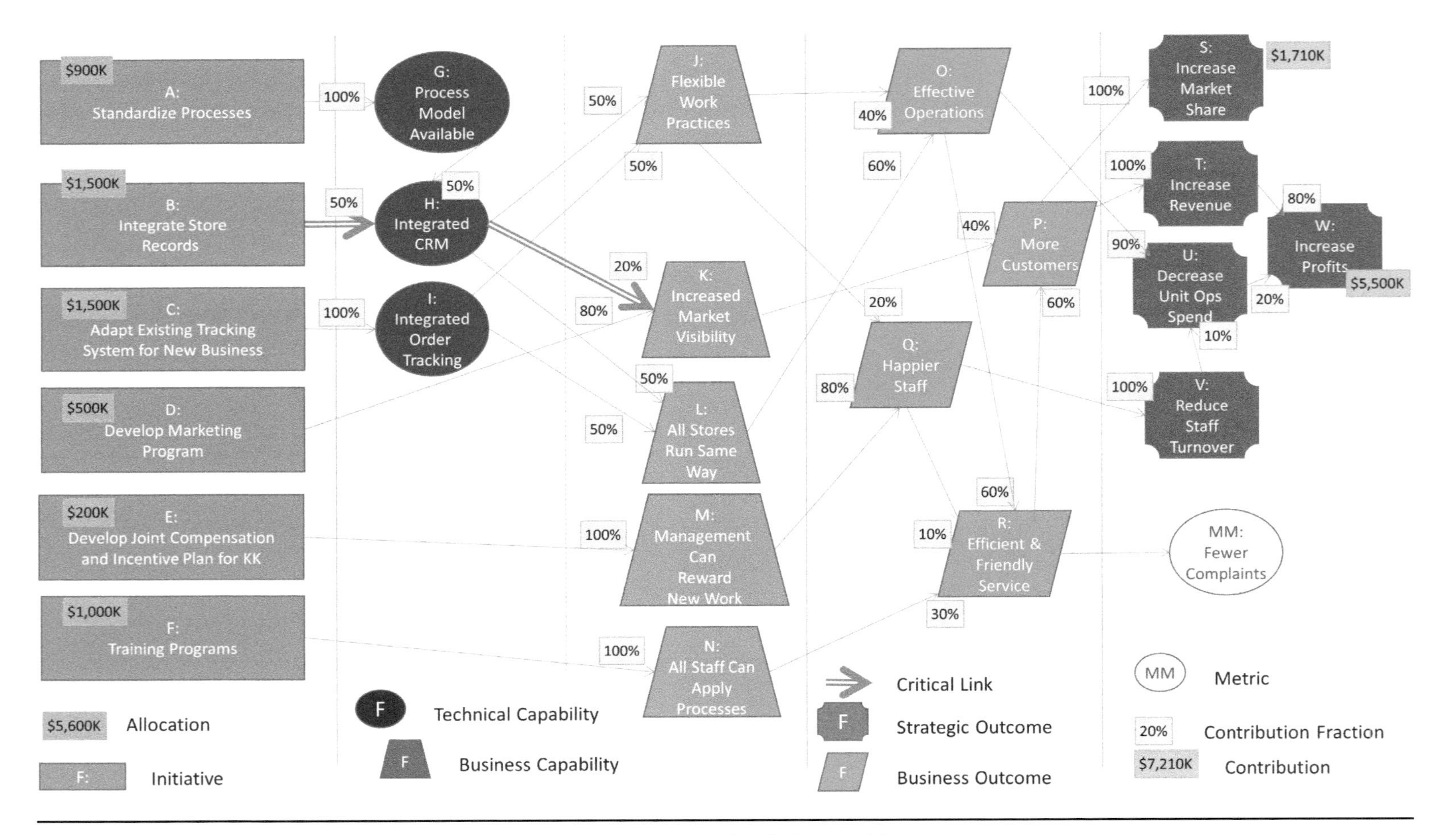

Figure 12.21 The QERTS Benefits Realization Map Indicating Essential Links by Double Arrows

Any value less than 1.0 in any of these index values indicates a slip in either cost or time—or both—which will need to be explained. If this is the case, the to-complete performance index and factor values (TCPI and TCPF) can be analyzed in order to evaluate the chances of recovering the situation.

12.12 Communicating the Information

At first sight, as for earned value, the concepts and formulae for earned benefit may seem hard to understand. However, it should be remembered that the approach provides a powerful and coherent link across the full life cycle and, as such, directs and adds value to all of the steps in the benefits realization process. The earned benefit concepts become easier to understand the more you apply them. That is not to say that the program manager should expect the management chain or the customer to understand the finer points of the technique. On the contrary, with a good understanding of what the numbers really indicate, the program manager will be able to translate them back into concepts linked to the relevant stakeholder's domain of interest. Communicating information is addressed in more detail in Chapter 14.

12.13 Conclusion

In line with the benefits realization life cycle that was defined earlier, along with the main processes for planning and tracking the benefits from inception to termination, the benefits-centric planning and tracking technique goes beyond the concepts of earned value to address the potentially complex contributions of the separate components and of essential links to the overall benefits of the program. The examples demonstrate that earned value and earned benefit can provide significantly different images of actual program status. Without a good knowledge of earned benefit, the program manager can be misled into making unwise decisions.

12.14 The Benefits of Earned Benefit

Earned benefit adds the genuine *value* dimension based on the corresponding benefits to the concepts of earned value, thereby allowing the key stakeholders to gain a better idea not only of progress with respect to the planned scope (earned value) but also with respect to the required outcomes (planned benefit). In this way, earned benefit provides a more complete and integrated value/cost/schedule control system (V/C/SCS) than the original C/SCS (cost/schedule control system) for earned value.

As shown in Table 12.12, there are many parameters and indicators that can be added to enhance still further the basic earned benefit and earned value formulae.

12.15 Earned Benefit Parameter Tables

Each parameter and indicator in Table 12.12 is shown as an abbreviation followed by its name; this is followed by an explanation of its significance in tracking and forecasting progress or, if relevant, its formula and the result of applying this to the Garden Services case study. Note that the tables address not only earned value management as extended to earned benefit but also support an expanded earned schedule approach that can be used in the context of earned value.

Table 12.12 Set of Earned Values and Earned Benefit Parameters and Indicators

Abbreviation	Indicator Name	Explanation	Calculated Value	Reference
AC	Actual Cost	Total spent to date	€370,000	12.5.1
AD	Actual Date	Date of data	July 1st	12.5.2
ADD	Actual Date Duration	Days since start	181 days	12.6.1
BAC	Budget at Completion	Total Budget	€470,000	12.5.7
BCPI	Benefit Cost Performance Index	EBEV / AC	86%	12.6.6
BPI	Benefit Performance Index	EB / PB	92%	12.6.7
BSPI	Benefit Schedule Performance Index	EBEDD / ADD	83%	12.6.12
BSV	Benefit Schedule Variance	EBEDD – ADD	–30 days	12.6.11
BV	Benefit Variance	EB – PB	–€34,000	12.6.10
BVPI	Benefit Value Performance Index	EDD – (VEDD – EBEDD) / EDD	66%	12.6.16
BVSV	Benefit Value Schedule Variance	EBEDD – VEDD	–61 days	12.6.15
CCB_i	Component Contribution to Benefit	Sum over columns j of {[CBM(i,j) * B_j}		12.8.18
CBM	Component–Benefit Matrix	See example		12.8.17
CEB_i	Component Earned Benefit	% complete[*component*(*i*)] * CCB_i		12.8.19
CPI	Cost Performance Index	(EVM Formula): EV / AC	93%	12.5.5
EB	Earned Benefit	The sum of all the CEB_i	€370,000	12.6.2
EBAC	Earned Benefit at Completion	Total forecast benefit value	€600,000	12.6.3
EBEC	Earned Benefit Equivalent Cost	The PV for a given EB	€320,000	12.6.20
EBED	Earned Benefit Equivalent Date	Planned date for current EB	June 1st	12.6.8
EBEDD	Earned Benefit Equivalent Date Duration	Days since start	151 days	12.6.9
EBEV	Earned Benefit Equivalent Value	The EV that corresponds to the EB	€320,000	12.6.2
EDD	Elapsed Duration	AD – PSD	181 days	12.8.16
EV	Earned Value	Sum of component Earned Values	€342,500	12.5.4
PB	Planned Benefit	The EB expected at a given date	€404,000	12.6.4
PPC	Program Percent Complete	EB / EBAC	62%	12.6.22
PED	Planned End Date	Program or stage completion date	September 1st	12.6.23
PEDD	Planned End Date Duration	Days since start	243 days	12.6.24
PSD	Program Start Date	Origin for calculating durations	January 1st	12.6.25
PV	Planned Value	Sum of component Planned Values	€345,000	12.5.4
SPI	Schedule Performance Index	(EVM formula): EV / PV	99%	12.5.6

(*Continued on following page*)

Table 12.12 Set of Earned Values and Earned Benefit Parameters and Indicators (*Continued*)

Abbreviation	Indicator Name	Explanation	Calculated Value	Reference
$TCPF_X$	To Complete Performance Factor	• Time: TCPIT / BSPI	178%	12.6.27
		• Cost: TCPIC / BCPI	173%	12.6.27
		• Benefit: TCPIB / BPI	128%	12.6.26
$TCPI_X$	To Complete Performance Index	• Time: (PEDD – EBEDD) / (PEDD – ADD)	148%	12.6.26
		• Cost: (BAC – EBEC) / (BAC – AC)	150%	12.6.26
		• Benefit: (EBAC – EB) / (EBAC – PB)	117%	12.6.26
VEB	Value Equivalent Benefit	The PB for a given AC	€437,500	12.6.21
VED	Value Equivalent Date	Planned date for spending AC	August 1st	12.6.13
VEDD	Value Equivalent Date Duration	Days since start	212 days	12.6.14

12.16 Summary

This chapter presented a broad extension of earned value management into the program management domain by the defining a generalized earned benefit method. It started by showing the link between initiatives and benefits by use of a component–benefit matrix.

It first presented the earned value concepts and then used the Garden Services case to show the added information provided by earned benefit at the business level. It linked earned benefit calculation with the BEER algorithm and showed the effect of essential links. It concluded by providing a table of all of the parameters and indicators as applied to the case study, along with their formulae and case study values.

12.17 References

Piney, C. "Chapter 7." In *Program Management, a Life Cycle Approach*, edited by G. Levin. Boca Raton, FL: CRC Press, 2013.

Project Management Institute. *Practice Standard for Earned Value Management.* Newtown Square, PA, USA: Project Management Institute, 2011a.

Chapter 13

Business Key Performance Indicators

In the previous chapter, we analyzed how to measure and track progress toward achieving business-oriented outcomes during the program implementation phase. Those indicators were relevant during the implementation of the program, but they cannot provide useful information once the program is complete. After completion, a new but compatible set of indicators is required to determine whether or not the completed program is actually delivering the planned benefits at the predicted rate. The results of this analysis can be used to adapt the original model for future use as well as to provide action-oriented information about any variance between the actual and the predicted figures.

13.1 Setting Key Performance Indicators

The objective in setting key performance indicators (KPIs) is to develop a set of measurements that will provide meaningful information as to progress with respect to specific objectives.

Bernard Marr (Marr 2010a) has explained the importance of ensuring that we determine the questions to be answered by these measurements before we define what to measure and how to make the measurements.

In the case of setting KPIs in the program management domain, we are not focusing on the classical balanced scorecard approach (Marr 2010b) because that is better suited for providing a holistic view of the overall organization. In our case, the focus should be on the strategic objectives that have been defined and planned through the benefits map. This benefits-focused approach is explained in this chapter, taking examples from the QERTS case study.

13.2 Operational KPIs

Although earned benefit is extremely useful for measuring progress based on work accomplished, business management also needs KPIs that are relevant before, during and after all of the implementation

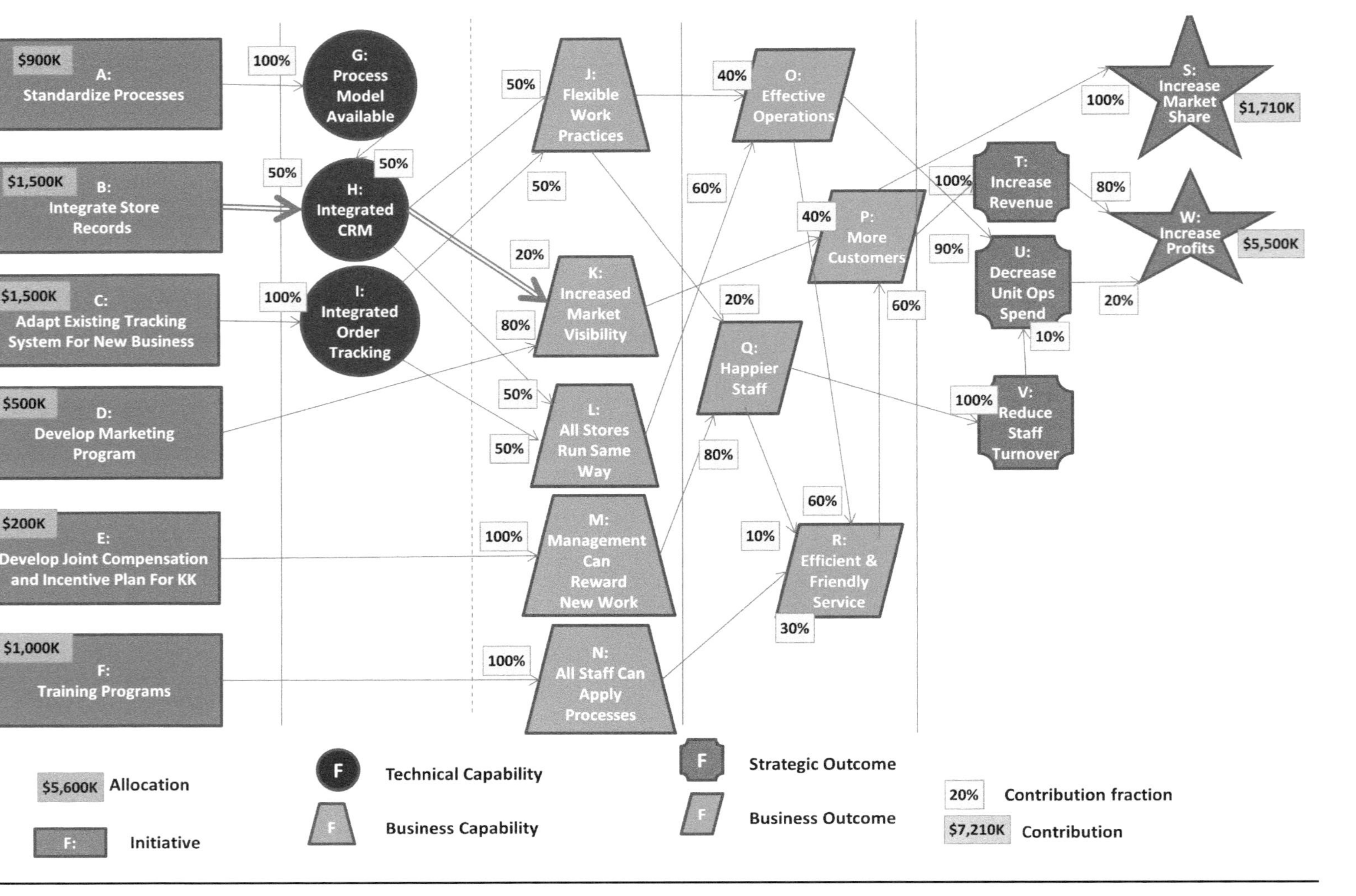

Figure 13.1 The Benefits Realization Map for the QERTS Example

work to track how the benefits are actually accruing. These KPIs allow management to assess the correctness of the underlying assumptions and improve their decision-making ability for the future. The benefits realization map (BRM) and its contribution fractions provide a good starting point for this analysis. The diagram for the QERTS case study is reproduced in Figure 13.1.

Since we are looking for KPIs of interest to the business, we should start from the right-hand side of the BRM to look at the strategic outcomes and then work back through business outcomes and business capabilities to assess which ones are relevant and which ones can be measured. The assessment should be done in that order (i.e., relevance and, only then, ability to measure) to avoid wasting effort and credibility in generating measurements that no one will use. In theory, however, all of the measurements should be of interest since they are shown in the BRM to contribute to the strategic objectives. Different stakeholders will be interested, to a greater or lesser extent, in different items, but overall, all of the metrics are expected to be of potential interest.

Some indicators may be less easy to measure directly than others. However, it is important in all cases to establish a consistent measurement system that can be used to provide absolute, or at least, relative values. This measurement system will then allow a direct link to be made between the value of the KPI and the contribution of the corresponding node. Quantifiable KPIs will also allow the plotting of these values over time, thereby helping to identify trends as well as showing the effect of specific actions that have been be taken to correct or enhance performance.

The types of indicators are explained in more detail in the following sections.

13.3 Agreeing on the Questions

Bernard Marr points out the importance of agreeing on the right questions prior to setting up the measurement model or process: key performance *questions* before key performance *indicators*. These questions will help define the indicators that should allow us to determine the answers to those questions in a clear and objective manner.

Based on the BRM, plus program management best practices, here are the main questions that need to be addressed:

- How well are we progressing toward our strategic objectives?
- How realistic is the original business case?
- For each capability or outcome: How close are we to achieving this result?
- For each capability or outcome: How confident are we in the value of the corresponding contribution?
- For each contribution fraction: How confident should we be in this value?

Earned benefit, as described in the previous chapter, represents the KPIs that address the first bullet above: How well are we progressing toward our strategic objectives? Note that this approach only provides KPIs that are valid up until completion of program implementation. Once the final outcomes are in place, new KPIs are required in order to answer the operational question: "Is the resulting situation delivering the benefits as defined in the signed-off business case?" This question addresses the operational realization of the benefits and provides a practical answer to the second question ("How realistic is the original business case?"), as will be explained later in the chapter, along with the other questions from the list.

13.4 Categories of KPIs

Two categories of KPIs are defined. They can be used not only to measure progress but also to assess the correctness of the model. These are:

- *Operational KPIs.* These measure the status directly at a given point of time.
  - Example: The node percent achieved, an estimate of how close the node is to achieving its full potential
- *Model-Related KPIs.* These measure values as predicted using the benefits map and the program schedule, as described in Chapter 8.
  - Example: The *model calculated node contribution,* giving the forecast contribution of a node

Operational KPIs can be analyzed alongside model-related KPIs to evaluate and analyze the accuracy of the model, as shown later.

13.5 Operational KPIs

Operational KPIs may be defined for each category of node.

13.5.1 Strategic Outcome KPIs

All of the strategic outcomes are by definition relevant. They can normally be directly measured: for example, market share, revenue, operational spend, or staff turnover.

Of direct interest to the program stakeholders is the way in which these strategic outcomes are influenced by the work of the program. Analysis of progress with respect to the strategic outcomes requires two sets of measurements:

- The value of the strategic outcomes themselves, called *business outcome KPIs,* and examined in the following section
- The KPIs that measure characteristics of the intermediate outcomes, called *model-related KPIs* in the following analysis (ideally, these should include KPIs related to the assumptions associated with the links in the BRM)

13.5.2 Business Outcome KPIs

Some of the business outcomes can be evaluated in terms of cardinal numbers—that is, numbers that measure quantity. These are called *quantifiable KPIs* in the following section. Some outcomes cannot be measured so precisely, although relationships between different measurements (such as larger, better, etc.) can be determined. These are called *non-quantifiable KPIs.*

13.5.3 Quantifiable Outcome KPIs in the QERTS Example

Effectiveness of Operations

The effectiveness of operations can be measured for example by evaluating the change in the effort expended by the operations group, normalized to the number of QERTS clients or transactions. If these figures are not available before the program is started, it will be necessary to add a program component to measure the preprogram state before the program goes live. This need for baseline data underlines the importance of deciding on KPIs and how to measure them in the initial stages of program planning and justification since, if the KPI definitions are left until later, this additional work will affect the values of cost and the schedule already planned.

More Customers

The change in customer numbers should be directly measurable—not just the number of customers, but their level of spending. The BRM does not make it clear that this spending pattern, rather than just the number of customers, must be included in the specification of the business outcome. Once again, this requirement shows the importance of defining the KPIs along with the model before completing the business analysis and developing the business case and the program plan.

13.5.4 Nonquantifiable Outcome KPIs in the QERTS Example

Staff Happiness

Staff happiness can be assessed with surveys to show increase or decrease, but numerical measures that would allow mathematical analysis (e.g., 1.75 times as happy as last year) cannot be defined as a valid quantification.

Efficient and Friendly Service

In a similar way to staff happiness, the friendliness and efficiency of the service can be assessed with surveys but does not provide a sound mathematical basis for precisely quantified progress reporting. However, this analysis suggests that modifying the BRM to split this outcome into Efficient Service (quantifiable) and Friendly Service (non-quantifiable) could be worthwhile.

13.6 Model-Related KPIs

Having KPIs associated with assumptions in the model will allow the accuracy of these assumptions to be assessed. This information will be of considerable value for adapting the model as delivery progresses, developing lessons learned, and creating future models.

Since we are aiming to compare the forecasts from the model with the operational measurements, the figures in question should be compatible with those generated by the cash flow analysis based on the model. These model-related KPIs are defined in the next section, and their relationship with the operational KPIs is used to analyze both the situation and the model.

13.7 Analyzing the KPIs

Taking these sets of KPIs—operational and model-related—together will provide information with which to analyze the accuracy of the assumptions used to develop the BRM.

Each of the indicators is listed in the following table (Table 13.1) and explained in more detail in the subsequent text.

Table 13.1 List of Key Indicators for Assessing the Model

Operational Indicators		
OAIP	Additional Improvement Potential	Assessed
OCNC	Current Node Contribution	KPI value
OFNC	Forecast Node Contribution	OCNC / ONPA
ONPA	Node Percent Achieved	100% – OAIP

Model-Related Indicators		
MCNC	Model Calculated Node Contribution	From the cash-flow analysis
MTNC	Model Total Node Contribution	From the BRM
MNPE	Model Node Percent Earned	MCNC / MTNC
MCFds	Model Contribution Fraction for (destination-source) link	

Analysis Indicators		
ALAV	Assessed Link Assumption Validity	See text
ANAV	Assessed Node Assumption Validity	See text
ARCFds	Assessed Revised Contribution Fraction for (destination-source) link	

13.7.1 Operational Values

The following approach will be taken.

- Each node will have a KPI based on an estimate of how close the node is to achieving its potential. This *operational node percent achieved* (ONPA) KPI may be the combination of a number of intermediate KPIs for the same node.
  - Each node in the BRM will have its own ONPA, which should be evaluated independently of the earned benefit or cash-flow model.
  - Ideally, the estimation of the ONPA should be independent of the financial numbers in the signed-off business case and based on the evolution of the program up to this point in time. This approach to estimating will allow progress to be measured independently of the original financial assumptions, to support an objective analysis of the correctness of the business case, and to avoid political pressure to modify the findings.
  - An independent approach to calculating ONPA could be obtained by asking the following question: "Given the current status, what additional improvement percentage (the *operational additional improvement potential,* or OAIP) do you think would be possible?" The ONPA can be calculated as (100% – OAIP).
- Similarly, it should be possible to define KPIs for most nodes to measure the actual value of the contribution at any point in time—the *operational current node contribution* (OCNC).

13.7.2 Using Operational and Model KPIs to Assess the Model

- The *operational forecast node contribution* (OFNC), based on the operational measurements defined above, is:

  - OFNC = OCNC / ONPA.
  - Any mismatch between this operational forecast of the node's contribution and the forecast value in the model (the *model calculated node contribution,* or MCNC) will indicate an error in the model.
- Each node will have a *model node percent earned* (MNPE) indicating how close that node was expected to be to having achieved its full contribution, based on the cash flow calculations. This parameter is similar in concept to the ONPA:
  - MNPE = MCNC / MTNC.
  - For initiatives (i.e., the left-hand side of the BRM), we will adopt the following convention:
    MNPE = ONPA = earned value "percent complete" of the initiative.
- Each node will have an *assessed node assumption validity* (ANAV) at any point in time, designed to measure how far the actual value achieved (i.e., the OCNC) differs from the values as calculated by the cash-flow model (i.e., the MCNC). We take the following convention: that the assumption is valid if the OCNC (based on the KPIs) is identical to the MCNC. If the two values are different, our confidence in the assumption depends on the difference between those two quantities; ANAV is normalized based on the average of their values, to give:
  - ANAV = 100% – 2 * {abs(MCNC – OCNC) / (MCNC + OCNC)}
- Each link in the model will have its *assessed link assumption variance* (ALAV) KPI. This number is closely related to the "real" value of the contribution fraction because, if the assumption on the link is incorrect, the source outcome will not contribute as expected to the destination node. Since the real situation can be a larger or smaller contribution fraction than initially assumed, the KPI can exceed 100 percent in the case of overachievement. The assumption validity will therefore be the difference between the *assessed revised contribution fraction* (ARCF—if we can assess it!) and the value of the contribution fraction in the original model (the *model contribution fraction,* or MCF):
  - ALAVds = 100% – (MCFds – ARCFds).

The original model holds if all of the node ANAVs and link ALAVs are 100 percent. This overall agreement will only be the case if the expected strategic objectives, the assumptions on the links as well as the contribution fractions are all correct. The model can therefore start to be reviewed as soon as the measurements begin to be taken. Let us examine in more detail how to carry out this analysis.

13.8 Analyzing the Model

Based on the measured (operational) numbers, the model may need to be adapted. This analysis must always be carried out at least once, at the end of the program, to provide an objective assessment of the original business case and the chosen solution. This updated model will be valuable for future programs in a similar environment. It should be noted that this capability of improving the accuracy of the model shows the potential value of carrying out small initial "pilot" programs in order to provide a baseline model on which to base the full program.

The starting point for assessing the model should be the operational KPIs, which can be evaluated quantitatively. The comparison between the operational KPIs and the forecast values from the model will indicate any mismatch. The other quantitative values as well as the qualitative ones can then be used to help identify the root cause of any difference.

The challenge is that BRMs have a large number of degrees of freedom: the values of the strategic outcomes, the links, the ramp-up times, and the contribution fractions. In addition, for some nodes, it will not be possible to define KPIs to assess the current contribution; this lack of quantifiable measures is generally the case for the nodes representing capabilities—even if a "percent achieved" may be measurable. However, as the program progresses, the trends in any mismatch between the measured and the calculated values can be used to help identify the causes. The approach to be taken is outlined below.

13.8.1 Left-to-Right Analysis

The basic approach is to work from left to right to determine the correct value to assign to each link contribution share at each status review, as the required KPIs become available. The algorithm to follow is to start from the initiatives and aim to obtain the model up to the first set of dependent outcomes that have quantifiable KPIs. These nodes will normally correspond to the set of business outcomes that are dependent on business capabilities. An initial model that ignores the nodes beyond this point can provide a basis for the initial analysis.

This initial analysis should be carried out by treating the terminal nodes on the right of the truncated diagram as the strategic outcomes of this model. The forecast value of these corresponding "strategic objectives" is the contribution of the node as calculated for the full model (i.e., the MCNCs).

The algorithm for pruning the model as closely as possible to the initiatives is as follows. Work from right to left:

1. Remove all nodes which have no successors.
   a. In the first pass, nodes with no successors will be the set of strategic objectives.
2. Check that the remaining nodes with no successors are measurable outcomes.
   a. If so, see whether there are any more nodes that can be removed by repeating step 1 above.
   b. If not, then reverse the previous pruning step to reinstate the terminal measurable outcome and exit.

An example is provided below for the QERTS example (Figures 13.2, 13.3, and 13.4). Once the model for the reduced map shown in Figure 13.4 has been validated, we will have a reliable basis for the rest of the model.

The choice was made to stop the pruning as shown in Figure 13.4 because the next step, as shown in Figure 13.5, would leave two unquantifiable nodes (K and R) exposed.

To obtain an overview of the part of the model on which we are focusing, it is useful to calculate the component–outcome matrix (COM) for this diagram. As described in Chapter 12, the COM provides the percentage contribution of each initiative to each of the chosen measurable outcomes. For the QERTS example shown in Figure 13.4, this initial simplification would give the following component–outcome table (Table 13.2). Note that the values of the calculated contributions for the end nodes, as taken from the full model, are also shown on the diagram in Figure 13.4.

But why is this COM relevant in validating the model in conjunction with the KPIs?

By providing the contribution fraction of each initiative to each of the outcomes that is being measured, the COM shows the relative impact of each initiative on each of the measured outcomes. If the KPIs indicate a mismatch between the predicted and the actual figures, the investigation as to the cause should start with the initiative with the highest contribution fraction.

It should be noted that this approach can be simplified by delaying the assessment until all of the nodes in the truncated model have reached (or are expected to have reached) their full benefit potential. This restriction serves to eliminate the uncertainty about the ramp-up times. However, delaying the assessment in this way can provide time for the program to drift out of control. The following example starts its analysis during the ramp-up period.

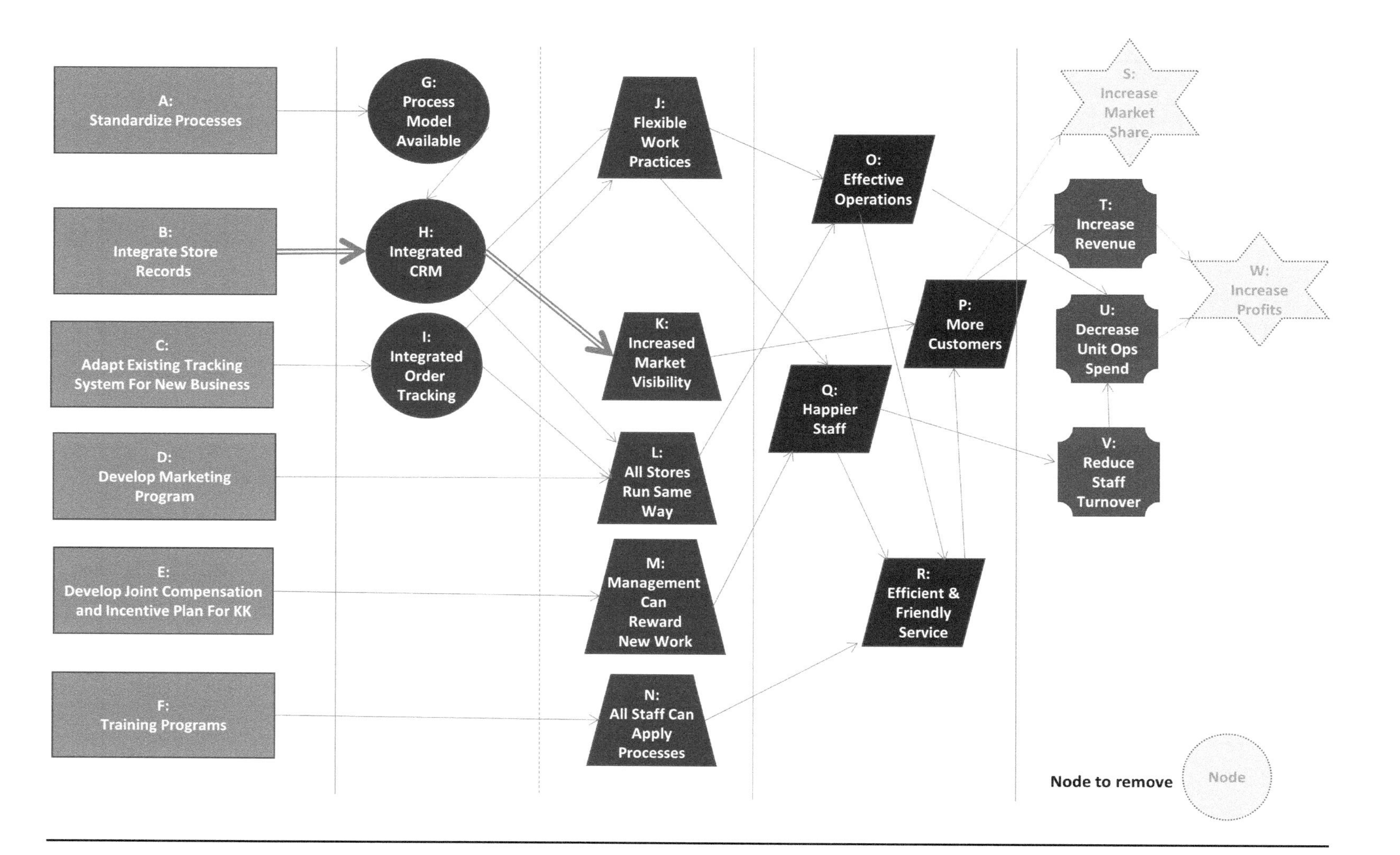

Figure 13.2 Initial Subset of the Results Chain for Analyzing Outcomes

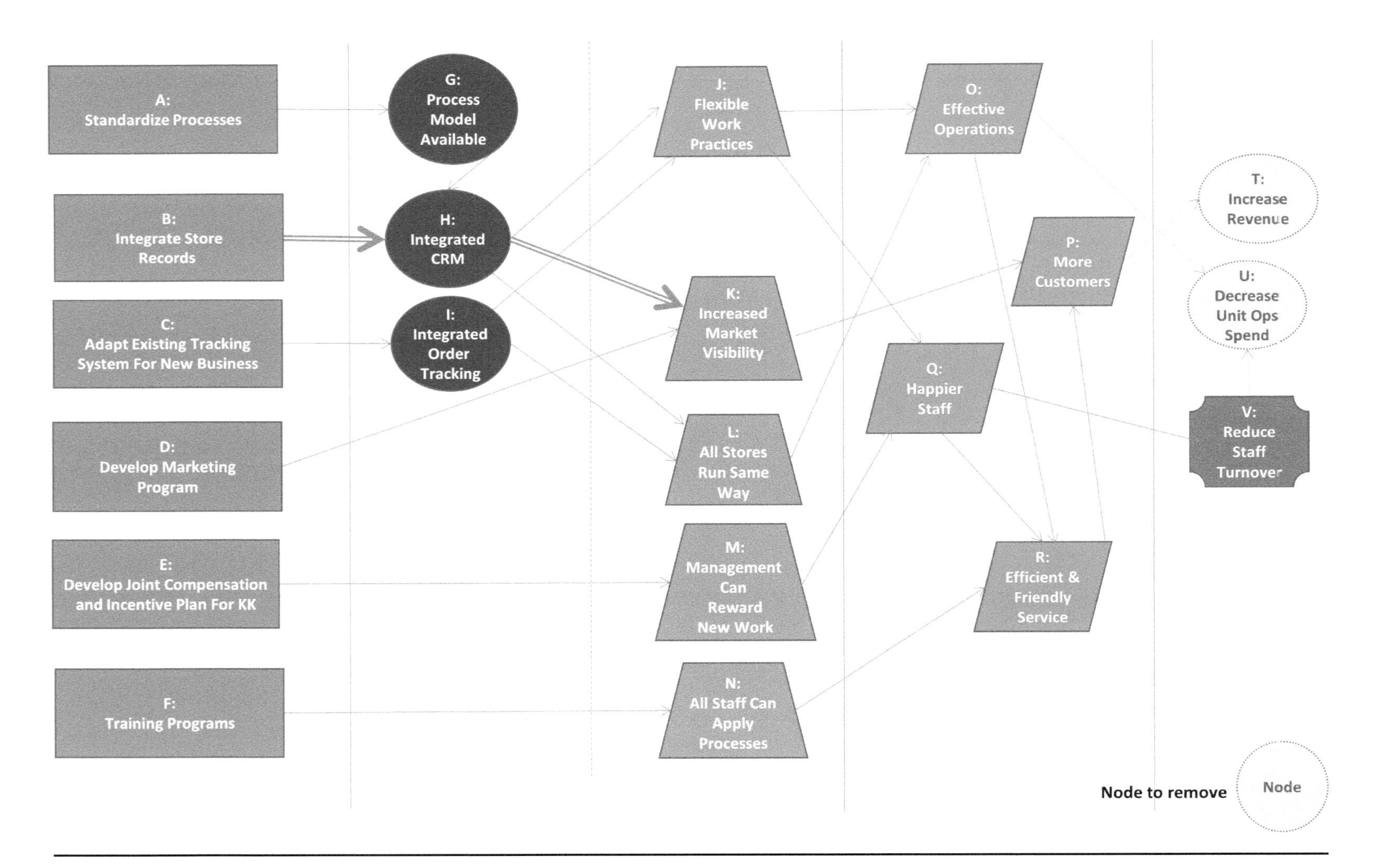

Figure 13.3 Second Step in Pruning the Results Chain for Analyzing Outcomes

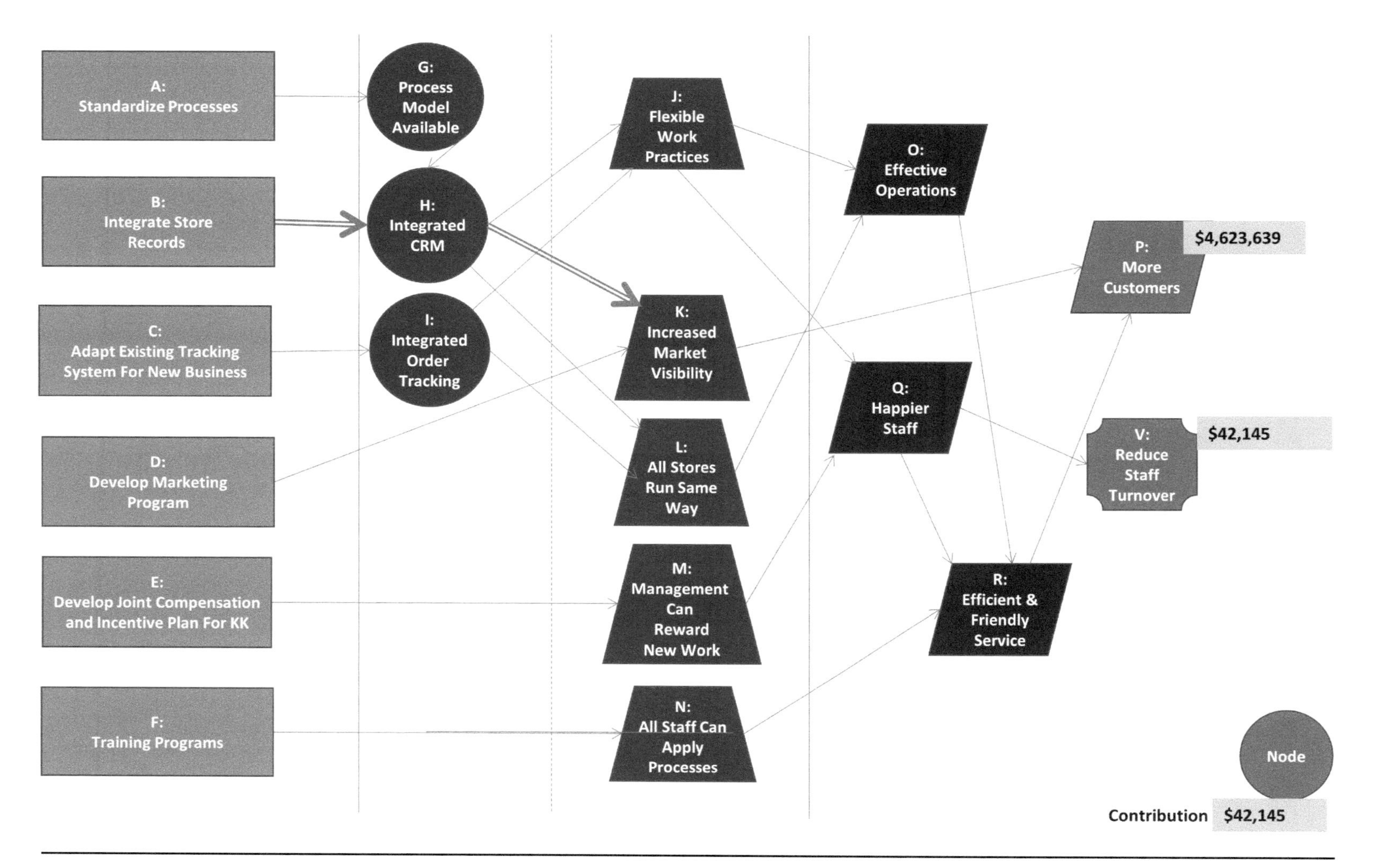

Figure 13.4 Final Step in Pruning the Results Chain for Analyzing Outcomes

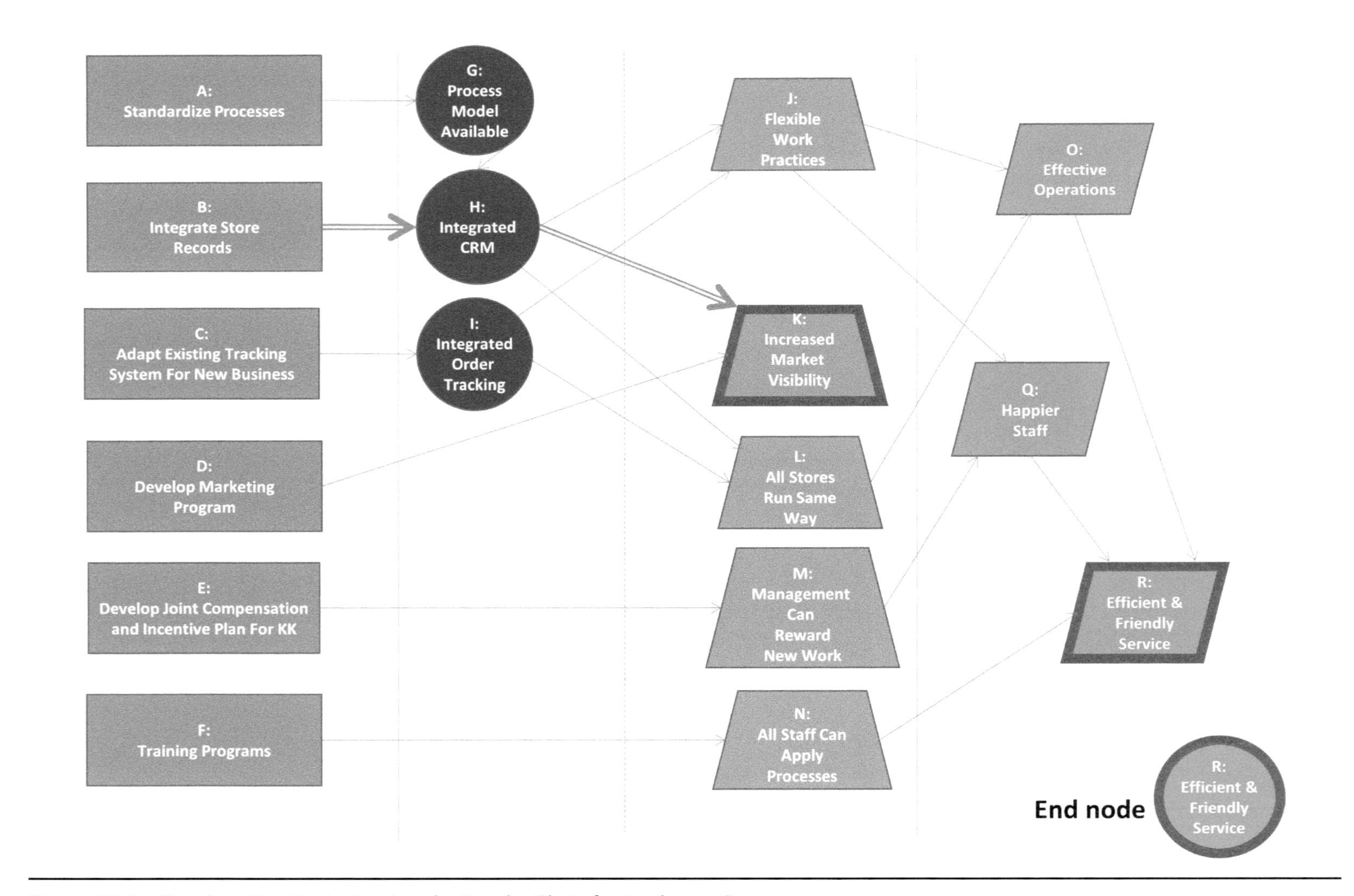

Figure 13.5 One Step Too Far in Pruning the Results Chain for Analyzing Outcomes

Table 13.2 Component–Outcome Matrix for the QERTS Example

	More Customers	Less Turnover
Standardize Processes	13%	5%
Integrate Store Records	13%	5%
Adapt Tracking System	22%	10%
Develop Joint Marketing	12%	0%
Align Compensation	14%	80%
Train Staff	26%	0%
	$4,623,639	$42,145

13.9 Example Based on QERTS Model

13.9.1 Developing the Component–Outcome Matrix

Using the terminology in Table 13.1, the COM allows the calculations to be carried out as far as the chosen outcomes, thereby giving the model-based value of the MNPE, based on the percent complete of each initiative. If the value of ONPA as calculated by the KPI differs from the MNPE calculated using the component–outcome matrix, then the model will need to be reviewed to understand the reasons for this mismatch.

The COM for the subset of the QERTS model shown in Figure 13.4 is given in Table 13.2.

13.9.2 QERTS Example KPIs

The following example is based on the results of the cash flow forecast of the contributions developed in Chapter 8.

The calculations provide the forecast values of the contribution at each month from the start of the program. As explained in the discussion above, some of the operational KPIs can only be evaluated as ordinal values (i.e., with numbers that allow you to evaluate rankings and trends but not precise values as a basis for mathematical calculations). However, even for these nodes, the trends can be compared with the trends from the corresponding cash flow calculations; they should both start at the same point in time and follow similar curves over time.

For the QERTS case, the forecast values at monthly intervals of each node from J = Flexible Practices onward are shown in Figure 13.6.

A diagram with all of the nodes is hard to follow even in the fairly simple QERTS case. This difficulty shows the value of the approach explained above of trimming the network to its proximal quantitative nodes and thereby reducing the number of nodes to consider. This simplification allows us, in this case, to concentrate initially on the pruned network in Figure 13.4 and focus on the two nodes shown in the COM in Table 13.2: nodes P = More Customers and V = Staff Turnover. The forecast curves for

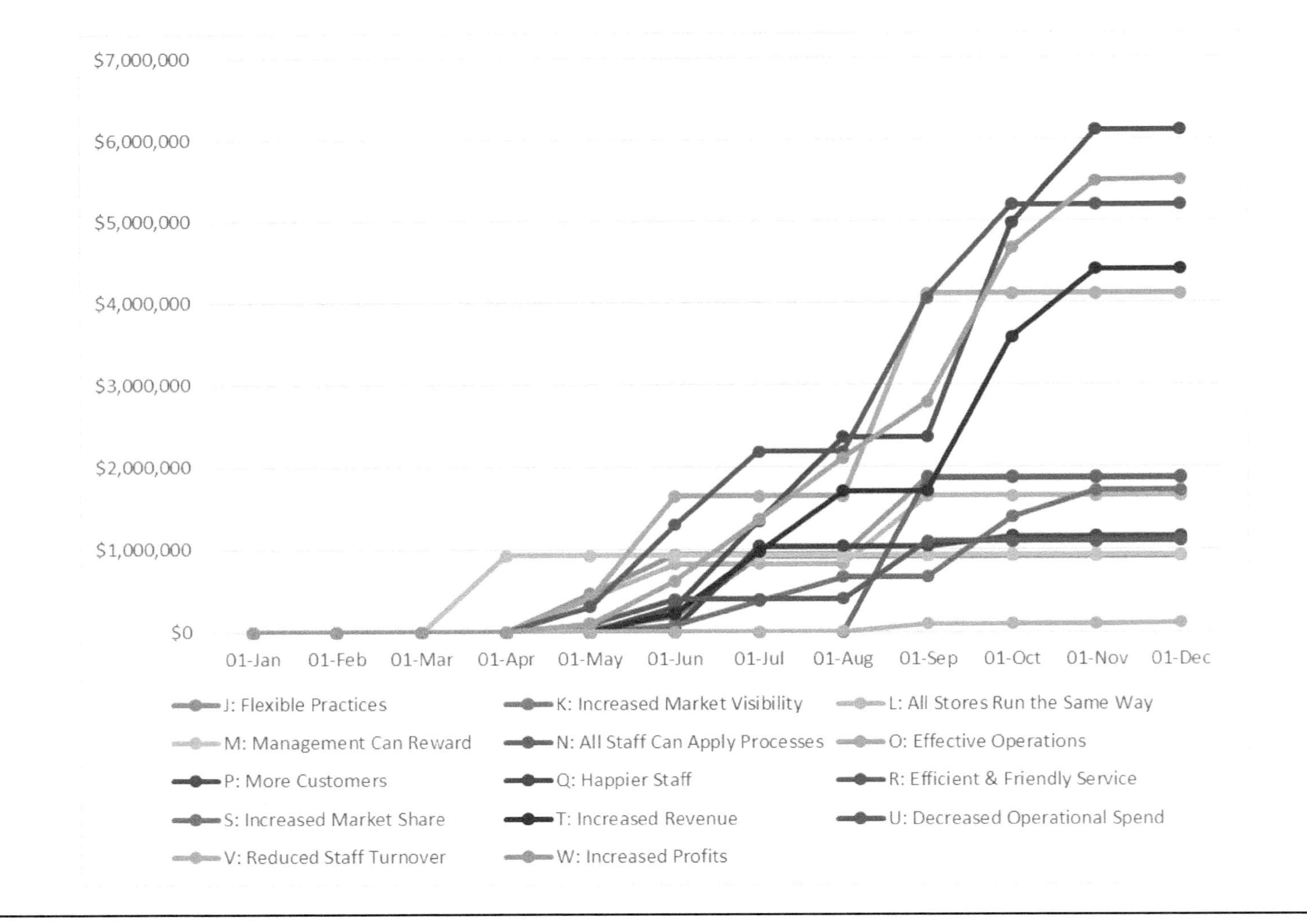

Figure 13.6 Forecast Values of the Nodes over Time

these two nodes are shown in Figure 13.7. These curves make it clear that the main KPI to focus on in this configuration is P = More Customers.

Based on the reasoning in Figure 13.4, we will now compare these predicted figures from the model with the actual values of the operational KPIs for nodes P = More Customers and V = Reduced Staff Turnover. For the purpose of this example, the predicted and actual figures for P = More Customers rounded to the nearest $1,000 are:

	Forecast	Actual
June	$312,000	$299,000
July	$1,344,000	$787,000

In June, there seemed to be nothing to worry about, as the percentage slip was of the order of 4 percent; that is, within the tolerance threshold. However, by July the percentage slip was ten times the size at 41 percent. More analysis is therefore not only justified but also, in fact, urgently required.

Using the key indicators from Table 13.1, the values for July are:

- MCNC (Model Calculated Node Contribution) = $1,344,000
- OCNC (Operational Calculated Node Contribution) = $787,000
- ANAV (Analysis Node Assumption Validity) =
  - ANAV = 100% – 2 * {abs(MCNC – OCNC) / (MCNC + OCNC)} = 48%

This analysis using the key indicators suggests that given the ANAV of about 50 percent, the likelihood of the node assumptions being valid could just as well be decided on the flip of a coin.

In addition, in this case study, the estimate of OAIP is stated to be in line with the predictions from the model (i.e., MNPE) of about 20 percent, indicating an ongoing trend for the OCNC over time. This mismatch between planned and actual values therefore needs to be investigated without delay.

The approach in this case is to look at the entire network of nodes that contribute to node P. They are as shown in Figure 13.8. That diagram takes into account the fact that, on July 1st, initiatives C = Adapt Existing Tracking System and F = Training Programs had not yet reached the point at which they could contribute to any beneficial outcomes. In the absence of these initiatives, the remaining links and nodes in the model that are assumed to contribute to node P = More Customers have been highlighted. The algorithm for choosing the nodes to highlight is as follows.

This analysis of contributing links and nodes is carried out from right to left, starting from the node in question, P:

1. Set P as the *focus node*.
2. Highlight all of the links going directly into the focus node.
3. If there are no new highlighted links, exit.
4. Set each of the source nodes of a highlighted link to be a highlighted node.
5. If there are no new highlighted nodes, exit.
6. Set each newly highlighted node as *focus node* in turn and carry on from step 2 above.

Non-highlighted nodes can be ignored in the subsequent analysis.

The numerical analysis is then carried out as follows:

1. Look at the direct predecessors of the node to be analyzed.
2. For each of these with quantitative operational KPIs, compare the OCNC with the MCNC to see whether the shortfall comes from there.

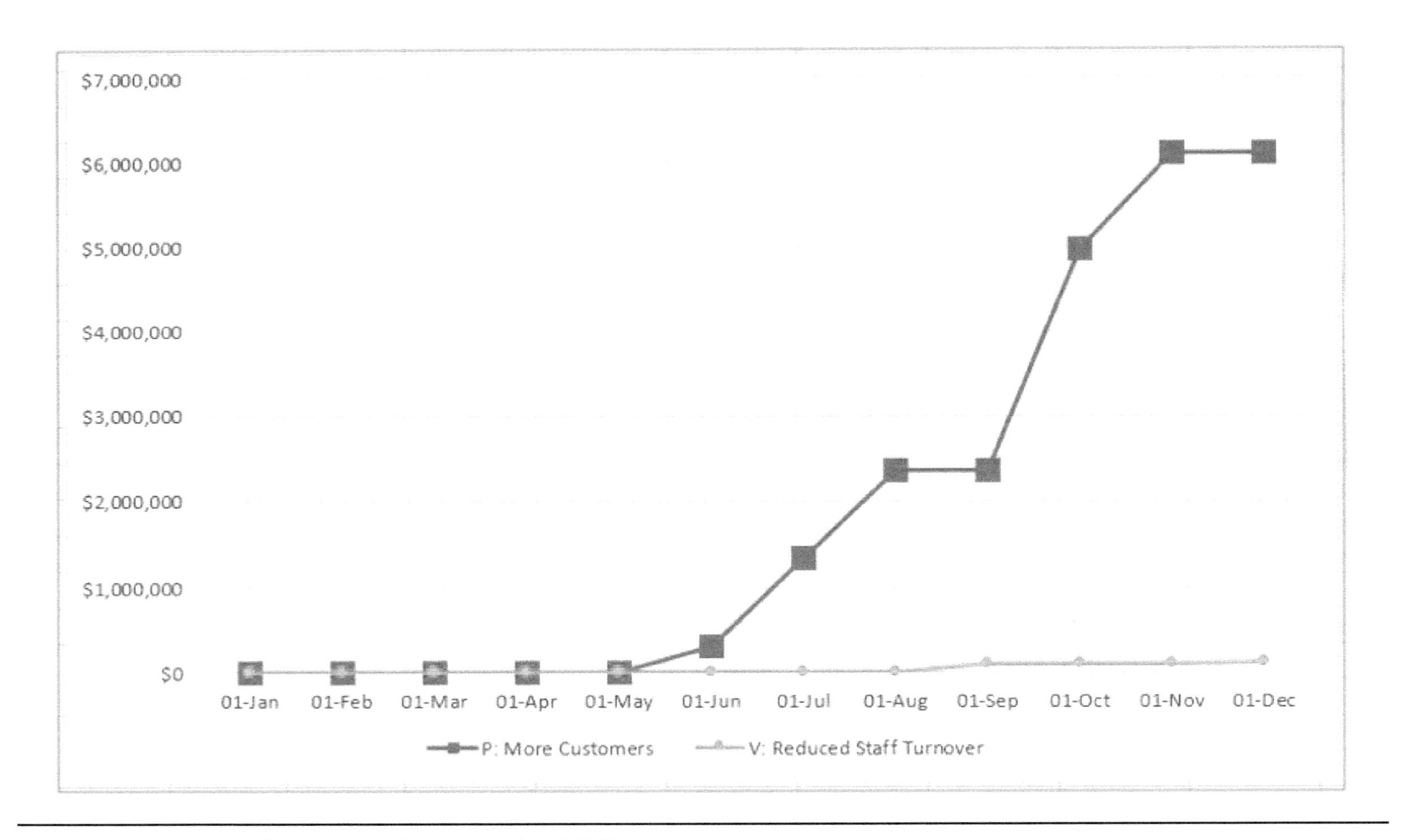

Figure 13.7 Forecast Values of Pruned Outcomes for QERTS

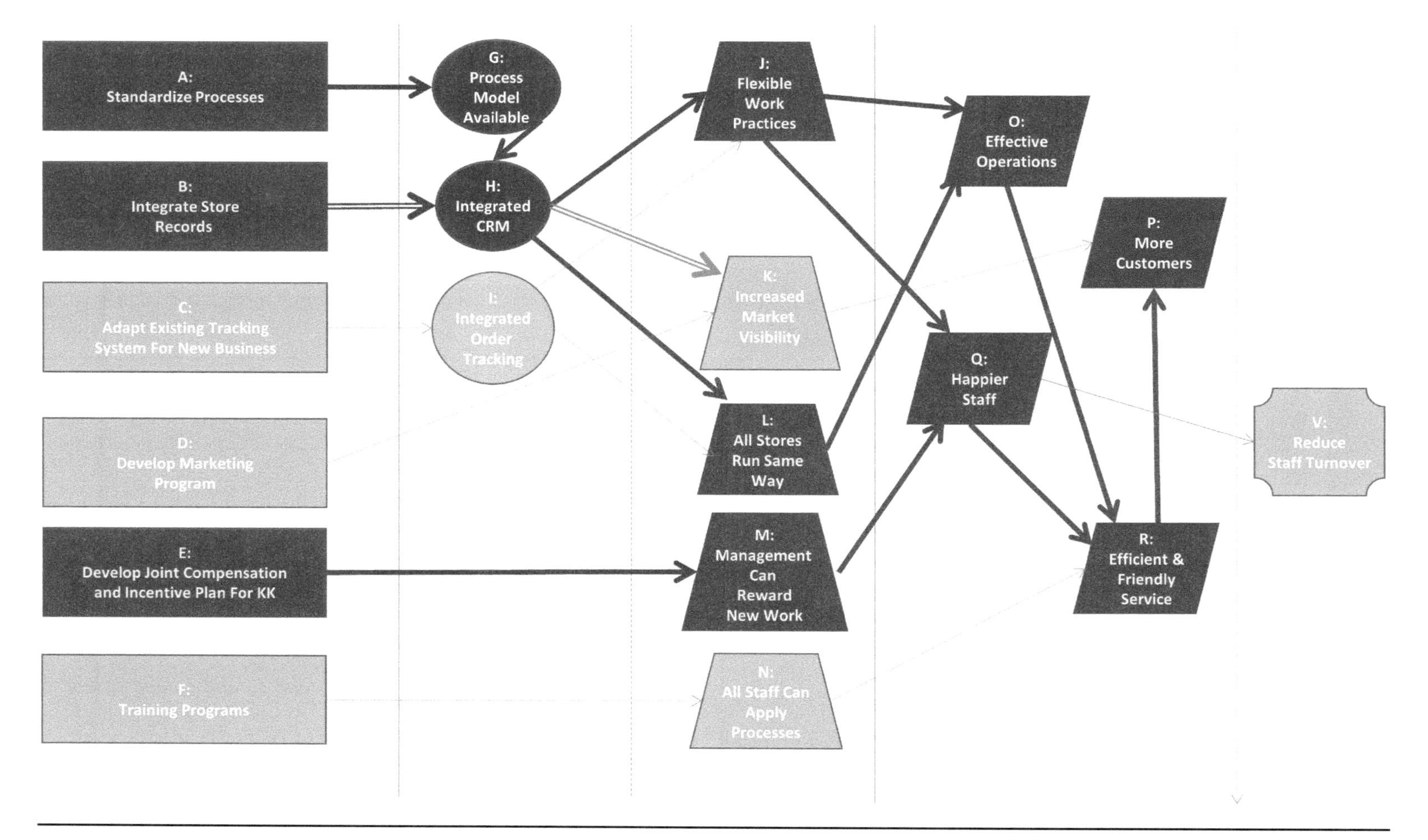

Figure 13.8 Analysis of Network of Contributors to P = More Customers

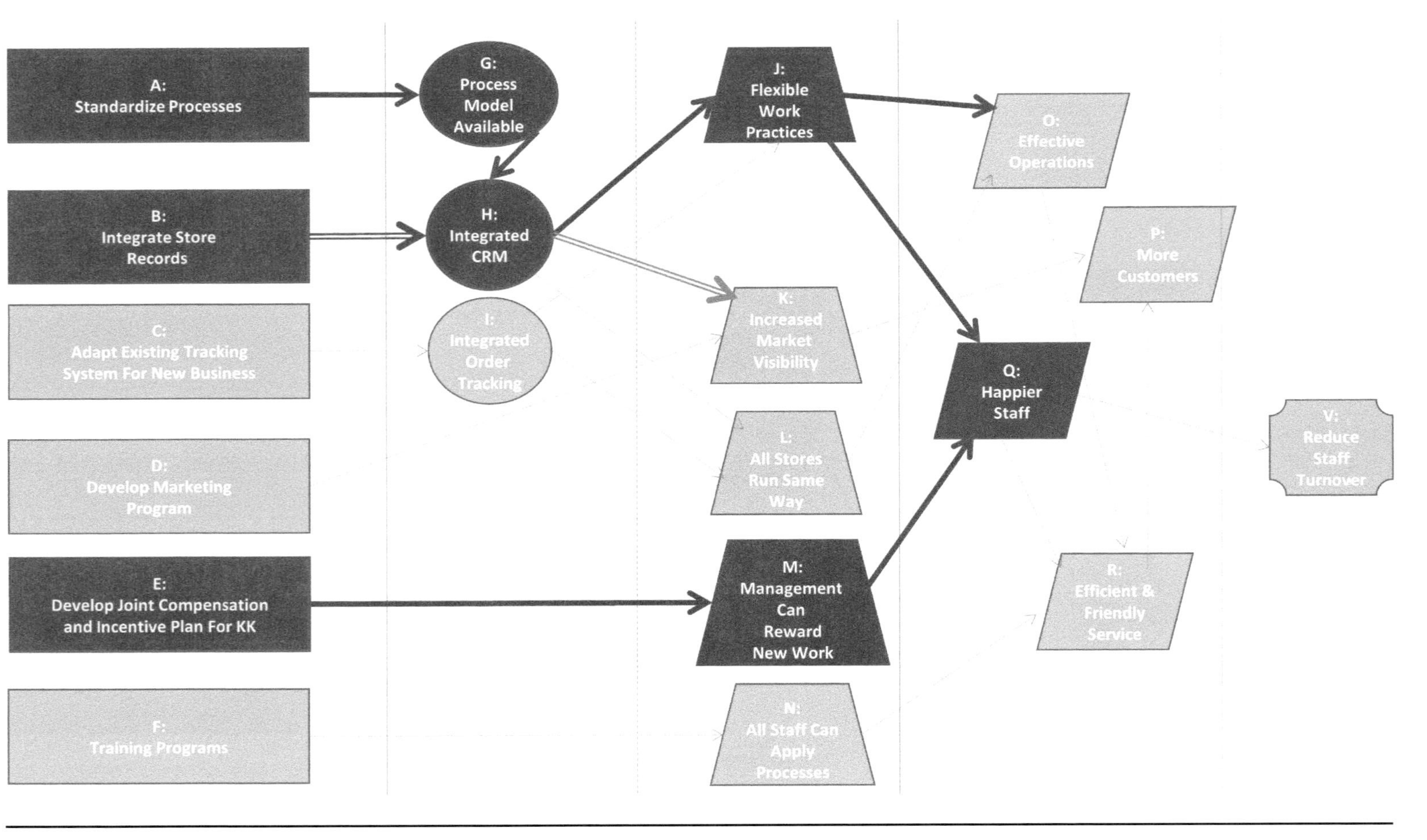

Figure 13.9 The Nodes in the Model that Contribute to Q = Happier Staff

a. In our example, the node in question is O = Efficient Operations, for which MCNC is $1,642,000 against an OCNC measured at the same value of $1,642,000. This result tells us that the problem must come from the only other contributing node R = Efficient and Friendly Service. However, R cannot be measured quantitatively.
b. Subjectively, however, we are told that the June level of customer satisfaction seemed to be rising as expected, whereas in July the feeling is that it should by now be rising 30 percent faster.
   - In addition, the OAIP for this node is similar to the MNPE so progress is not just late; it is not expected to recover.
c. Since we know that the problem does not come from O = Efficient Operations, we need to examine Q = Happier Staff and evaluate the degree of improvement.
d. In our model, the level of happiness of the staff (node Q) as forecast by the MCNC should be extremely high by now. In the current extreme, we find that none of the staff feel that the actions taken to date have contributed in the slightest to the enjoyment of their work. The problem is clearly focussed around node Q.

3. We now need to analyze the network of nodes in the model that contribute to node Q = Happier Staff. As on the previous step, we can prune the network back from right to left to leave only the nodes that contribute to Q = Happier Staff. The result is shown in Figure 13.9.
   a. We need to examine nodes J = Flexible Work Practices and M = Management Can Reward Work and base a staff satisfaction survey around these two capabilities in order to assess the accuracy of the assumptions on the corresponding links.
   b. For the purpose of this example, it turns out that the satisfaction survey does not identify any particular problems that are due to flexible work practices or staff rewards. However, an extra question was included on the survey asking: "What is the most annoying or disruptive feature of the way you are now expected to carry out your daily work?" The answer turned out to be the lack of challenge and novelty with all of the stores being run the same way. This was earlier identified as a "disbenefit" link and discussed in Chapter 7. However, in our example, it is clear that no action was taken to resolve what started out as a threat and has now shown itself to be an actual and costly problem.

Since the program is only seven months old, there is still time to address this disbenefits issue and carry out the actions identified in Chapter 7. The program should then recover to reach its expected level of benefits delivery, but with a delay of well over seven months.

In this way the use of operational and model-based KPIs can serve to identify problems at an early date, thereby giving the program team a chance to understand the situation and to take effective action to resolve it.

All of these tools provide the program manager and the business analyst with a large amount of information. However, one key role of the program manager is not only to know what is going on and act on it appropriately, but also to be able to communicate this information in a way that is meaningful to the various stakeholders. Effective communication is addressed in the next chapter.

13.10 Summary

This chapter explained the basis for developing valid KPIs. The link between earned benefit progress tracking during implementation and KPIs once benefits start to accrue was developed. An example based on the earlier case study using the various indicators was then analyzed. The way in which the KPIs can be used to review and adapt the model was demonstrated on the same model.

13.11 References

Marr, B. *What Are Key Performance Questions*? Management Case Study, The Advanced Performance Institute (www.ap-institute.com), 2010a.

Marr, B. *What Is a Modern Balanced Scorecard*? Management Case Study, The Advanced Performance Institute (www.ap-institute.com), 2010b.

Chapter 14

Stakeholder Analysis

Although a stakeholder's immediate focus may be on a specific element of the program, the synergy between program elements can be used to identify other related areas. This chapter defines an analysis technique using the benefits realization map to provide an original way of applying the power–interest grid to determine the range of interests of each stakeholder. This technique provides a broad stakeholder analysis that remains consistent with the overall program approach.

However good the strategy, planning, and general governance, a program can only succeed with the active support of key stakeholders. Even when the key stakeholders have been correctly identified, one common mistake is to focus only on their principal or most obvious areas of interest. Although this narrow focus may be a valid approach for projects, it is insufficient for programs due to the complex interactions between the program entities.

14.1 Stakeholder Interest and Power

The goal of the stakeholder analysis is to evaluate each stakeholder's potential engagement with respect to each node and link of the benefits realization map (BRM).

The potential level of engagement depends on two separate characteristics of stakeholders:

- *Level of interest:* How strongly stakeholders think that the program—or a subset of the program—affects their objectives. The values range from highly positive (*ideal*) through slightly positive (*supportive*) to slightly negative (*resistant*) and up to highly negative (*destructive*).
- *Level of power:* The amount of political or other form of power the stakeholder could exert on the program or on a subset of the program. The values range from high (*powerful*) to low (*ineffective*).

14.1.1 Three Categories of Stakeholders

For the purpose of the detailed analysis using the BRM, three categories of stakeholders need to be defined:

- Stakeholders directly involved in providing a deliverable (*producers*)
  - Producers are involved in program implementation at the technical level.
  - They affect the entities downstream—that is, to the right in the BRM.
- Stakeholders directly involved in transforming a deliverable into an outcome (*transformers*)
  - Transformers are involved in the program at the operational level.
  - Their focus is on the internode links and affects downstream entities.
- Stakeholders impacted by an outcome (*receivers*)
  - Receivers define what they expect from the program and are therefore involved at the strategic level.
  - Their involvement affects the upstream entities—that is, to the left.

These roles are mapped on to a typical BRM in Figure 14.1. The figure shows the role of stakeholders involved in each entity in the map:

- A = Standardize Processes is a Producer.
- B = Integrate Store Records is a Producer.
- J = Flexible Work Practices is both a Receiver and a Producer.
- K = Increased Market Visibility is both a Receiver and a Producer.
- O = Effective Operations is both a Receiver and a Producer.
- S = Increased Market Share is a Receiver.
- T = Increased Revenue is both a Receiver and a Producer.
- U = Decrease Unit Ops Spend is both a Receiver and a Producer.
- W = Increase Market Share is a Receiver.
- All of the links, such as G ⇨ H, H ⇨ J, H ⇨ K, etc., are Transformers.

Each stakeholder belongs to one, two, or all three categories. For each category, the level of interest and power of a given stakeholder can be different.

The goal of stakeholder relationship management is to enable each positive stakeholder to achieve their maximum level of interest and power while attempting to ensure that disruption due to negative opinions and influence is kept to a minimum.

14.1.2 Level of Power

The level of power has a direct effect on the potential result of any operation.

- A *producer* with no power will not be able to produce the deliverable.
  - Lack of producer power will reduce or prevent the output of that deliverable.
- A *transformer* with no level of power will not be able to contribute to the outcome.
  - Lack of transformer power will negate the assumption associated with the link from the source to the destination in the BRM.
- *Receivers* with no level of power will not be able to have an impact on the overall result of the program while reducing the value of all contributing nodes.
  - Receivers with no power will disrupt one of more of the benefits of the realization strategy.

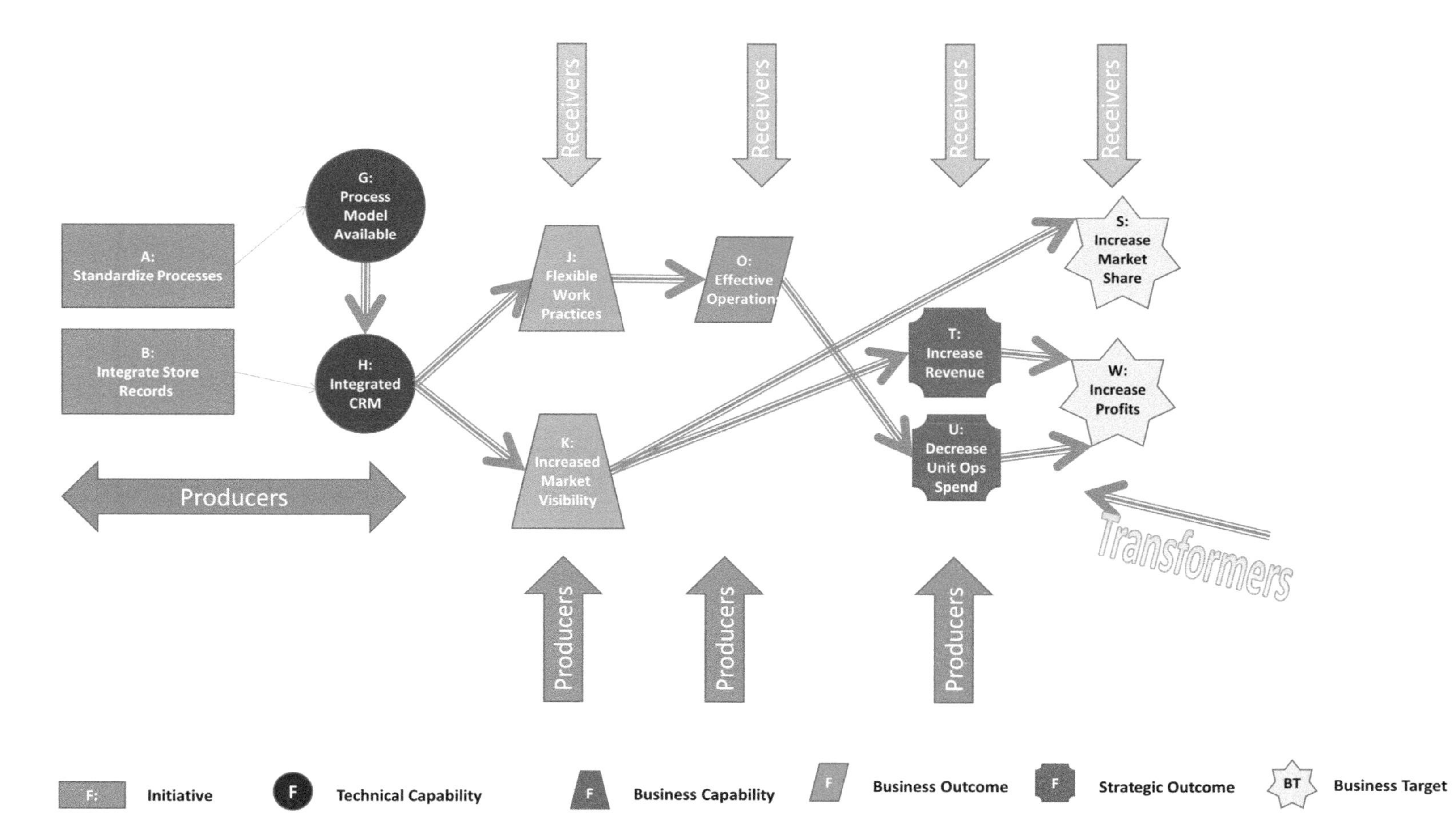

Figure 14.1 Subset of BRM Showing the Categories of Stakeholders

The level of power of a stakeholder needs to be evaluated for each node and link in the BRM. The effects build up due to the interactions between the entities in the BRM.

14.1.3 Level of Interest

The level of interest affects how likely we are to be able to engage the stakeholder in the corresponding activity. A low level of interest can reduce the commitment of the stakeholder and, therefore, the efficiency, effectiveness, and quality of the work that is done.

14.1.4 Stakeholder Impact

Although ideally we would engage all stakeholders involved in all of the program steps, this is never possible in reality. To be able to identify where best to place the effort of engaging stakeholders, it is therefore necessary to be able to determine which stakeholders would have the greatest potential impact on a given entity in the BRM and, by extension, those that would have the greatest effect on the overall program.

The goal of the BRM-based analysis is to identify the areas of focus most closely associated with each stakeholder's maximum potential impact based on their levels of power and interest on the entities of the BRM. Stakeholder impact on a given program depends on the business value of the program entity involved as well as the stakeholder's level of power and interest with respect to that entity.

The analysis makes use of the results of the break-even everywhere requirement (BEER) algorithm developed in Chapter 5. This technique developed the rules for evaluating allocation fractions in such a way that left- to-right calculations using these allocation fractions provide values identical to those based on right-to left calculations using the contribution fractions from the business case. The same approach is used for evaluating how the stakeholder power and influence diffuse across the BRM. The analysis technique is defined next.

14.1.5 Stakeholder Impact Analysis

The way in which each stakeholder can affect the overall program is based on the BRM for the program. Use of the BRM ensures that this analysis remains consistent with the business model used to direct the program as a whole.

As mentioned earlier, the levels of power and interest of each stakeholder can be different for each entity of the BRM. In addition, the impact of one entity in the BRM can also have an effect on the impact other entities. These effects are additive.

The effect of each stakeholder on a node depends on the corresponding power and influence of that stakeholder in each of the three categories. Each stakeholder can have several such effects—one per category—as explained next.

14.2 Mapping the Analysis onto the Model

14.2.1 Mapping Stakeholder Power

The actual power of a stakeholder on the overall program depends not only on the stakeholder's level of power but also on the business value of the entity being affected. For simplicity, the product of the level of power (0 to 3) by the value of the entity being affected will be taken as a measure of the size of the power (simply known as "the power" in contrast to "level of power"). The power diffuses in a different way for each category of stakeholder. To calculate the power, start by setting the power of each stakeholder on each entity in the model to zero. The power of a given stakeholder is then calculated as follows:

- Producers:
  - The total producer power of the stakeholder on the relevant node is increased by the level of power of the producer as initially assessed.
  - The effect of the producer power of this stakeholder on the successor nodes in the BRM diffuses from left to right across the relevant links based on the allocation fractions and adds to the total producer power in the corresponding nodes.
- Transformers:
  - The power of the transforming stakeholder on a destination node depends on the contribution that the corresponding link makes to the overall benefit. The transformer power of the destination node for this stakeholder is increased by the product of the level of power (0 to 3) by the power delivered by the producer.
  - The sum of these values over all incoming links gives the basic transformer power.
  - The effect of the transformer power of this stakeholder on the successor nodes in the BRM diffuses from left to right across the relevant links based on the allocation fractions and adds to the total transformer power of the corresponding nodes.
- Receivers:
  - The total receiver power of a node is increased by the product of the receiver's level of power by the transformer-delivered value of that node.
  - In contrast to producers and transformers whose power diffuses across the BRM from left to right, the effect of receivers travels back from right to left toward the initiatives.
  - The effect of the receiver power of this stakeholder on the predecessor nodes in the BRM diffuses across the links of the BRM based on the contribution fractions.

Once this analysis is complete, the power of each stakeholder on each node can be ranked in each category of stakeholder to indicate for each node whose power is most important and identify the reason for it.

A similar approach then needs to be carried out for the stakeholder interest.

14.2.2 Mapping Stakeholder Interest

Whereas each category of power is measured differently (e.g., based on the business value), the categories of interest all depend only on the level of interest. For this reason, the overall level of interest of each stakeholder can be calculated for each entity, as well as separately when considered by category as producers, receivers, or transformers. The overall level of interest will allow the program manager to determine an order of priority for stakeholder engagement based on the total interest. The communications can then be tailored to take into account the interest of each chosen stakeholder in each category.

The total interest in each category is calculated as follows:

- Producers:
  - Set the initial value as the interest level of the node.
  - The interest now needs to diffuse from left to right based on the allocation fractions.
- Transformers:
  - The interest value of a link is set to the product of the result of the producer interest for the source node by the initial interest level of the link.

- The sum of these values over all outgoing links gives the initial interest of the stakeholder as a transformer.
- The effect of the transformer interest of this stakeholder on the predecessor nodes in the BRM diffuses from right to left across the relevant links based on the allocation fractions.

- Receivers:
  - Set the initial value as the interest level from the transformer for that node.
  - The interest now needs to diffuse from right to left, using the original contribution fractions.

14.2.3 Normalizing the Results

Once these calculations have been carried out, the result for each BRM node should be normalized as a percentage of the category total. This normalization is carried out separately on the results of the producer, receiver, and transformer calculations for each of power and interest, to provide six sets of normalized data.

14.2.4 Using Mendelow's Grid

A useful model for evaluating how to assess stakeholders' potential involvement in a program is the use of Power–Interest mapping (Mendelow 1991) as shown in Figure 14.2.

The standard approach for using this grid is to map the stakeholders or stakeholder groups onto the grid based on their levels of power and interest. The grid allows the program team to prioritize the stakeholders and their interests. In our case, in decreasing order of priority, the quadrants are: High Power + High Interest, High Power + Low Interest, Low Power + High Interest, Low Power + Low Interest. Once this prioritization has been carried out, the key stakeholders can be selected and the specific per-stakeholder analysis based on the BRM as explained above can be applied to each chosen stakeholder.

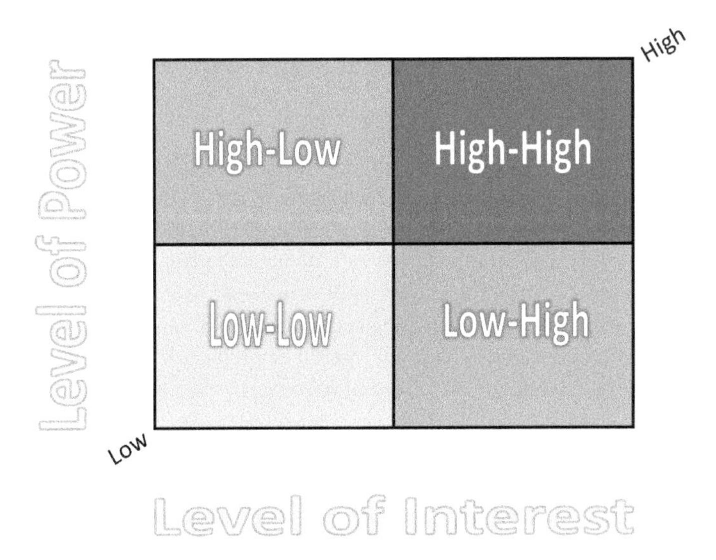

Figure 14.2 Mendelow Power–Interest Grid

To link the initial stakeholder analysis to the BRM, in addition to the standard grid showing all of the relevant stakeholders, one additional version of the grid is produced for each key stakeholder. This additional analysis will provide a basis for developing a specific engagement plan for each key stakeholder or stakeholder group. Each per-stakeholder grid is used to link the entities that compose the BRM to the power and interest of that specific stakeholder. The result is shown by mapping the BRM entities onto the quadrants based on the specific stakeholder's priorities as evaluated in the BRM-based analysis.

The results of analysis can be mapped onto the per-stakeholder grid as follows:

- Generate a single power score for each node of the BRM by taking the average of the producer, receiver, and transformer scores.
- Define threshold values to separate *high* from *low* separately for power and interest.

It is important to carry out the mapping twice for each chosen stakeholder. The first set of numbers (the *first level*) corresponds to the Power–Interest values before processing through the BRM. The second set (the *second level*) comprises the values after processing. The first-level analysis provides the "obvious" areas in which to concentrate. The second-level analysis identifies the additional areas that can help enhance and broaden the stakeholder relationship.

The QERTS case study shows how these ideas work in practice.

14.3 QERTS Example

This example is based on the full QERTS example. The corresponding BRM from earlier chapters is shown in Figure 14.3.

To compare the different areas of focus of different stakeholders, the analysis will initially be carried out on two contrasting groups of stakeholders: the IT group and the Marketing group.

14.3.1 Marketing Group

Marketing Group Power

- As a producer, the Marketing group is involved as follows:
  - D = Develop Marketing Program: High (3)
  - K = Increase Market Visibility: High (3)
  - R = Efficient and Friendly Service: Medium (2)
    - Marketing will provide support to staff to assist in enhancing customer relations.
- As a transformer, the Marketing group is involved on the following links:
  - D = Develop Marketing Program to K = Increased Market Visibility: High (3)
  - F = Training Programs to N = All Staff Can Apply Processes: Low (1)
    - Marketing can become involved in training to explain the value of the joint processes.
  - K = Increased Market Visibility to P = More Customers: High (3)
    - Only with the help of marketing can visibility be transformed into actual business.
  - P = More Customers to T = Increased Revenue: Medium (2)
    - Marketing can help by encouraging upselling to the customers.
- As a receiver, the Marketing group is involved as follows:
  - J = Flexible Work Practices: Low (1)
    - Marketing is involved in understanding how the work practices can affect the service.

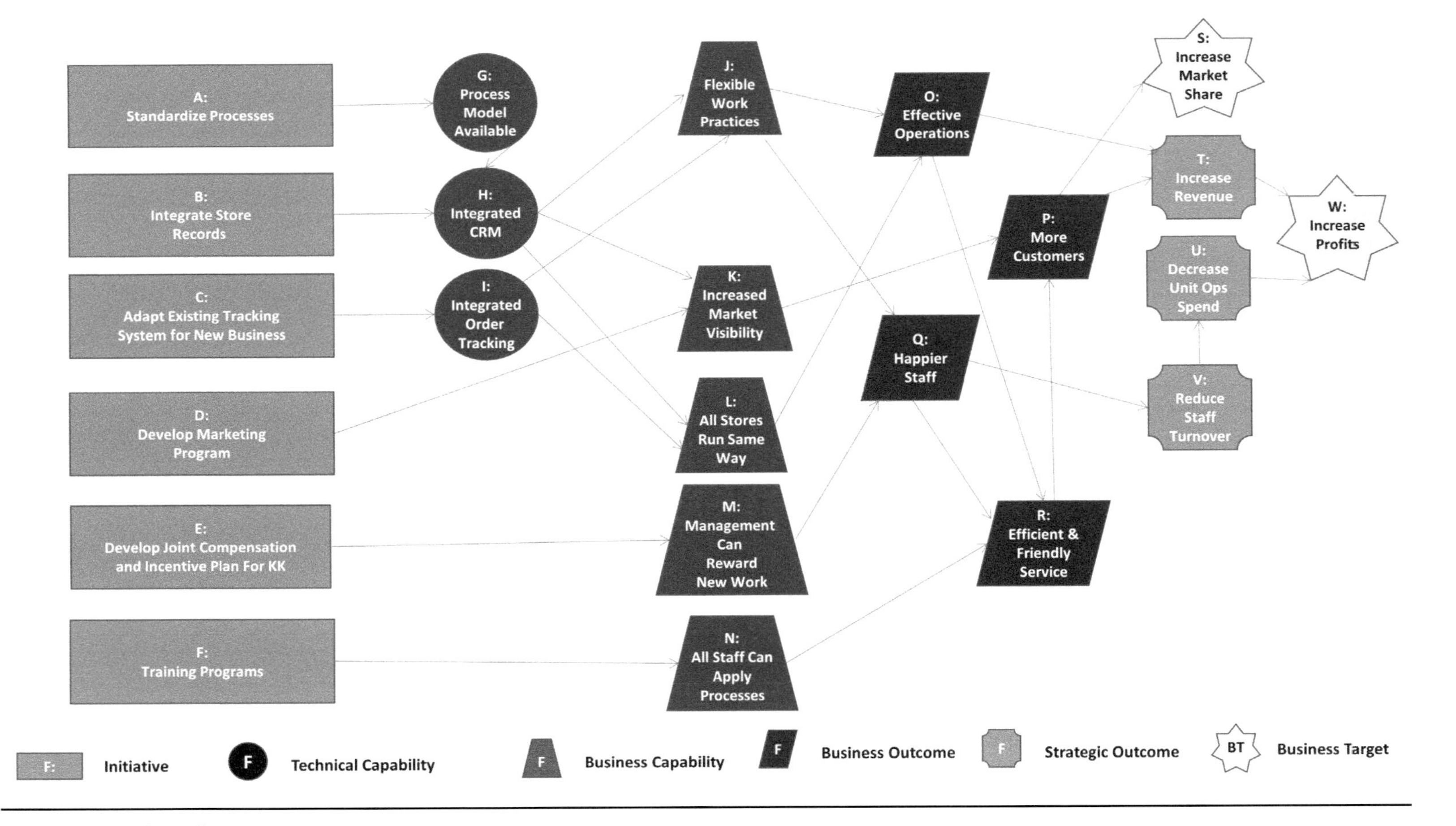

Figure 14.3 QERTS BRM

    - K =Increased Market Visibility: High (3)
    - N = All Staff Can Apply Processes: Low (1)
        - Marketing can build a message on the perceived excellence of the staff.
    - R= Efficient and Friendly Service: Low (1)
        - Marketing can build a message on the attitude of the staff.

Marketing Group Interest

- As a producer, the Marketing group is interested as follows:
    - D = Develop Marketing Program: High (3)
    - K = Increase Market Visibility: High (3)
    - P = More Customers: Medium (2)
    - R = Friendly and Efficient Service: Low (1)
    - S = Increase Market Share: High (3)
    - W = Increase Profits: High (3)
- As a transformer, the Marketing group is interested in the following links:
    - D = Develop Marketing Program to K = Increase Market Visibility: High (3)
    - H = Integrated CRM to K = Increase Market Visibility: High (3)
    - K = Increased Market Visibility to P = More Customers: High (3)
    - P = More Customers to S = Increase Market Share: High (3)
    - P = More Customers to T = Increase Revenue: High (3)
- As a receiver, the Marketing group is interested as follows:
    - H = Integrated CRM: High (3)
        - Marketing needs knowledge of all existing customers.
    - K = Increase Market Visibility: High (3)
    - P = More Customers: High (3)
    - S = Increase Market Share: High (3)

14.3.2 IT Group

IT Group Power

- As a producer, the IT group is involved as follows:
    - B = Integrate Store Records: High (3)
    - C = Adapt Tracking System: High (3)
    - F = Training Programs: Low (1)
    - H = Integrated CRM: High (3)
    - I = Integrated Order Tracking: High (3)
- As a transformer, the IT group is involved as follows:
    - B = Integrate Store Records to H = Integrated CRM: High (3)
    - C = Adapt Tracking Systems to I = Integrated Order Tracking: High (3)
    - F = Training Programs to N = All Staff Can Apply Processes: Low (1)
        - IT may be involved in providing training.
    - G = Process Model Available to H = Integrated CRM: High (3)
    - I = Integrated Order Tracking to L = All Stores Run the Same Way: High (3)
        - IT must be involved in the system deployment.
    - L = All Stores Run the Same Way to O = Effective Operations: High (3)
        - IT is potentially involved in adapting the system to the needs of operations.

- As a receiver, the IT group is involved as follows:
  - B = Integrate Store Records: High (3)
  - C = Adapt Tracking System: High (3)
  - G = Process Model Available: High (3)

IT Group Interest

- As a producer, IT is interested in the following:
  - A = Standardize Processes: High (3)
  - B = Integrate Store Records: High (3)
  - C = Adapt Existing Tracking System: High (3)
  - G = Process Model Available: Low (1)
  - H = Integrated CRM: High (3)
  - I = Integrated Order Tracking: High (3)
  - O = Effective Operations: Medium (2)
- As a transformer, IT is interested in the following:
  - G = Process Model Available to H = Integrated CRM: High (3)
  - H = Integrated CRM to L = All Stores Run the Same Way: Medium (2)
  - I = Integrated CRM to J = Flexible Work Practices: Low (1)
  - I = Integrated Order Tracking to L = All Stores Run the Same Way: High (3)
  - L = All Stores Run the Same Way to O = Effective Operations: Medium (2)
- As a receiver IT is interested in the following:
  - A = Standardize Processes: Medium (2)
  - B = Integrate Store Records: High (3)
  - C = Adapt Tracking System: High (3)
  - G = Process Model Available: High (3)
  - I = Integrate Order Tracking: Low (1)

14.3.3 Numerical View

The following table (Table 14.1) shows the results of the analysis based on the preceding values (first level) and the values based on the BRM algorithm described earlier (second level) for the IT group. Although this table is instructive, it is not immediately easy to interpret. The Mendelow grid can provide a valuable synthesis of the information and a serve as a basis for prioritizing action.

14.3.4 Mendelow's Grid for QERTS

In Tables 14.2, 14.3, and 14.4, the nodes are indicated by their identification letter. Lower-case letters are used to indicate a first-level involvement. Capital letters are used for nodes with a second-level involvement. The position of the letter identifying the BRM node within a quadrant is unrelated to its actual score. Only the membership of the quadrant is relevant.

Table 14.1 Numerical Analysis of the IT Group's Power and Interest

		IT Group							First Level				Second Level			
		Power			Interest				Low P= 0 - 4, I=0-1				Low P = 0 - 7%; Low I = 0 - 10%			
		Producer	Transformer	Receiver	Producer	Transformer	Receiver	Overall Interest	Power Value	Interest Value	Power Level	Interest Level	Power Value	Interest Value	Power Level	Interest Level
A	Standardize Processes	3%	31%	8%	3%				6	6	H	H	14%		H	
B	Integrate Store Records	3%			3%				6	6	H	H				
C	Adapt Tracking System															
D	Develop Marketing Program															
E	Align Compensation															
F	Train Staff		19%	5%		6%	100%	35%	3	4	L	H	8%	35%	H	H
G	Process Model Available	7%	31%	8%	6%			2%	3	3	L	H	15%	2%	H	L
H	Integrated CRM	14%		8%	12%			4%	3	4	L	H	7%	4%	L	L
I	Integrated Tracking								1		L					
J	Flexible Working Practices	6%		4%	5%	2%		2%	3		L		3%	2%	L	L
K	Increased Market Visibility	1%			1%	8%		3%						3%		L
L	All Stores Same	7%		4%	5%			2%					3%	2%	L	L
M	Management Can Reward Staff								1		L					
N	All Staff Can Apply Processes		19%	5%	3%	9%		4%		2		H	8%	4%	H	L
O	Effective Operations	12%		10%	11%	5%		5%					7%	5%	L	L
P	More Customers	11%		11%	12%	16%		9%					7%	9%	L	L
Q	Happy Staff								1		L					
R	Efficient and Friendly Service	10%		10%	11%	8%		6%					7%	6%	L	L
S	Increase Market Share	3%		3%	3%	4%		3%					2%	3%	L	L
T	Increase Revenue	8%		8%	9%	11%		7%					5%	7%	L	L
U	Decreased Unit Ops Spend	3%		2%	3%						L		2%		L	
V	Reduce Staff Turnover										L					
W	Increased Profits	11%		10%	11%	30%		14%			L		7%	14%	L	H

Mapping the IT Group

The approach explained previously for mapping the results onto Mendelow's grid has been applied to the numbers for the IT group in Table 14.1. The result is shown in Table 14.2. The nodes and the level relating to the P-I score are indicated as explained above. The conventions for the Mendelow grid in this case are as follows:

- First level: "low power" < 5, "low interest" < 2
- Second level: "low power" < 8%, "low interest" < 11%

Table 14.2 Mendelow's Grid for the IT Group

A	Standardize Processes
B	Integrate Store Records
C	Adapt Tracking System
D	Develop Marketing Program
E	Align Compensation
F	Train Staff
G	Process Model Available
H	Integrated CRM
I	Integrated Tracking
J	Flexible Working Practices
K	Increased Market Visibility
L	All Stores Same
M	Management Can Reward Staff
N	All Staff Can Apply Processes
O	Effective Operations
P	More Customers
Q	Happy Staff
R	Efficient and Friendly Service
S	Increase Market Share
T	Increase Revenue
U	Decreased Unit Ops Spend
V	Reduce Staff Turnover
W	Increased Profits

IT Group

Level of Power

A G N	a F b
P L H R S J O T	f g h W n

Level of Interest

Table 14.2 provides the following information:

- Highest quadrant showing the main priority topics for the IT group:
  - a = Standardize Processes (first level only)
  - b = Integrate Store records (first level only)
  - F = Train Staff (second level only)
- Including the next level of priority (high power, low interest):
  - A = Standardize Processes (second level only)
  - G = Process Model Available (second level only)
  - N = All Staff Can Apply Processes (second level only)
- In addition, the IT group does, however, also have a high level of interest in the following entities:
  - h = Integrated CRM (first level only)
  - W = Increased Profits (second level only)

This analysis shows that the priorities for the IT group have a purely a technical focus at the highest level. However, IT's interest in the effect of these technical areas on the business itself, as represented by the increase in profits, is important from the point of view of managing the relationship with these stakeholders.

Mapping the Marketing Group

The grid for Marketing can be developed in a similar way to the IT group grid and is shown in Table 14.3:

Table 14.3 Mendelow's Grid for the Marketing Group

A	Standardize Processes
B	Integrate Store Records
C	Adapt Tracking System
D	Develop Marketing Program
E	Align Compensation
F	Train Staff
G	Process Model Available
H	Integrated CRM
I	Integrated Tracking
J	Flexible Working Practices
K	Increased Market Visibility
L	All Stores Same
M	Management Can Reward Staff
N	All Staff Can Apply Processes
O	Effective Operations
P	More Customers
Q	Happy Staff
R	Efficient and Friendly Service
S	Increase Market Share
T	Increase Revenue
U	Decreased Unit Ops Spend
V	Reduce Staff Turnover
W	Increased Profits

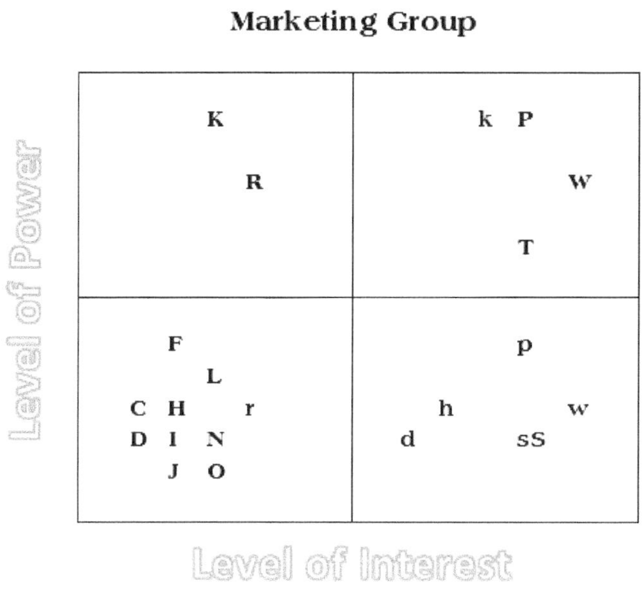

Marketing Group

Level of Power		
High	K R	k P W T
Low	F L C H r D I N J O	p h w d sS
	Low	High

Level of Interest

- Highest quadrant showing the main priority topics for the Marketing group:
  - k = Increased Market Visibility (first level only)
  - P = More Customers (second level only)
  - T = Increased revenue (second level only)
  - W = Increased Profits (second level only)
- Including the next level of priority (high power, low interest). These are all second-level effects:
  - K = Increased Market Visibility (second level—after first level in top quadrant)
  - R = Efficient and Friendly Service (second level only)

As expected, Marketing has a different focus from the IT group in the highest quadrant. The only common focus of the two groups is for the overall profitability.

This two-level analysis shows the various levels of relationship management that a given group requires in order to satisfy the stakeholders' needs and to enhance their engagement.

A similar analysis can be carried out when a potentially negative stakeholder is identified, in order to tailor arguments and thereby to improve the relationship. An example is provided as follows.

Mapping a Negative Stakeholder

For this example, we have determined that the group of customer-facing QE staff are resentful of the takeover because it will force a degree of standardization (L = All Stores Run the Same Way) that is contrary to their approach to delivering service. We have carried out the following analysis based on their interest and power to resist the merger strategy.

QE Staff Power

- As a producer, QE staff can affect:
  - J = Flexible Work Practices: Low (1)
  - L = All Stores Run the Same Way: Medium (3)
  - N = All Staff Can Apply Processes: Medium (2)
  - P = More Customers: Medium (2)
  - Q = Happier Staff: High (3)
  - R = Efficient and Friendly Service: High (3)
- As a transformer, QE staff can affect:
  - F = Training Programs to N = All Staff Can Apply Processes: High (3)
  - H = Integrated CRM to L = All Stores Run the Same Way: Low (1)
  - I = Integrated Order Tracking to J = Flexible Work Practices: Low (1)
  - I = Integrated Order Tracking to L = All Stores Run the Same Way: High (3)
  - J = Flexible Work Practices to O = Effective Operations: High (3)
  - M = Management Can Reward New Work to Q = Happier Staff: High (3)
  - N = All Staff Can Apply Processes to R = Efficient and Friendly Service: High (3)
  - Q = Happier Staff to R = Efficient and Friendly Service: Medium (2)
  - Q = Happier Staff to V = Reduce Staff Turnover: High (3)
- As a receiver, QE staff can affect:
  - I = Integrated Order Tracking: Low (1)
  - J = Flexible Work Practices: Low (1)
  - L = All Stores Run the Same Way: Medium (2)
  - N = All Staff Can Apply Processes: Medium (2)
  - Q = Happier Staff: High (3)
  - R = Efficient and Friendly Service: High (3)

QE Staff Interest

- As a producer, QE staff are interested in:
  - J = Flexible Work Practices: Medium (2)
  - N = All Staff Can Apply Processes: Medium (2)
  - O = Effective Operations: High (3)
  - Q = Happier Staff: High (3)
  - R = Efficient and Friendly Service: Medium (2)
- As a transformer, QE staff are interested in:
  - F = Training Programs to N = All Staff Can Apply Processes: Low (1)

- H = Integrated CRM to L = All Stores Run the Same Way: High (3)
- I = Integrated Order Tracking to J = Flexible Work Practices: High (3)
- I = Integrated Order Tracking to L = All Stores Run the Same Way: High (3)
- J = Flexible Work Practices to O = Effective Operations: High (3)
- L = All Stores Run Same Way to O = Effective Operations: High (3)
- M = Management Can Reward New Work to Q = Happier Staff: High (3)
- N = All Staff Can Apply Processes to R = Efficient and Friendly Service: Low (1)
- Q = Happier Staff to R = Efficient and Friendly Service: Medium (2)
- Q = Happier Staff to V = Reduce Staff Turnover: Medium (2)

- As a receiver, QE staff are interested in:
  - I = Integrated Order Tracking: High (3)
  - J = Flexible Work Practices: High (3)
  - L = All Stores Run the Same Way: High (3)
  - M = Management Can Reward New Work: High (3)
  - O = Effective Operations: High (3)

The resulting Mendelow's grid is shown in Table 14.4. In addition to "All Stores Run the Same Way," it shows the key nodes:

- o = Effective Operations (at first level)
- q = Happier Staff (at first level)
- W = Increased Profits (at second level)

Table 14.4 Mendelow's Grid for QE Staff

A	Standardize Processes
B	Integrate Store Records
C	Adapt Tracking System
D	Develop Marketing Program
E	Align Compensation
F	Train Staff
G	Process Model Available
H	Integrated CRM
I	Integrated Tracking
J	Flexible Working Practices
K	Increased Market Visibility
L	All Stores Same
M	Management Can Reward Staff
N	All Staff Can Apply Processes
O	Effective Operations
P	More Customers
Q	Happy Staff
R	Efficient and Friendly Service
S	Increase Market Share
T	Increase Revenue
U	Decreased Unit Ops Spend
V	Reduce Staff Turnover
W	Increased Profits

QE Staff

Level of Power

P rR T	l q W O
F U L Q V C H I nN S J	m i j o

Level of Interest

Analysis of these results raises some interesting points:

- It is clear that the primary focus of the QE staff is on l = All Stores Same and q = Happier Staff because they find these two outcomes incompatible with each other.
- At the second level, these staff have power and interest in a number of additional areas that are key to the strategic success: o = Effective Operations and W = Increased Profits.

Table 14.5 Mendelow's Grids for the Dual Interests of QE Staff

A	Standardize Processes
B	Integrate Store Records
C	Adapt Tracking System
D	Develop Marketing Program
E	Align Compensation
F	Train Staff
G	Process Model Available
H	Integrated CRM
I	Integrated Tracking
J	Flexible Working Practices
K	Increased Market Visibility
L	All Stores Same
M	Management Can Reward Staff
N	All Staff Can Apply Processes
O	Effective Operations
P	More Customers
Q	Happy Staff
R	Efficient and Friendly Service
S	Increase Market Share
T	Increase Revenue
U	Decreased Unit Ops Spend
V	Reduce Staff Turnover
W	Increased Profits

Positive QE Staff

Level of Power		
High	P l R O T	q r W
Low	U Q V N S J	m i n j o

Level of Interest

Negative QE Staff

Level of Power		
High	q rR W O T	P l
Low	U L Q S	

These insights can all serve as arguments that may help to win over the dissatisfied staff. These messages can be further enhanced with additional analysis.

The analysis can be taken still further because stakeholders with reservations with respect to one or more of the program entities may feel positively about the others. The positive and negative aspects can be analyzed separately by focusing on the corresponding nodes of the BRM. In the QERTS case, at the first level, the QE customer-facing staff only have negative feelings about L = All Stores Run the Same Way. The analysis can therefore be carried out as above, except that the positive analysis should omit this node from the interest list for the producer and receiver categories, whereas the negative interest analysis should similarly include only this node. No changes are needed to the transformer settings because the changes to the producer and receiver settings provide the required focus on node L. The corresponding results are shown in Table 14.5.

The main conclusion from the negative analysis in Table 14.5 is as follows: As was stated, the staff do not want L = All Stores Run the Same Way. However, the positive analysis shows that they have the potential for q = Happy Staff and r = Efficient and Friendly Service leading to W = Increased Profits. In the negative analysis, their feelings and their power are a major threat to P = More Customers. This supports the analysis in Chapter 7 on the importance of addressing the situation.

The additional results of the positive analysis show a high level of interest of the QE staff in the following areas: Integrated Tracking, Flexible Working Practices, Management Can Reward Staff, All Staff Can Apply Processes, and Effective Operations. This positive–negative analysis therefore helps to provide arguments to mitigate the corresponding resistance.

14.4 Summary

This chapter took a classic Power–Interest model for stakeholder analysis as a starting point. The way in which the BRM can then be used to extend the understanding of stakeholder power and interest throughout the map was explained. This approach was applied to two supportive and one negative groups of stakeholders in the existing case study. This analysis demonstrated the contrasting positions and priorities of different groups of stakeholders and suggested strategies for enhancing their engagement in the program.

14.5 From Analysis to Communications

Once the program management team understands the key points on which to focus, it can use this knowledge to optimize the way in which stakeholder relationships are handled. One key feature of stakeholder relationship management is understanding and use of effective communications. This topic is addressed in the next chapter.

14.6 References

Mendelow, A. *Stakeholder Mapping*, Proceedings of the 2nd International Conference on Information Systems, Cambridge, MA, 1991. Cited by Johnson, G. and Scholes, K. in *Exploring Corporate Strategy*. Harlow, UK: Financial Times/Prentice Hall, 1998.

Chapter 15

Communication—Why, How, and What

This chapter addresses the importance of effective communication in the success and potential failure of programs. It explains the data–information–knowledge–wisdom hierarchy and proposes a model that maps data and information flows onto the typical organizational structures. It shows how these ideas can be linked to the benefits realization approach in order to provide the correct level and volume of communication to the various stakeholders.

The saddest aspect of life right now is that science gathers knowledge faster than society gathers wisdom.

—Isaac Asimov

Any fool can know. The point is to understand.

—Anonymous

Data and information are not useful on their own and cannot be turned into knowledge or wisdom unless we have questions we want to answer. Once we have got a question we can then use data to turn it into knowledge and learning. Without questions there can be no learning, and without learning there can be no improvement.

—Bernard Marr

This chapter was written in collaboration with Max Wideman. His input was invaluable. Any inconsistencies come from my misunderstanding and certainly not from any lack of wisdom on his part.

15.1 Introduction

Communication is the foundation of every project environment. How so? Although this principle is poorly explained in most project management literature about project success and failure, the logic is simple: A project only exists when there is an objective and an intent to get there. The objective needs to be defined and communicated. You can only get a project to its objective by doing work. Work requires people. But those people can only progress the project if they know what to do. Knowing what to do requires some sort of a plan; developing the plan requires teamwork, and the plan must be communicated before it can be implemented. Therefore, communication is the foundation of every project! This is shown in Figure 15.1.

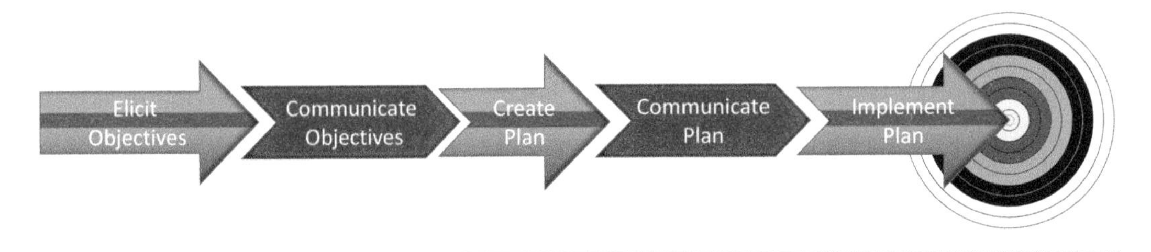

Figure 15.1 The Essential Role of Communication

Programs and projects succeed or fail depending on the decisions taken along the way: Why should we do the program? How should we do the project? What should we do if . . . ? And so on. *The Standard for Portfolio Management*—Third Edition (Project Management Institute 2013b) uses the word "decision" 131 times, as in "The PMO supports the program manager by providing the information needed to make *decisions* that guide the program" (p. 19). The *PMBOK® Guide*—Fifth Edition (Project Management Institute 2013a) uses the words "communications" or "communicate" just as much. This chapter investigates the relationship between data, information, and decision-making: Where does the information come from, how is it communicated, and what is required to achieve success?

The starting position is the reference in the *PMBOK® Guide* in Appendix X1.5. This refers to the data, information, knowledge, and wisdom (DIKW) model that is applied consistently to the inputs and outputs for the relevant processes. D, I, K, and W can each be considered as a level in a hierarchy: The *data* level supports the *information* level, *information* supports *knowledge*, and *wisdom* is based on *knowledge*.

15.2 Understanding Hierarchies

15.2.1 Definition of Hierarchy

In the context of this chapter, the term *hierarchy* is taken to mean a set of groupings or concepts arranged so that each level depends on the content of the layer directly below it. Alternatively, a hierarchy can be seen as a set of groupings or concepts in which the higher level provides a more general view or synthesis of the concepts at the level directly below it.

15.2.2 Examples of Hierarchies

This section is taken from Max Wideman (Wideman, *Hierarchies*, 2015), in which he gives examples of hierarchies of concepts from highest to lowest level in a number of different areas.

Management Responsibility Hierarchy

Typical documentation:

- *Governance*
- *Purpose*
- *Principles*
- *Practice*
- *Policies*
- *Procedures*
- *Processes*
- *Guidelines*
- *Recommendations*
- *Explanations or Notes*
- *Structure of the Organization*

Project Family Lifetimes

From longest to shortest:

- *Portfolios*
- *Programs*
- *Projects*
- *Periods or Macro Phases*
- *Phases*
- *Stages or Tranches*
- *Milestones*
- *Activities*
- *Tasks*

Management Approach

- *Organization*
- *Strategy*
- *Tactics*
- *Roles*
- *Responsibilities*
- *Style*
- *Systems*
- *Resources*
- *Application*
- *Techniques*
- *Tools*
- *Results*

Product Life Cycle Management

- *White Paper*
- *Opportunity Statement*

- *Decision Case (Programs)*
- *Business Case (Projects)*
- *Project Charter, Project Brief*
- *Project Implementation Plan*
- *Work Breakdown Structure*
- *Detailed Design Briefs*
- *Production Contracts*
- *Change Orders*
- *Product Delivery Acceptances*
- *Usage and Benefit Delivery Reports*
- *Retirement and Disposal*

15.3 Addressing Data, Knowledge, Information, and Wisdom

But what are data, information, and knowledge; how do we acquire them; and how does wisdom fit in? And, even with valid knowledge, what can get in the way of making the "right" decision?

This is a vast subject, so this chapter will focus on its relevance to the management of projects, programs, and portfolios. It will not address the classical sender/receiver models of communication. Instead, it explains how to align the content and quantity of information to the needs and capabilities of the relevant stakeholders.

This chapter will use the term *intel* (an abbreviation of *intelligence*, as used for example in the name of specialized government agencies) to encompass each and all of D, I, K, and W.

15.3.1 Background to DIKW

The basis of the DIKW concept has been around for a long time, and most authors refer to two main starting points:

The first to address the problem directly was T. S. Eliot in his 1934 poem "The Rock":

Where is the Life we have lost in living?
Where is the wisdom we have lost in knowledge?
Where is the knowledge we have lost in information?

Were it not poetically arrogant, I would add to T. S. Eliot's poem

. . . and where is the information we have drowned in data?

The second classical reference is R. L. Ackoff (1999): He presented the ideas in a more formal manner in his paper, "From Data to Wisdom." In this paper, Ackoff links information, knowledge, and understanding to efficiency. He then explains his view that wisdom provides the ability to transform efficiency into effectiveness.

15.3.2 Definitions

The following definitions have been adopted from Ackoff's paper to align more directly with the program management environment and to relate to ideas already addressed in earlier chapters:

- *Data* are the product of observation or measurement. They are represented by statements of fact or by numbers. They are independent of the context and are neither analyzed nor processed. Of themselves, they mean very little, and only represent potential. They have no value at all unless they are relevant.
    - The raw numbers in the benefits realization map (BRM)—allocations to the initiatives, contributions from the strategic outcomes—are data. The expected contributions of the strategic outcomes are data available to the program's initiation phase; estimates of the required allocations are data developed for the planning phase.
- *Information* tells you something about the status or content of whatever is being described. Information is normally inferred from data by some form of analysis. Data can be integrated with existing information to create additional information. In addition, the same data can be applied in different ways in order to provide different types of information. As opposed to data, which has only potential, information has intrinsic value.
    - The BRM, after applying the algorithms to evaluate all of the parameters associated with the nodes, provides the information required for a business case.
- *Knowledge* could be defined as *information in context*. It is the result of integrating information into a cohesive and consistent whole with respect to a specific subject. Knowledge provides the basis for developing skills and for assessing truth or falseness of a statement. For example, the *PMBOK® Guide* specifies a number of Knowledge Areas, such as Cost, Time, and Risk, to define specific relevant contexts. Knowledge, therefore, has relevance.
    - Knowledge is gained by evaluating the viability of the business case and the likelihood of success based on multiple organizational features. The features can include risk levels, organizational maturity, reliability of capacity planning, and the like.
- *Wisdom* is related to applying knowledge and experience to a situation. It supports judgement, advice, or decision making. It is associated not only with the information and knowledge, but also with the person making the judgement. It is involved in the effective transformation of data into information (e.g., "are the data valid?"), of information into knowledge (e.g., "is the information relevant?"), and of knowledge to expand wisdom (e.g., "given what I know now, how should that change my attitude?"). Wisdom delivers success.
    - Wisdom needs to be applied to determine the best decisions to be made based on the earlier modeling of the program (the BRM), the business case, and ongoing progress as measured using earned benefit and strategic key performance indicators (KPIs), as described in Chapter 13.

15.3.3 Overview of the DIKW Process

The process of gathering intel begins by collecting *data*, at the first level. The data are then processed to form *information* at the second level. When this information is processed for a specific purpose, it takes the form of *knowledge*, at the third level. The creation of knowledge when integrated into the consistent worldview of the context in question leads to acquiring *wisdom* at the fourth or topmost level.

This DIKW process can be represented as a hierarchy. *Data* form the base of the hierarchy, and value is added to these items by interpreting and converting them into a relevant and meaningful form termed *information*. When information is applied in a particular situation and converted to into an action-oriented form that supports decision making, it is defined as *knowledge*. *Wisdom* is found at the pinnacle of the DIKW hierarchy. Wisdom differs from data and information, both of which are based solely on facts. Wisdom requires adding meaning to the accumulated knowledge. The worldview is more abstract, as it is something intrinsic and is typically an accumulation of values, judgements, prior experience, or interpretations.

Seen another way, *data* are neutral and may be totally irrelevant to the topic in question, *information* is relative to the situation, *knowledge* is relative to the broader picture, and *wisdom* involves your personal experience and model of reality.

15.3.4 Negotiating the DIKW Ladder

Although the DIKW hierarchy could be pictured as the rungs of a ladder, it is not sufficient to recognize the rungs. To climb a ladder, you need to understand how to move from one rung to the next (i.e., what needs to happen to data to turn it into information, etc.). This is shown schematically in Figure 15.2 and addressed as follows.

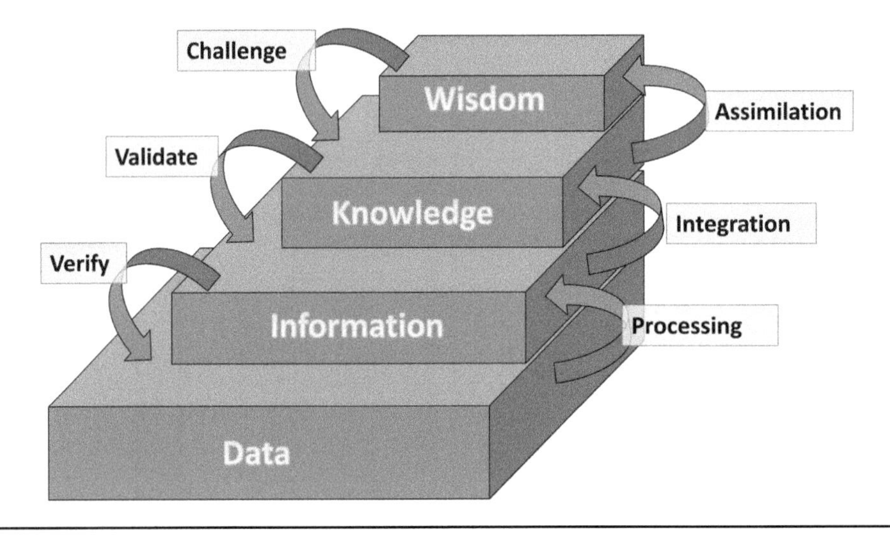

Figure 15.2 Moving Up and Down the DIKW Model

Each higher level is represented in Figure 15.2 as smaller than the preceding one to indicate the following points:

- The process of translating from one level to the next higher one entails selecting relevant content and, as necessary, discarding some of the content of the lower layer.
- The content of the higher layer is more concentrated and as such is destined to be more easily assimilated or applied.
- The process of translation from one layer to the next carries the risk of discarding useful and valid content.

From Data to Information: Processing

The conversion of data into information is where the term *data processing* needs to be applied. The reason for processing raw data is to extract and align the key elements in such a way as to be visibly relevant to the situation that needs to be analyzed.

Based on Bernard Marr's concept of "key performance questions" (2010a) referred to in the quotation at the start of this chapter, the step from data to information entails aligning and pruning the data to map it onto a specific key performance questions.

From Information to Knowledge: Integration

Information is transformed into knowledge by building an interconnected set of relevant facts to create a consistent whole—a *body of knowledge*. Note that the transformation of information to knowledge is where Ackoff's concept of "Understanding" is integrated into the DIKW hierarchy as defined in this chapter.

The application of the concept of key performance questions linked to the information from the previous step is used to provide relevant answers to specific questions—that is to say, to generate additional knowledge.

From Knowledge to Wisdom: Assimilation

Moving from knowledge to wisdom implies updating your worldview based on the knowledge in such a way that the combination can be used for informed decision making.

At this point, of course, you may realize from the result of your analysis that you were asking the wrong questions; wisdom resides not only in knowing how to apply new information, but also in rejecting or reassessing received ideas!

And Looking Down

From each rung of the ladder above the lowest, you can benefit from a broader perspective:

- Wisdom needs to be repeatedly challenged by reexamining the knowledge base.
  - Wisdom can also be used to assess the relevance of information and data.
- The credibility of knowledge can be reassessed by validating the information.
  - Knowledge also can be used to structure the information and data.
- The accuracy of information can benefit from verifying the data.

15.3.5 A DIKW Example from Earned Value

Details of earned value and earned benefit were provided in Chapter 12.

EVM Data

The raw data to be used in earned value management (EVM) include task status data such as *percent complete* and *amount spent*.

Background EVM Information

The project information required for EVM includes the schedule and the rules for evaluating the earned value of each task.

Processed EVM Information

The EVM variables such as schedule performance index (SPI) and cost performance index (CPI) can be calculated by applying the project information to the EVM data.

EVM Knowledge

The CPI and SPI values allow you to know how the project is performing with respect to planned schedule and budget.

Applying Wisdom

Assume that EVM tells you that you are behind schedule. The choice of action to be taken may well need additional information, such as critical path analysis, and the application of wisdom in proposing and evaluating options.

15.3.6 Applying DIKW to Communications and Stakeholders

As explained at the start of this chapter (see Figure 15.1), the objective of complete project and program communications is to ensure that the intel is developed, structured, and exchanged in such a way as to achieve the objectives of the project or program.

Data processing skills are therefore required for information management, whereas emotional intelligence skills are required to apply the communication and to support stakeholder relationship management.

15.4 Effective Communications

15.4.1 The Scope of Communications Management for Projects and Programs

The challenge is to identify which subset of the overall area of communications management is specific to project and program management. Although it is reasonable to start from general principles, the goal of this chapter is to provide a program management focus—especially since the field of information and communications is so broad.

For projects and programs, the focus of communications management is on the relevant documents and their content—that is, what is communicated. This is clearly necessary because these documents provide the basis for the data flows that form the links between the communications actions and processes. However, this narrow focus for communications leaves a gap that needs to be filled: interpersonal communication, or how the intel is communicated.

15.5 Information and Communications

15.5.1 Planning the Work

This section is based closely on *Issacons*© by Max Wideman (2006).

Whether the plan is formal or informal, three components are required:

1. Communications with *internal* stakeholders: those directly associated with program activities
2. Communications with *external* stakeholders: those with some interest in the program—for example, the public
3. Storage and retrieval of information, including confidentiality, integrity, and accessibility features

15.5.2 Communication and the Program Manager

Stakeholder relationship management was dealt with in detail in the previous chapter in relation to the structural details of the program. However, there are general communications requirements in all programs that the program manager needs to understand. The principal communication types are shown in Figure 15.3, adapted from Wideman (2006) to describe the role of the program manager rather than the project manager.

The program manager acts as the central hub for all of the flows of data and information. Without this coordination, the data would stay where they are and never be transformed into information; no knowledge would be gained as to the status of the program; and no decisions based on reality could be made—an unwise manner in which to proceed!

As shown in Figure 15.3, there are four main groups of stakeholders plus the corresponding communication flows between them. The peripheral communications shown as dotted lines in the diagram are communication flows that support the work of the program manager. They are not intended to bypass the program manager completely. The four sets of direct communication channels are used by the program manager to ensure that these four groups of stakeholders and the expectations, strategy, program plans, and activities remain aligned.

The Stakeholder Groups

The four stakeholder groups shown in Figure 15.3 are as follows:

- Senior management:
  - This group is responsible for setting the overall strategy to which the program needs to align. Senior management provides executive-level governance by addressing issues that require greater vision or authority than the program manager is formally authorized to apply.
- Client, sponsor, or portfolio manager:
  - This group of stakeholders will receive the results of the program. They need to be involved at all stages of the program. Their level of satisfaction has a major effect on how the success of the program is ultimately perceived.
- Line managers and other program managers:
  - These other managers are both colleagues and rivals at the same time. Discussion and collaboration are required in order to achieve an effective working relationship.
- Program team members, component managers, and contractors:
  - A program can engage both internal and external resources. The challenge for the program manager is to develop these resources into an effective and efficient team.

Direct Communications Channels

The four channels are between the program manager and:

- Senior management:
  - This channel is used at key points in the program and is also most useful in case of serious issues. The main program communication with senior management involves the peripheral channel to the sponsor.
- Client, sponsor, or portfolio manager:
  - This channel serves to keep the client, sponsor, or portfolio manager aware of the plans and progress of the program. The format and content of the information will normally change

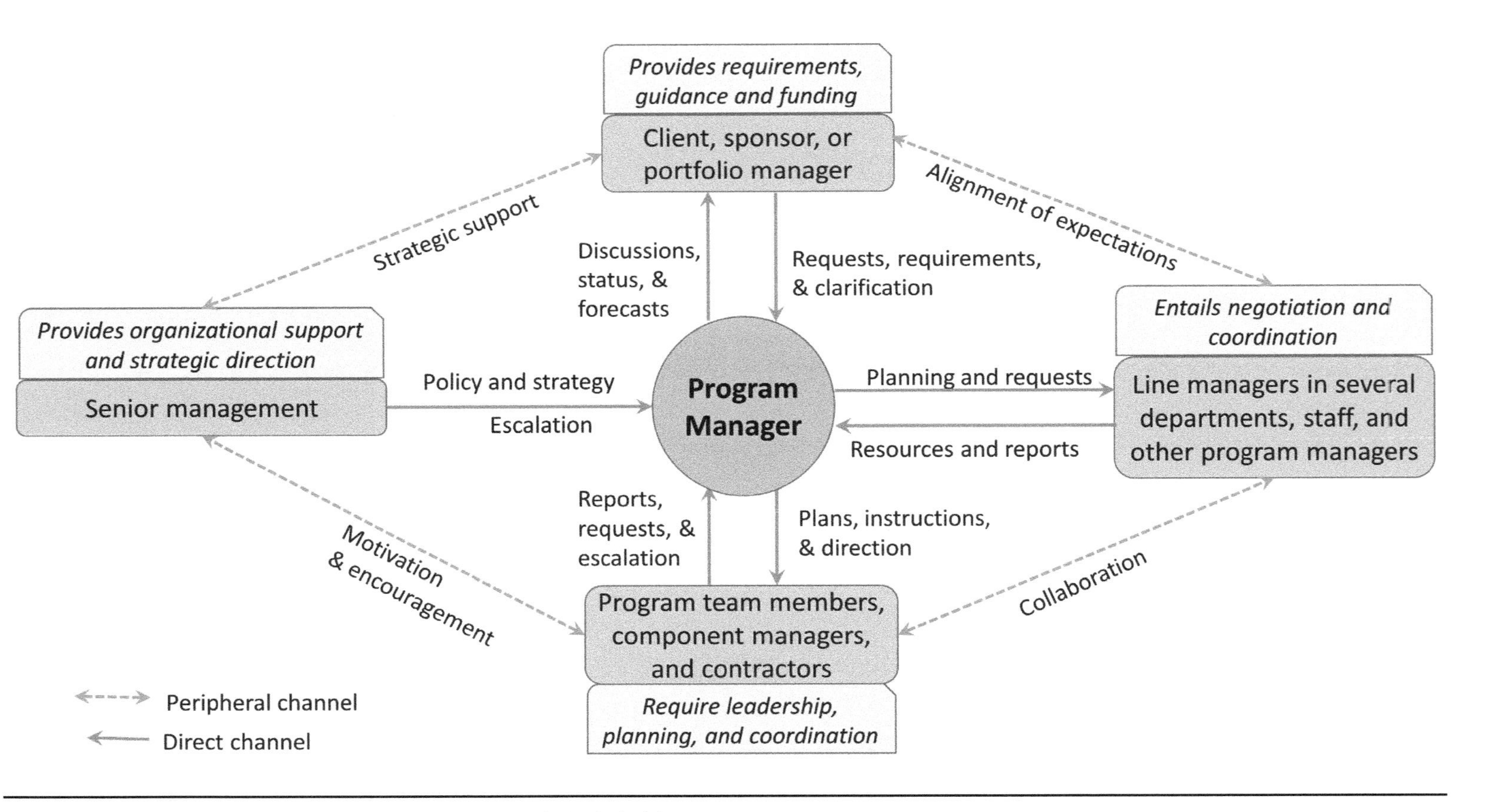

Figure 15.3 Program Manager Communications with Stakeholders

depending on the stakeholder. For example, some information may need to be kept confidential from the client.

- Line managers and other program managers:
  - This channel will be used to discuss and negotiate on common issues, such as access to—or shared use of—resources.
- Program team members and contractors:
  - This channel carries all of the interactions involved in providing guidance and leadership to the people most directly involved in delivering the required outcomes.

Peripheral Communication Channels

- Senior management ⇔ client, sponsor, or portfolio manager:
  - The channel serves to establish a strategic level relationship between the two groups.
- Client, sponsor, or portfolio manager ⇔ line managers and other program managers:
  - This channel serves to ensure that the line manager group understands any strategic decisions that affect them. It also allows the line manager group to feed tactical information back to the client group.
- Line managers and other program managers ⇔ program team members, component managers, and contractors:
  - This channel provides for hierarchical interactions such as, for example, staff member to line manager. It can be a source of confusion for program team members if they receive information across this channel that conflicts with the direct information from the program manager. The program manager, therefore, needs to use the direct channels with each of these two groups (line managers and team members) to maintain alignment and avoid confusion.
- Program team members, component managers, and contractors ⇔ senior management:
  - This communication is almost entirely a one-way path from senior management to the team members and contractors. It serves to increase understanding, motivation, and a feeling of belonging to the wider organization. The path from team members to senior management should be used only rarely, such as escalation in case of major problems that cannot be resolved by any other means.

All of these paths contribute to providing the program manager and key stakeholders with the information required to direct and control the program. Good intel is the key ingredient of good decisions.

Given the increasing complexity of the project, program, and portfolio environments, an additional challenge is to guard against information overload while still providing everything that is required to create relevant knowledge. The viable system model (Beer 2003) described next provides a means for addressing this challenge.

15.6 The Viable System Approach

In Stafford Beer's viable system model (VSM), the roles of information and of communication channels in controlling the inherent complexity of the overall system are addressed. The relevance of this theoretical, systemic approach can be applied to increase understanding of the overall structure of a given domain. The following definitions form the basis of the viable system approach:

- A system is *viable* if it is able to respond appropriately to the various threats and opportunities presented by its environment, and it is able to survive.

- The term *variety* is used to describe the variability of the environment, as well as the capability of the system and its components to cope with, and to change appropriately as required by, the structure of the system. This system variety is closely allied to the current concept of agility, which appeared much later than the viable system ideas and with considerable hubris.
- *Requisite variety* is the level of variety of a system that is compatible with the environment in which the system has to exist, adapt, and survive.

This VSM approach can be applied to communications and information management in projects, programs, and portfolios, supported by the channels in the VSM, as explained next.

15.6.1 The Viable System Model

The overall objective of the VSM is to provide a mechanism by which the variety of the business environment and that of the solutions can be balanced and jointly managed. The optimum solution is one that maintains a maximum diversity and flexibility while providing sufficient control to avoid unmanageable variability. As stated above, a viable system is defined as a system that is able to survive in a specific environment. *Survival* here means maintaining many of its important features over the short term and some fundamental (identifying) features over the long term. This link between the short term (projects) and long term (programs), with a focus on maintaining an identity (strategic direction) in a changing environment, makes the model particularly suitable for integrated project, program, and portfolio management. The links (or *channels*) between the systems in the VSM correspond to the communication paths that support adaptation and alignment between the systems.

15.6.2 The Components of the VSM

All of the VSM systems can be considered recursive or re-entrant, with each level able to be split into a set of similar systems. Here we give a brief introduction to the description of the organization encapsulated in a single level of the VSM.

The VSM is composed of five interacting subsystems (see Figure 15.4), which may be mapped onto aspects of organizational structure. In broad terms, Systems 1–3 are concerned with the "here and now" of the organization's activities, and System 4 is concerned with the "there and then"—strategic responses to the effects of external, environmental, and future demands on the organization. System 5 is concerned with balancing the "here and now" with the "there and then" to give policy directives that maintain the organization as a viable entity. Communication is represented by links, called *channels* in the VSM, between the systems. To support the overall viability, these communications channels must have sufficient capacity for the data or information exchanges without overloading the communicating entities.

The general role of each system is defined next, followed by a description of the channels. The brief names in brackets for each system have been provided as an aid to understanding and are not part of the formal VSM terminology. The way program management can be mapped onto these concepts is described after the general model shown in Figure 15.4 has been explained.

System 1 (*Transformation*) in a viable system, System 1 contains several primary activities concerned with performing a function that implements at least part of the key transformation of the organization. Each System 1 primary activity is itself a viable system because of the recursive nature of systems as described above.

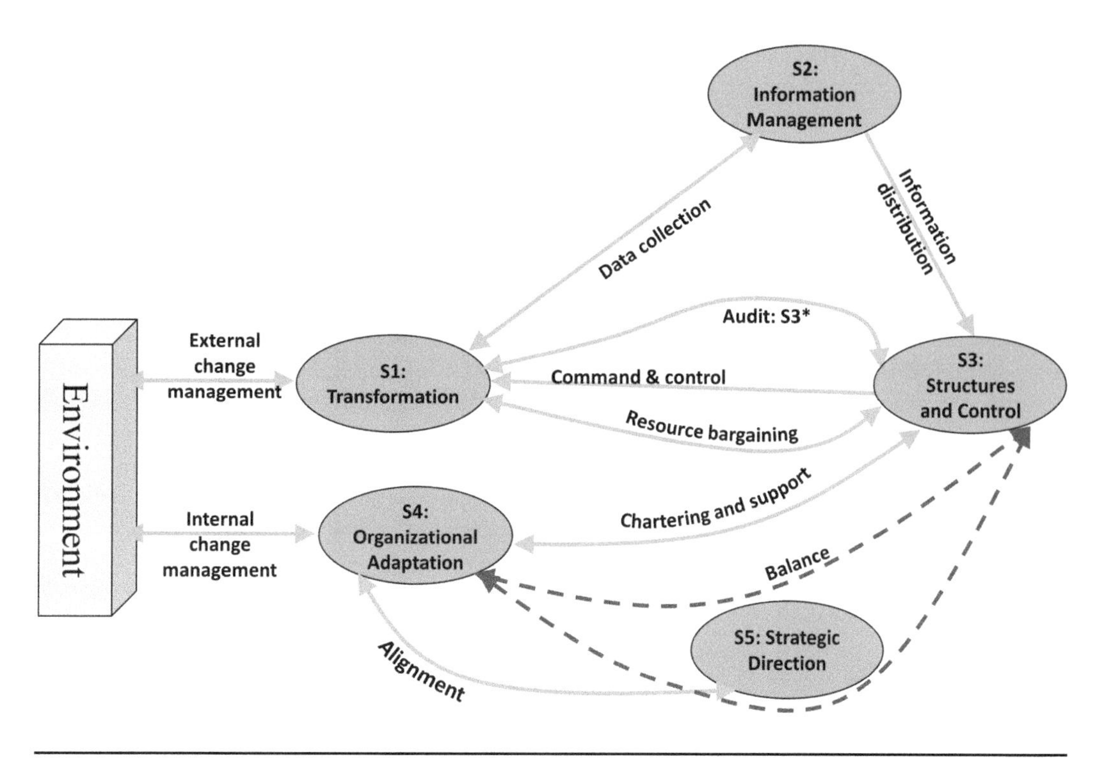

Figure 15.4 The VSM and Its Channels

System 2 (*Information Management*) represents the information channels and bodies that allow the primary activities in System 1 to communicate between each other. These System 2 components also allow System 3 components to monitor and coordinate the activities within System 1.

System 3 (*Structures and Control*) represents the structures and controls that are put in place to establish the rules, resources, rights, and responsibilities of System 1. These structures and controls also provide an interface with Systems 4 and 5.

System 4 (*Organizational Adaptation*) is responsible for looking outward to the environment to monitor how the organization needs to adapt to remain viable.

System 5 (*Strategic Direction*) is responsible for policy decisions within the overall organization to balance demands from different parts of the organization and to steer the organization as a whole.

The business environment is also represented in the model. The presence of the business environment is necessary because it is the domain of action of the system. Without it, there would be no feedback between the organization and the solution, and no reason for the solution to exist.

Among the key points to be retained from the VSM are the following (as shown in Figure 15.4):

- The changes and adaptation are mediated through internal communication *channels* that interconnect the components of the viable system as well as those between the viable system and the business environment.
- The concept of *requisite variety* applies to all elements of the model, including the channels. Requisite variety specifies that these elements must have sufficient bandwidth, acceptable

signal-to-noise ratio, and processing power to carry the required information without excessive risk of misunderstanding between the sender and the receiver—and without overloading either of them.

Communications and Management within the VSM

The VSM addresses the variety implicit in a management situation from two points of view. The first is the horizontal relationship depicted as the links between the environment of customers and other external stakeholders with the overall organization and senior management. The second comprises the vertical links, which connect the management of the project with that of the larger organization. Beer has indicated that these two sets of links tend to interact, although not always optimally.

Vertical communications reduce the autonomy of component parts by constraining their conversations and transactions or overriding the application of lower-level "on the ground knowledge" for the good of the larger system. Some common purposes are satisfied if everyone is moving in the same general direction in their own way. Others require close cooperation and tight scheduling in order to succeed. Still others are subject to higher levels of control, which may be required because of direct safety and security considerations or the risk to the larger system or the environment if things go wrong. Both horizontal and vertical communications channels are used for formal and informal conversations and transactions.

Vertical Communications

The VSM serves to show the role of communications as a key factor for structuring and control. It also shows the different types of communication needed to hold the systems together. Each type of communication deserves its own specific approach at the interpersonal level. For example:

- S1 ⇔ S2: *Data collection.* The data will be collected and exchanged based on well-defined rules and formats.
- S2 ⇨ S3: *Information distribution.* The data from S1 are processed in S2. The processed data in the form of information allows analysis of status and trends. The data also allow scenario analysis and decision making. The resulting information about the system is sent to S3 to update the level of knowledge and allow for informed decision-making. (Program monitoring was addressed in Chapters 12 and 13.)
- S3 ⇨ S1: *Command and control.* This channel carries the results of decision making based on the information processed in system S2.
- S1 ⇔ S3: *Audit.* The accuracy and credibility of data with respect to its relationship to the situation and the overall knowledge are reviewed. The audit requires a question–answer data-focused approach based on openness, honesty, and a no-blame culture.
- S1 ⇔ S3: *Resource bargaining.* As the planning and implementation in system S1 progress, the need for resources becomes clearer, and a dialogue is set up to address the options and requirements. (Capacity planning was addressed in Chapter 10.)
- S4 ⇔ S3: *Chartering and support.* There is the need to link rules for structuring and controlling the program to the overall needs of the organization. The communications involve the skills of relationship building, negotiation, issue management, and the like as required by stakeholder relationship management. (See also Chapter 14 for determining the prime interests of key stakeholders.)

- S4 ⇔ S5: *Alignment.* An ongoing conversation is required in order to ensure alignment between the changes to the organization and the overall organizational strategy. (The tools for ensuring this alignment are built into the BRM described throughout this book, but especially in Chapters 5–7.)
- S3 ⇔ S4 ⇔ S5: *Balance.* One of the challenges for organizations is to balance two knowledge-related activities:
  - Transferring knowledge from the staff of System 4, where it originates, to the line managers who will apply it using System 3
  - Providing the higher-level strategic management (System 5) with the capability to track progress and intervene (only) in cases where their involvement is required for strategic or organizational reasons

Horizontal Communications

Horizontal communications provide the links between the environment of customers and other external stakeholders with the overall organization and senior management.

- Environment ⇔ S1: *External change management*

  There is a lot more variety in the environment than a program manager and team could cope with or need to know about, and a lot more variety in project and program management than senior management needs to address. Efficient program members and senior managers develop skills to select the information they need and ignore the rest, while remaining alert to signs of change and incipient instability.

  Customer (or client) knowledge is an important aspect of the knowledge that an individual practitioner, a small group, or an organization must integrate and manage. It includes general knowledge based on demographics, markets, and preferences as well as specific confidential information about who bought what for how much and what they liked or complained about the most frequently. This information is crucial to attract and retain customers. Customer knowledge also contributes to improving the speed to market of new or modified products—a major competitive advantage.

  Individual practitioners and private, public, and voluntary operations all engage in iterative exchanges with their environments, seeking information from contractors, suppliers, competitors, outside experts, and others as well as from their past, current, and potential customers. They adapt their activities on the basis of this information and evaluate the result to assess whether the desired results were obtained.
- Environment ⇔ S4: *Internal change management*

  System 4 brings together all the functions that look to the future for the organization and must be in communication with each other.

  Individuals, functions, and organizations must maintain a balance between their activities in the present and those oriented toward the needs of the future business environment. In cutting-edge technology businesses, this balance is likely to be more tilted more toward the future than it is within manufacturing organizations.

By its nature and design, the VSM is well suited to analyzing the activities and communications flows in programs.

15.6.3 The VSM and the Program Environment

Mapping the Systems

The five VSM systems in Figure 15.4 can be mapped into the program environment as follows (see also Figure 15.5):

System 1 (*Program Management,* corresponding to *Transformation*) covers the development and maintenance of the Program Management Plan with all of its subsidiary plans. These subsidiary plans give rise to one class of system-within-system recursion. Another class of recursion is generated by the fact that programs are decomposed into components. Components can be subprograms, projects, or operational activities—all of which need also to be viable systems.

System 2 (*Program Reporting,* corresponding to *Information Management*) includes activity management, tracking, and performance reporting.

System 3 (*Program Control,* corresponding to *Structures and Control*) is provided by monitoring and control in general, and project integration management in particular.

System 4 (*Program Sponsorship,* corresponding to *Organizational Adaptation*) is provided by executive management support.

System 5 (*Steering Committee,* corresponding to *Strategic Direction*) encompasses Portfolio Management and provides governance as well as a link to senior management.

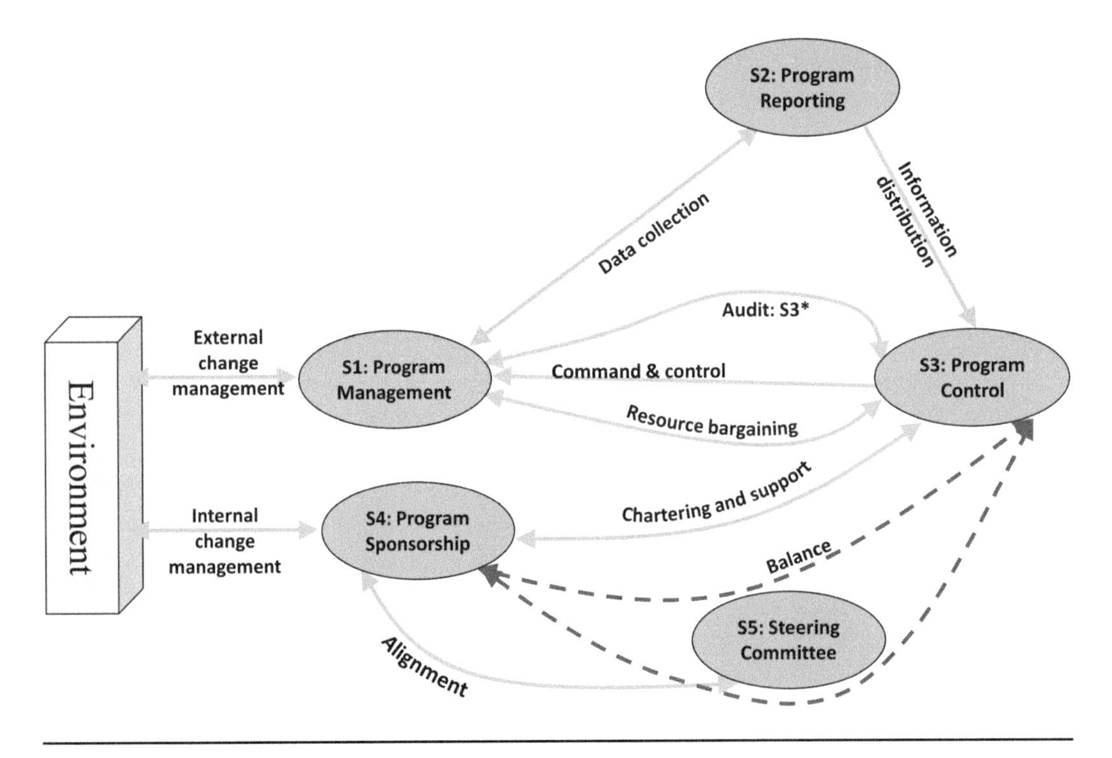

Figure 15.5 The VSM Applied to Program Management

Mapping the Channels: Vertical Communications Paths

Each of the vertical channels involves communications within the organization. Some carry data, others carry information in order to enhance the level of knowledge of the participants and consequently leverage the wisdom of the decision-makers. Each path serves a particular purpose and may require its own communication strategy.

- S1 ⇔ S2: *Data collection*
  During implementation, this channel will report earned benefit data (Chapter 12). As the implementation progresses and once some or all of the outcomes are available, program KPIs will be included in the reports.
- S2 ⇨ S3: *Information distribution*
  The data from S2 are processed, and the corresponding information is sent to S3 to be used for action planning—such as the additional data-gathering exercise for QERTS described in Chapter 13 to analyze the shortfall between actual and planned progress.
- S3 ⇨ S1: *Command and control*
  This channel carries the results of decision making based on the information processed in system S2. In the QERTS example from Chapter 13, this channel would be used to inform the program manager of the need to collect the specific data required for additional analysis.
- S1 ⇔ S3: *Audit*
  Externally funded programs and major internal programs are likely to be audited periodically to ensure that all stakeholders are confident in the accuracy of the data and the wisdom of the decisions that have been taken.
- S1 ⇔ S3: *Resource bargaining*
  The sets of required data and the approach for tailoring the organization to meet resource requirements were addressed in detail in Chapter 10.
- S4 ⇔ S3: *Chartering and support*
  Program sponsorship is normally provided from the portfolio level. The relevant documents formalizing the results of initial discussions and official negotiations include the program mandate and the program charter. Formal structures and reporting mechanisms are defined at the sponsorship level, and the corresponding paths are established. The information provided from the program includes status reports and escalation requests. Information to the program will include formal organizational decisions and updates on strategy.
- S4 ⇔ S5: *Alignment*
  Portfolio management relies on ongoing communication with a designated management team to exchange information, thus ensuring that the corresponding programs and strategic management are working toward the same objectives. Any relevant information from the environment as identified by the sponsor must also be communicated to senior management across this path.
- S3 ⇔ S4 ⇔ S5: *Balance*
  One criticism that is often heard in program teams is that senior management seems to become disconnected from the reality of the program environment. This can occur if the program sponsorship (S4) becomes a bottleneck for information exchanges between the top strategic level (S5—the Steering Committee) and program operations (S3—Program Control). The three-way communication involving all three of these systems is designed to address this risk.

Mapping the Channels: Horizontal Communications

- Environment ⇔ S1: *External change management*
  The link between the program activities and the external environment provides a window through which the program team can view the effect of their activities for creating the planned outcomes. This is also the channel through which the team and the external stakeholders interact. The importance of these interactions in establishing valuable stakeholder relationships is key to successful program management. Michel Thiry (2015) has underlined the importance of defining an explicit activity within the program management plan for marketing your program both beyond the organization—that is, via this channel—and inside the organization. The link to external stakeholders is also one of the basic requirements to achieve effective agility and adapt to changing needs and expectations. The relevant data can arrive in many different formats. The program team needs to be adept at interpreting the incoming data and taking the appropriate action in a timely manner. Treatment of the varied external data is an area in which many teams find that they do not have the *requisite variety*. They might therefore overlook important data that could otherwise, for example, have been used to avoid serious problems. This specific need for requisite variety should be addressed from the start when structuring the program management team.
- Environment ⇔ S4: *Internal change management*
  Because of the size, reach, and duration of programs, the conditions in which they perform will change during the lifetime of the program. The sponsor will need to capture the relevant information from the environment. The sponsor will then need to determine changes to the strategic direction and priorities of the program and communicate them to the program team (via the chartering channel to System S3). These changes will normally need prior discussion with senior management (via the alignment channel to System S5).

15.6.4 VSM Conclusion

The recursive nature of the VSM is mirrored in portfolios, programs, and projects by the fact that a portfolio can be composed of projects, programs, and operational activities, and programs can be composed of subprograms, projects, and nonproject (e.g., operational) activities.

The VSM provides a practical framework on which to base an efficient and effective communications strategy. The VSM concept of *requisite variety* could be paraphrased as "Goldilocks communications"—not too much, not too little, just enough. The inherent variability of the program environment is well matched to the concept of variety in general, while the broad range of program stakeholders argues strongly in favor of a formal model on which to base communications.

15.6.5 Synergy Between VSM and DIKW

The DIKW model explains the role and relationship between data, information, knowledge, and wisdom. However, it is not sufficient to know how each of these relates to the others. To manage a program efficiently, you need to know when and where to use data, or information, or knowledge, and how to apply wisdom to ensure effective communication in order to achieve the required outcomes. The VSM provides the basis for structuring control and communication of data, information, and knowledge.

15.7 Concluding Remarks

15.7.1 Vision and Intuition

Where do vision and intuition fit in? They can be smooth extrapolations or violent tectonic shifts from the knowledge and certainties of today: data, information, and knowledge aim for value-added certainties. Wisdom allows you to add the power of uncertainty to your knowledge of the situation.

For example, the go/no-go decision on product development may be based on consumer research, but it must always be balanced by experience and vision. Henry Ford is quoted as saying, "If I'd asked my customers what they wanted, they'd have said, 'a faster horse.'"

15.7.2 Attaining Wisdom

Quand bien nous pourrions être savants du savoir d'autrui, au moins sages ne pouvons-nous être que de notre propre sagesse.
(Though we might become learned through the learning of others, we can only become wise through our own wisdom.)

(de Montaigne, *Essais* [reprinted 2009])

We have seen how the DIKW hierarchy provides a basis for structuring communications and control throughout the project domains. We have discussed the optimal way of analyzing the capabilities of the various program areas and stakeholders in order to provide *requisite variety* in the exchanges. We have seen that you can communicate data and information as well as provide knowledge about explicit areas of interest. One challenging question remains: Can you communicate wisdom, and if so, how?

Whereas you *transmit* data, you *communicate* information and *convey* knowledge You might possibly *impart* wisdom but, in reality, wisdom has an intimate nature, and only *you* can *acquire* it.

This is analyzed as a personal account by Max Wideman in "Legacy" (2007). Two key points are:

- Providing advice based on your "wisdom" will frequently be resisted by the recipient as a direct attack on their personal values.
- A more indirect (and "ancestral") approach, such as storytelling, can provide examples in a way that is easier for the recipient to assimilate.

Since wisdom is the result of knowledge and personal experience, it is clear that wisdom itself is a faculty that takes time to develop. Data, information, and knowledge can be shared, whereas each person's wisdom is, in the final analysis, personal.

15.8 Summary

This chapter has addressed data, information, knowledge, and wisdom, and the relationship between them. The way in which they form a hierarchical basis for analysis and communication was explained, and examples based on earlier chapters were presented. The techniques for using the DIKW hierarchy to communicate effectively in programs have been explained and analyzed. The chapter concluded by adapting a formal systems model to address the challenge to communications caused by the size and complexity of the program organizational environment and the amount of data and information that need to be exchanged and processed in order to support effective decisions. The special position of wisdom in the hierarchy was discussed.

15.9 And Finally

This concludes the study of the various aspects of Earned Benefit Program Management. It is hoped that the information presented was sufficiently varied to have added to the readers' knowledge in a number of related areas, without creating an overload. Only time will tell if it will lead to enhanced organizational wisdom and to improving the success rate of programs in general.

15.10 References

Ackoff, R. L. *Ackoff's Best: His Classic Writings on Management.* UK: John Wiley & Sons, 1999.
Beer, S. *Diagnosing the System for Organizations.* UK: John Wiley & Sons, 2003.
de Montaigne, M. E. *Essais, I, 25.* Paris, France: Éditions Pocket, 2009.
Eliot, T. S. "The Rock" (originally published 1934). In *Collected Poems.* Orlando, FL: Harcourt Brace Jovanovich, 1991.
Marr, B. "What Are Key Performance Questions?" Management Case Study: The Advanced Performance Institute (www.ap-institute.com), 2010a.
Project Management Institute. *A Guide to the Project Management Body of Knowledge (PMBOK® Guide)—Fifth Edition.* Newtown Square, PA: Project Management Institute, 2013a.
Project Management Institute. *The Standard for Portfolio Management—Third Edition.* Newtown Square, PA: Project Management Institute. 2013b.
Thiry, M. *Program Management.* Surrey, UK: Gower Publishing Limited, 2015.
Wideman, M. "Hierarchies." *Max's Project Management Wisdom* (website), 2015. http://www.maxwideman.com/musings/hierarchies2.htm
Wideman, M. "Issacons© (Issues and Considerations): Project Management Guidelines for Program Managers." *Max's Project Management Wisdom* (website), IAC #1439, 2006. http://www.maxwideman.com/issacons/1030.htm
Wideman M. "Legacy." *Max's Project Management Wisdom* (website), 2007. http://www.maxwideman.com/musings/retire.htm

Finale

Benefit Mapping the Book

This finale maps Chapters 1 to 15 into the set of benefits that the book intends to provide.

F.1 Mapping the Book

The mapping is based on the approach defined for the benefits realization map (BRM) technique developed for programs in Chapters 4 through 6. For a book, however, the following conventions have been adopted:

- The initiatives are replaced by the chapters of the book.
- The concept of strategic outcomes has been retained.
- Chapters provide learning capabilities.
- Learning capabilities provide learning outcomes.
- Learning outcomes lead to benefits of the strategic outcomes.

This model is called the book review model, and this new approach has been used to generate the following diagram as shown in Figure F.1 and described here.

F.2 The Book's Strategic Benefits

F.2.1 The Benefits

There are two complementary benefits:

1. Enhancing the practice of project, program, and portfolio management
2. Improving the success of programs

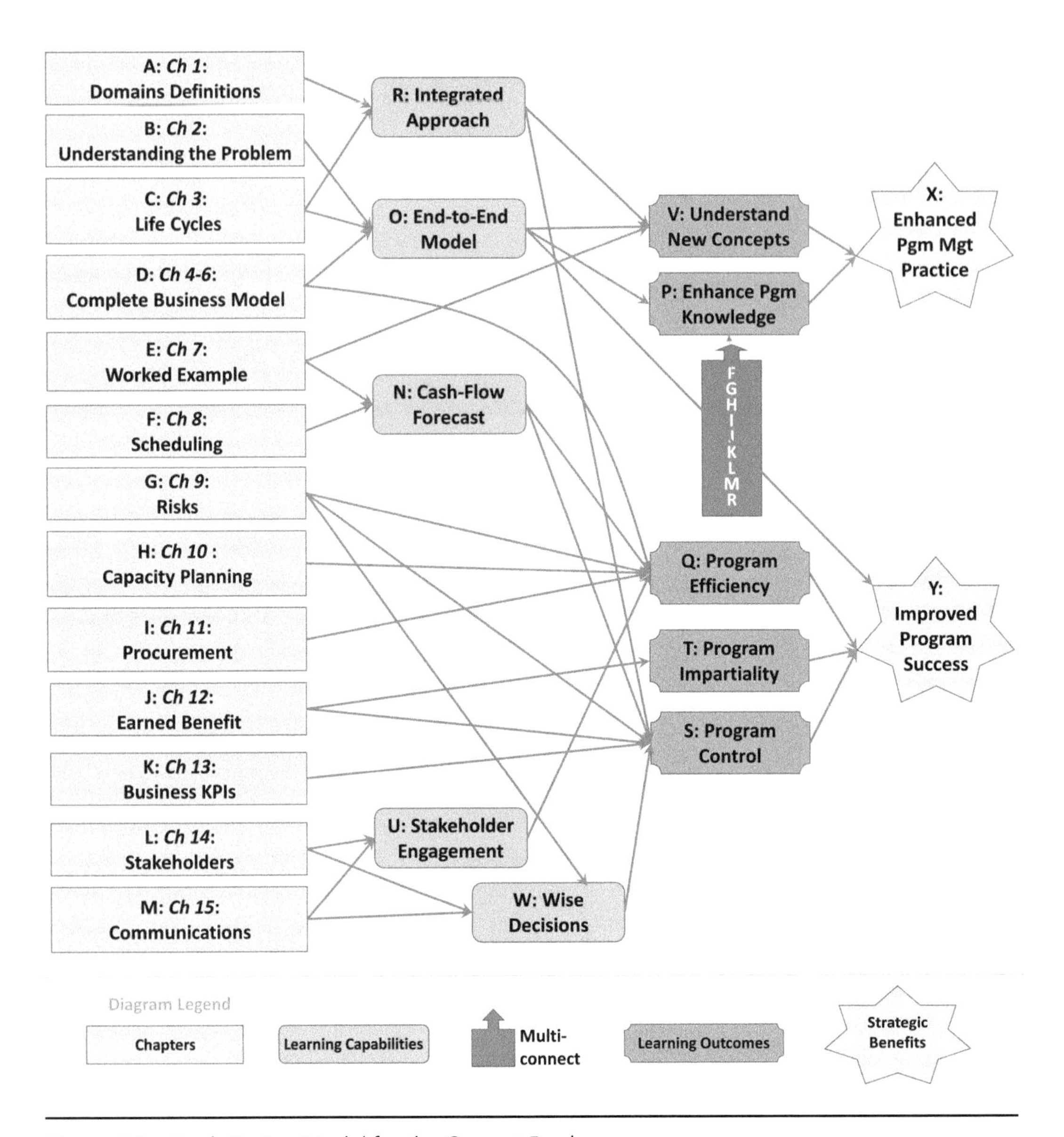

Figure F.1 Book Review Model for the Current Book

F.2.2 The Contributions

The value of the contributions of the strategic benefits have been evaluated for a theoretical business and determined as follows:

- The business has 30 program managers. In salaries and other benefits, these program managers cost the company $3 million per year. Enhancing their knowledge adds 10 percent to their value to the organization. The contribution of *enhancing the practice* is therefore $300,000.

- The business spends $50 million per year on programs. Based on the findings in *The Pulse of the Profession®* (Project Management Institute, 2015), the difference between poor program management and *improved success* is on the order of 20 percent of the total spend. Assuming that concepts in the book increase success by only 10 percent of the potential total improvement, the contribution of *improving success* is therefore 10 percent of 20 percent of $50 million = $1 million.

The total contribution is therefore $1,300,000.

Now that the contributions have been defined, the allocations to the initiatives need to be evaluated.

F.3 The Book's Initiatives

F.3.1 The Initiatives

I initially considered equating each chapter of the book, from Chapter 1 to Chapter 15, to an initiative of the BRM. However, because of their inherent interdependence, Chapters 4, 5, and 6 are grouped together in the same way as subprojects that contribute together to a project.

F.3.2 The Allocations

To gain value from the book, you need to invest time and effort. Assuming that the time and effort are proportional to the amount of reading involved, the investment for each learning initiative is proportional to the number of words in the chapter or group of chapters. The allocation value will therefore be defined in units of word. To avoid excess precision, the word counts are rounded to the nearest 500 words.

This approach shows that, in general, the benefits mapping technique can be adapted to undertakings in which the contributions and allocations are not measured in the same units. This is the case, for example, in nonprofit enterprises such as charities, in which the allocations are measured in monetary terms but not the contributions.

In our specific case, the book review model maps the network that connects Chapters 1 to 15 to the strategic benefits.

F.4 The Book Review Model

F.4.1 Learning Outcomes

The learning outcomes are derived by determining what needs to occur for the benefits from the book to be achieved.

Enhanced Program Management Practice

The benefit of *enhanced program management practice* requires the following learning outcomes:

- Understanding new concepts
- Enhanced program management knowledge

Improved Program Success

The benefit of *improved program success* depends on the following learning outcomes:

- Program efficiency
- Program impartiality: protecting the program business model from political or other bias
- Program control

F.4.2 Learning Capabilities

The learning capabilities are derived by clustering and categorizing the requirements of the learning outcomes, based on overarching program management concepts as follows:

Integrated Approach

The *integrated approach* ensures consistency across all the concepts.

End-to-End Model

An *end-to-end model* provides an understanding of how all of the concepts fit together from start to end of a program.

Cash-Flow Forecast

The capability of forecasting the cash flow links together the three key criteria for success: time, cost, and return.

Stakeholder Engagement

In planning to comply with the defined methodology and create a detailed technical solution, the program manager must ensure the support of all of the people who need to be involved for the program to succeed. All too often, this is not the case.

Wise Decisions

A program manager needs to understand how to use and deliver the information required for decision making at all of the levels involved in the program: senior management, program management, line management, technical and operational staff, and so on.

F.4.3 Chapters

The chapters are the building blocks of the book. The scope and content of each chapter is described in the Preface at the very start of this book.

F.4.4 The Links

The links are shown in Figure F.1. In the case of the links leading into *enhance program management knowledge*, the diagram has been simplified by listing the source nodes rather than drawing all of the links.

The values of the contribution fractions have been assigned in a fairly subjective manner. For this reason, and to avoid overloading the figure, they are not shown.

F.5 Analysis of the Book Review Model

The standard algorithm implementing the break-even everywhere requirement (BEER) has been applied to the book review model shown in Figure F.1. The result is shown in Table F.1.

Table F.1 Analysis of the Book Review Model

		Allocation	Contribution	Value per Word
A	Chapter 1: Domains Definitions	5,500 Words	$26,140	$4.75
B	Chapter 2: Understanding the Problem	4,500 Words	$108,000	$24.00
C	Chapter 3: Life Cycles	7,500 Words	$80,140	$10.69
D	Chapter 4-6: Complete Business Model	15,000 Words	$147,120	$9.81
E	Chapter 7: Worked Examples	5,000 Words	$48,500	$9.70
F	Chapter 8: Scheduling	6,000 Words	$84,800	$14.13
G	Chapter 9: Risks	15,500 Words	$140,000	$9.03
H	Chapter 10: Capacity Planning	3,000 Words	$61,800	$20.60
I	Chapter 11: Procurement	2,500 Words	$61,800	$24.72
J	Chapter 12: Earned Benefit	9,000 Words	$231,500	$25.72
K	Chapter 13: Business KPIs	5,000 Words	$94,700	$18.94
L	Chapter 14: Stakeholders	5,000 Words	$106,700	$21.34
M	Chapter 15: Communications	7,500 Words	$108,800	$14.51
N	Cash Flow Forecast	7,000 Words	$85,000	$12.14
O	End-to-End Model	14,000 Words	$180,000	$12.86
P	Enhance Pgm Knowledge	13,500 Words	$210,000	$15.56
Q	Program Efficiency	19,500 Words	$300,000	$15.38
R	Integrated Approach	16,000 Words	$130,700	$8.17
S	Program Control	30,500 Words	$400,000	$13.11
T	Program Impartiality	18,500 Words	$300,000	$16.22
U	Stakeholder Engagement	7,000 Words	$120,000	$17.14
V	Understand New Concepts	10,000 Words	$90,000	$9.00
W	Wise Decisions	5,500 Words	$80,000	$14.55
X	Enhanced Pgm Mgt Practice	23,500 Words	$300,000	$12.77
Y	Improved Program Success	68,000 Words	$1,000,000	$14.71
	Overall Result	91,000 Words	$1,300,000	$14.29

The analysis provides an estimate of the value of the contribution of each chapter or group of chapters to the total value of the book's benefits in the sample case used in this Finale. The numbers can be interpreted in many different ways—value per chapter and value per word are provided in Table F.1. However, in different organizational environments, these numbers may be weighted based on the perceived strategic importance each topic, capability, or benefit.

F.6 Final Summary

This Finale has shown that the BRM and the associated algorithms can be applied in areas beyond classical program management. The BRM approach is valid even when the contributions and the allocations are measured in different units. The results from the BRM analysis can provide a mapping or equivalence between the different units used for quantifying the model.

Based on the assumptions at the start of this Finale, each word in each chapter from Chapter 1 to Chapter 15 provides a value between $8 and $19, with an average of $15 per word. It is to be hoped that the reader will recognize the personal value and benefit that this book can provide.

F.7 References

Program Management Institute. *The Pulse of the Profession®*. Newtown Square, PA: Project Management Institute, 2016.

Appendix

Carrying Out the Calculations

This appendix explains how to implement the benefits realization map algorithm in Excel. People with access to more sophisticated capabilities can obviously program the algorithms directly using their favorite programming language instead and cross-check simple numerical examples using Excel. A suggested set of requirements for a complete benefits realization map tool is proposed.

The relevant logic, formulae, and rules for calculating contributions and allocations were explained and described in Chapters 5 and 6.

A.1 Calculating the Contributions

The process if you wish to carry it out is as follows (assuming use of Excel):

1. List the nodes, one per row.
   a. Enter the known contributions for the strategic outcomes.
   b. Enter the known allocations for the initiatives.
   c. Leave the other fields blank.
      i. For the Garden Centre case, there will be no blank fields.
      ii. For the QERTS case, all nodes except A to F (initiatives) and S and W (strategic outcomes) have no value assigned initially.
2. Build the contribution table.
   a. Each column represents a node, and each cell contains the corresponding contribution fraction.
      i. The sum in each row therefore adds to 100% (sum of contributions fractions into the corresponding node).
   b. The Garden Centre example from Chapter 12 (see Figure A.1) is shown (Table A.1).
3. Create a transposed copy of the same table to make the following calculations easier to set up.
4. Calculate the corresponding contribution shares from the known values.

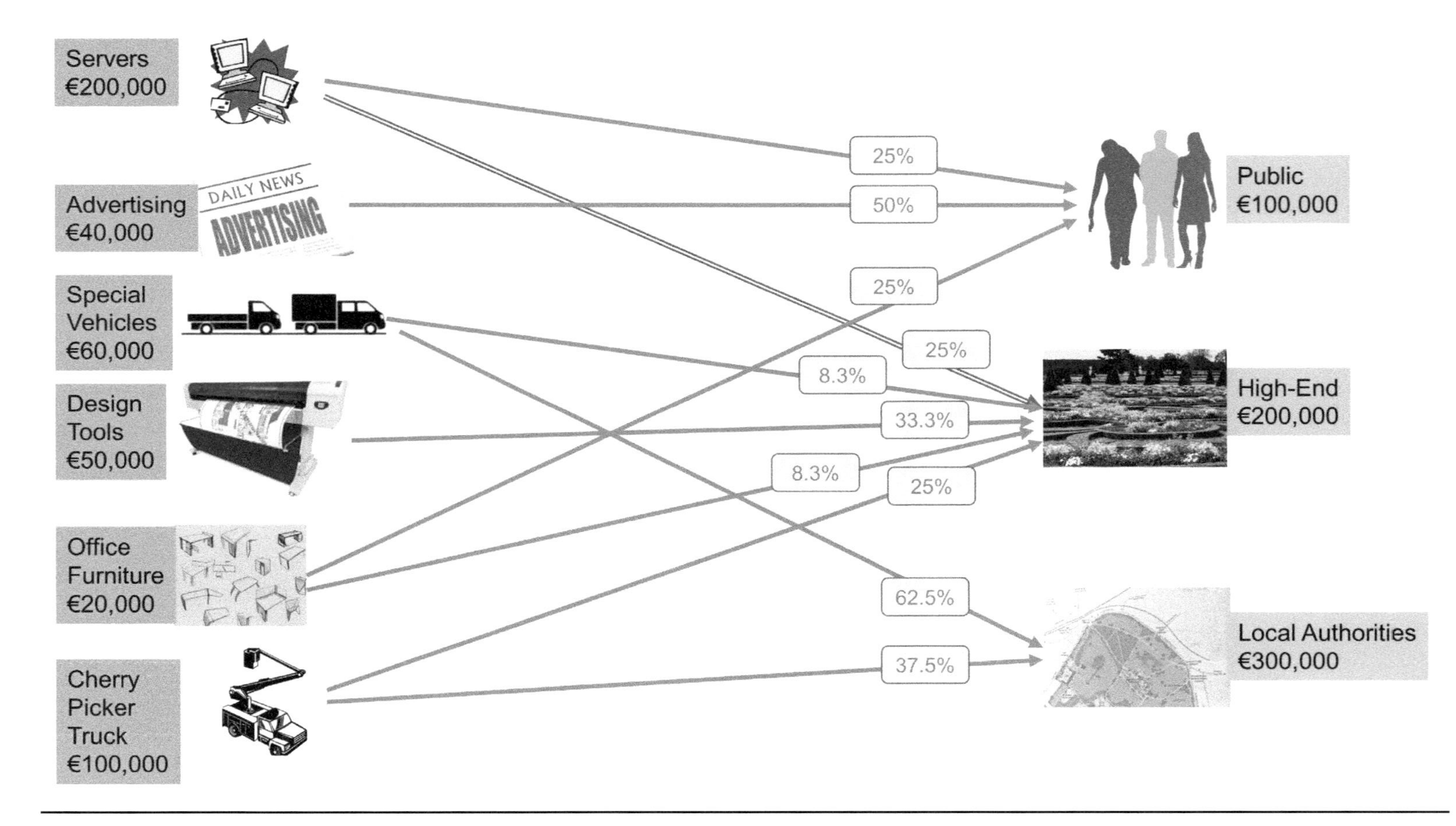

Figure A.1 The Garden Centre Benefits Realization Map

Table A.1 Contribution Table for the Garden Centre Example

		Estimated cost / value	initiative?	end-node?	Servers A	Advertising B	Special Vehicles C	Design Tools D	Office Furniture E	Cherry Picker Van F
A	Servers	€200,000	TRUE	FALSE						
B	Advertising	€40,000	TRUE	FALSE						
C	Special Vehicles	€60,000	TRUE	FALSE						
D	Design Tools	€50,000	TRUE	FALSE						
E	Office Furniture	€20,000	TRUE	FALSE						
F	Cherry Picker Van	€100,000	TRUE	FALSE						
G	Public	€100,000	FALSE	TRUE	25%	50%			25%	
H	High-End	€200,000	FALSE	TRUE	25%		8%	33%	8%	25%
I	Local Authorities	€300,000	FALSE	TRUE			63%			38%

a. Enter the known contributions.
b. Each of the unknown contributions is calculated as a SUMPRODUCT of the two following vectors: (all of the contributions [known and unknown]) by (the transpose of the corresponding row in the transposed table).
    i. This approach is taken so that each formula can be copied down to the relevant cells below, and Excel will increment the row number automatically.
    ii. If necessary, look up how to calculate vector operations (using ctrl-shift-enter to enter a SUMPRODUCT() formula).
    iii. Tell Excel that the circular reference is okay (it is circular because each contribution calculation uses the entire column of the contributions that contains itself!).
        - Go to File/Options/Formulas and enable iterative calculations.

A.2 Calculating the Allocations

5. To get the allocation fractions, calculate the normalizing factor. Remember: The allocation fraction is based on the contribution fraction and the contribution share.
    a. Start by calculating the *normalization factor*—the total contribution share of each node. This is calculated as a SUMPRODUCT of (the set of contributions for the node = the column of the Contribution Table) by (the set of contributions we just calculated).
        i. If it has a zero value, set it to 1 to avoid Excel identifying errors when dividing zero by zero.
6. Now work on the allocations: Create the allocation table of allocation fractions. Use the normalizing factors as follows:
    a. Each cell of the allocation table is the product of the contribution by the corresponding fraction (i.e., its contribution share) divided by the normalizing factor.
7. To the left of the allocation table, list the names of the nodes as well as the allocations that correspond to initiatives (in our case, the values for A to F are known). Set the other allocations to zero.
8. In a similar way to the calculation of the contributions, we can now evaluate the unknown allocations from the known ones as SUMPRODUCT of the (column of allocations) by (the transpose of the corresponding row of the allocation table).
9. Now for the essential links: In a similar way to the contribution table, create a matrix in which the cell is set to 1 if the column-to-row (source to destination) link is essential, otherwise zero.
    a. Call this the *essential link table.*
10. Make a copy of the calculation of the contributions based on the allocation table (i.e., starting from the contributions of the initiatives and calculating left to right).
    a. This will be the *essential contribution list.*
11. Make an *essential contribution zero list* from the essential contribution list:
    a. The cell is 1 if the corresponding cell in the essential contrition list is zero.
    b. Otherwise, it will be zero.
12. Change values in the essential contribution list as follows:
    a. Set the value to zero for the initiative(s) you might wish to cancel.
    b. Conditionally set to zero any other value in this list (at position *p*).
        i. If any source to it is zero (e.g., there is a 1 on "its" row in the essential link table corresponding to a zero in the essential contribution list), set it to zero if the product of the *essential contribution zero list* and "its" row in the essential contribution table is non-zero.

A.3 Completing the Calculations

The final calculation is in two steps. The first step (13) is to renormalize the contribution fractions: Wherever a source node has a zero essential contribution, the corresponding contribution fraction should be set to zero and the remaining ones to the destination renormalized to sum to 100 percent (unless their sum is 0 percent, of course), Identify these using the letter N instead of C. The second step (14) is to apply the reworked contribution fractions as normal from right to left, based on the essential values of the strategic outcomes.

13. Make two successive copies of the transposed contribution fraction table.
    a. Conditionally set the fractions corresponding to zero-contributions to zero.
    b. Set the renormalized table based on the column-sums.
14. Recalculate the contributions as above.
    a. Use the SUMPRODUCT() calculation.

A.4 Requirements from a Tool

This section consolidates the set of capabilities required from a tool.

- Diagramming:
  - Ability to create the results chain model and check the validity of linkages
    - Give instructive hints and error messages to ensure that the rules for connections are respected.
- Support for all of the defined entities:
  - Initiatives
  - Technical capabilities
    - Both prerequisite and resultant ones
  - Business capabilities
    - Both prerequisite and resultant ones
  - Business outcomes
  - Strategic outcomes
- Entity parameters:
  - Initiatives
    - Costs
    - Durations
    - Link to full description (e.g., project requirements document)
  - Capabilities (automatic calculation)
    - Contribution value
    - Allocated cost
    - Link to full description (e.g., vision statement, mission statement)
  - Business outcomes (automatic calculation)
    - Contribution value
    - Allocated cost
    - Link to full description (e.g., vision statement, mission statement)
  - Strategic outcomes
    - Value of the outcome
    - Allocated cost (automatic calculation)
    - Link to full description (e.g., vision statement, mission statement)

- Links:
  - Intelligent routing of lines
  - Indicators (selectively) for link characteristics
    - Normal or essential contributions for example
- Dependency parameters:
  - Contributions
  - Assumptions
  - Confidence (likelihood) levels
  - Delay (lag)
- Automatic calculations:
  - Contribution values [calculated right-to-left (top-down)]
    - Operational and essential values
  - Allocation percentages
    - Allocations [calculated left-to-right (bottom-up)]
  - Optional subsets of the program, taking into account the essential links
  - Gantt chart or automatic link for providing the requisite parameters to a standard scheduling tool
  - Probability distributions
    - Time-based
      - For total work
      - For various milestones (e.g., intermediate outcomes)
    - Cost-based
      - For total implementation
      - Per entity from allocations
    - Value-based
      - Per entity from contributions
    - Value-cost ratios
  - Likelihood-based sensitivity analysis
    - Progressive left-to-right (bottom-up) focus
  - Creation of the component matrices
    - Creation of the component–benefit matrix
    - Creation of the component–outcome matrix
    - Creation of the component–allocation matrix
    - Application of each of these matrixes for performance calculations
      - Based on status figures from the components
  - Calculations for earned benefit (Chapter 12) and the business key performance indicators (Chapter 13)
  - Calculations to analyze stakeholder power and interest

A.5 Final Comment

Although this appendix describes how to use a spreadsheet to implement the calculations, this approach should soon be superseded. A leading software company, experienced in this domain, is currently integrating the break-even everywhere requirement (BEER) algorithm into its existing product. The planned upgrade of this product will provide a graphical front-end for developing the results chain, automatic calculation, and additional visual indicators, as well as artificial intelligence–based decision support features.

Acronyms

See also Chapter 8 (Scheduling): 8.7 Additional Definitions, and Chapter 12: 12.7 Earned Value Abbreviations, Values, and Indicators and 12.15 Earned Benefit Parameter Tables.

ALAV	Assessed link assumption variance (Chapter 13)
ALE	Algorithm for link evaluation; takes into account the effect of *essential links* for evaluating the cash flow (Chapter 8)
ANAV	Assessed node assumption validity (Chapter 13)
ARCF	Assessed revised contribution fraction (Chapter 13)
BAKA	Business analysis knowledge area (Chapter 3)
BAU	Business as usual; the current operational environment (Chapter 1)
BCSU	Bank Station Capacity Upgrade Project (Chapter 11)
BEER	Break-even everywhere requirement: the basis for the algorithm for calculating the *allocation* parameters based on the *contribution* details of the model (Chapter 5)
BRM	Benefits realization map; the *results chain* plus all of the details concerning the *entities* and the implementation approach (Chapter 4)
CAD/CAM	Computer-aided design/computer-aided manufacturing (Chapter 4)
CAF	Current allocation fraction (Chapter 8)
CCPM	Critical chain project management (Chapter 8)
CDC	Current date contribution (Chapter 8)
COM	Component–outcome matrix (Chapter 12)
DIKW	Data, information, knowledge, and wisdom (Chapter 15)
EDF	Effective date factor (Chapter 8)
ENA	Essential node adjustment (Chapter 8)
EVM	Earned value management (Chapter 12)
ICE	Innovative contractor engagement (Chapter 11)
IPM	Integrated project management (Chapter 1)
IT	Information technology (Chapter 1)

ITT Invitation to tender (Chapter 11)

JIT Just-in-time (see also *Kanban* in the Glossary) (Chapter 4)

KBR Keep the business running: ongoing operations (Chapter 10)

KPI Key performance indicator (Chapter 13)

LAGER Lag evaluation rules; applied to calculating the dates at which *benefits* actually start to appear (Chapter 8)

LAR Loss avoidance ratio; the relative loss of value to a node that would occur if a given *essential contribution* were missing (Chapter 6)

MCF Model contribution fraction (Chapter 13)

MCNC Model calculated node contribution (Chapter 13)

MEAT Most economically advantageous tender; the evaluation choice that takes into account value-added options in a procurement (Chapter 11)

MNPE Model node percent earned (Chapter 13)

NC New contribution (Chapter 8)

OAIP Operational additional improvement potential (Chapter 13)

OCNC Operational current node contribution (Chapter 13)

OFNC Operational forecast node contribution (Chapter 13)

ONPA Operational node percent achieved (Chapter 13)

PDM Precedence diagramming method (Chapter 8)

PERT Program evaluation and review technique (Chapter 8)

PM Project manager

PMKA PMI® knowledge area (Chapter 13)

ROI Return on investment (Chapter 7)

SME Subject matter expert (Chapter 3)

SOA Scheduled operational activities (Chapter 10)

SONC Sum of new contributions (Chapter 8)

SOW Statement of work (Chapter 1)

SW Southwest Airlines (Chapter 2)

TRIM Total risk and issue management (Chapter 9)

URA Unplanned remedial actions (Chapter 10)

VEST Vectorized event scheduling technique (Chapter 8)

VFM Value for money (Chapter 11)

VSM Viable system model (Chapter 15)

WBS Work breakdown structure (Chapter 1)

Glossary

Advantage A positive *issue*.

Allocation The total cost for realizing a given *node*. The allocation of an *initiative* is given by its estimated budget.

Allocation Fraction The percentage of the *allocation* that the source "charges" the destination for the result of the work carried out by the source. The sum of all allocation fractions originating from any node is 100 percent.

Allocation Share The amount of the *allocation* of a source that the source "charges" the destination for the result of the work carried out by the source.

Benefit A change in internal or external business conditions that has a positive effect with respect to the business objectives of the organization concerned.

Benefits Realization Map The *results chain* plus all of the details concerning the *entities* and the implementation approach.

Business Capability A means of acting on or within the current business environment.

Business Outcome A change to conditions or options in the business environment that has an effect with respect to business objectives.

Capability A means, such as a product, tool, technique, or knowledge element, that can be applied for achieving a result. There are two categories of capabilities: technical and financial.

Contribution The total value of a *node* measured in the same units as the projected value of the strategic *outcomes*.

Contribution Fraction The percentage of the destination value or *capability* that depends on the source. The sum of all contribution fractions entering any node is normally 100 percent.

Contribution Share The amount of *contribution* that a *node* inherits from a given destination in exchange for the *capability* or *outcome* provided by this node to that destination.

Critical Chain A scheduling method that takes resources, team accountability, and natural variability into account.

Dependency A relationship between the source and destination nodes. Examples of these relationships concern *contributions* and *allocations*.

Disbenefit A side effect of an action or *outcome* that generates a *problem* or causes a reduction in a planned *benefit*.

Earned Benefit Method A method for tracking program progress with respect to the business case. It entails translating work completed into the potential value of the work with respect to the program's quantified objectives.

Earned Value Management A method for tracking execution progress with respect to a program or project plan. It entails mapping work completed onto the program or project's baselined budget.

Entity Any component of the *benefits realization map*. It can be represented pictorially by a *node* in the *results chain*.

Essential Contribution A *contribution* whose presence is required before the destination can acquire value from any of the other contributions.

Essential Link The link between the source and the destination of an *essential contribution*.

Financial Outcome A change to conditions or options in the environment that has an effect with respect to financial objectives. Generalized as *strategic outcome* to cover nonprofit organizations.

Hierarchy A set of groupings or concepts arranged so that each level depends on the content of the layer directly below it. Alternatively, a set of groupings or concepts in which the higher level provides a more general view or synthesis of the concepts at the level directly below it.

Initiative A specific action carried out to create a given *capability*. This is normally implemented as a project or a *program*.

Issue A situation that, if not addressed effectively, will affect success. An issue can have negative consequences (a problem) or positive results (an advantage).

Kanban A scheduling technique to reduce the overheads caused by developing intermediate or final deliverables too long before they are required. Kanban is linked to the just-in-time supply chain approach.

Link A connection between two *nodes* (the source node and the destination node) indicating that the source contributes to the destination node.

Node Any one of the following: an *initiative*, a *capability*, or an *outcome*.

Outcome A change to conditions or options in the environment. There are two categories of outcomes: technical, and strategic.

Portfolio A collection of projects, *programs*, and, possibly, subportfolios that share critical resources and contribute in a given strategic area.

Portfolio Management The application of limited resources to a *portfolio* in such a way as to optimize the overall return on the use of these resources, measured with respect to a set of strategic *objectives*.

Problem A negative *issue*.

Program A set of projects, *subprograms*, and, possibly, nonproject activities for which a joint management approach provides positive synergies towards a defined business goal.

Program Management	Coordinated management of all *program* components in such a way as to optimize the overall business *outcome*.
Program Success	Achieving the objectives and delivering the scope while complying with relevant constraints.
Results Chain	The full diagram of *links* and *nodes*.
Risk	A potential *issue*. The combination of an uncertain situation, the likelihood of occurrence of the situation, and the impacts (positive or negative) that the occurrence of the situation would have on the success of the endeavor.
Risk Attitude	A chosen response, based on perception, to an uncertain situation, the uncertainty of the situation, and the impacts (positive or negative) that the occurrence of the situation would have on the success of the endeavor.
Risk Management	Proactive issue management.
Strategic Outcome	A change to conditions or options in the environment that has an effect with respect to strategic *objectives*.
Subprogram	A separable subset of a *program*. This is defined in more detail in Chapter 7.
Technical Capability	A means of performing a task, such as a tool or a skill.

Bibliography

Ackoff, R. L. *Ackoff's Best: His Classic Writings on Management.* UK: John Wiley & Sons, 1999.
Beer, S. *Diagnosing the System for Organizations.* UK: John Wiley & Sons, 2003.
Bernoulli, D. "Exposition of a New Theory on the Measurement of Risk." *Econometrica* 22, no. 1 (1954): 23–36.
Cabinet Office, Strategy Unit. "Improving Government's Capability to Handle Risk and Uncertainty." 2002. Retrieved from http://www.rdec.gov.tw/DO/DownloadControllerNDO.asp?CuAttachID=6279
Center for Security Studies (CSS). "Fukushima and the Limits of Risk Analysis." 2001. Retrieved from http://www.css.ethz.ch/publications/pdfs/CSS-Analysis-104-EN.pdf
CMMI Institute. "Capability Maturity Model Integration." 2010. Retrieved from http://cmmiinstitute.com/cmmi-models
Ciano, G. (Count). *The Ciano Diaries, 1939–1943*, Vol. 2. Portsmouth, NH: Heinemann, 1947.
de Montaigne, M. E. *Essais, I, 25.* Paris, France: Éditions Pocket, 2009.
Eliot, T. S. "The Rock," in *Collected Poems.* Orlando, FL: Harcourt Brace Jovanovich, 1991.
Erlang, A. K. *The Theory of Probabilities and Telephone Conversations* [*Nyt Tidsskrift for Matematik B*], Vol 20, 1909.
Fahey, L., and Randall, R. M. *Learning from the Future: Competitive Foresight Scenarios.* Hoboken, NJ: John Wiley & Sons, Inc., 1997.
Fleming, Q. W., and Koppelman, J. M. *Earned Value Project Management.* Newtown Square, PA: Project Management Institute, 2000.
Full House Committee on Oversight and Government Reform. "Wasting Information Technology Dollars: How Can the Federal Government Reform Its Investment Strategy?" Full House Committee on Oversight and Government Reform Hearing (2013). Available from https://oversight.house.gov/hearing/wasting-information-technology-dollars-how-can-the-federal-government-reform-its-it-investment-strategy/
Hillson, D. A. *Exploiting Future Uncertainty.* Farnham, UK: Gower, 2010.
Hynuk, S., and Benoît, R. "Measuring Portfolio Strategic Performance Using Key Performance Indicators." *Project Management Journal* 41, no. 5 (2010): 64–73.
International Institute of Business Analysis™. *A Guide to the Business Analysis Body of Knowledge® (BABOK® Guide), version 3.* Toronto, Ontario, Canada: International Institute of Business Analysis™, 2015.
International Organization for Standardization (ISO®). *Risk Management—Principles and Guidelines* [ISO 31000:2009]. Geneva, CH: International Organization for Standardization, 2009.
Kahneman, D. *Thinking, Fast and Slow.* New York, NY: Farrar, Strauss and Giroux, 2013.
Kahneman, D., Slovic, P., and Tversky, A. *Judgement under Uncertainty: Heuristics and Biases.* Cambridge, UK: Cambridge University Press, 1982.
Kahneman, D., and Tversky, A. "Prospect Theory: An Analysis of Decision under Risk." *Econometrica* XLVII (1979): 263–291.
Kaplan, R. S., and Norton D. P. "The Office of Strategy Management." *Harvard Business Review* (October 2015). Available from https://hbr.org/2005/10/the-office-of-strategy-management
KoolToolz, Call Center Calculator. http://www.kooltoolz.com/ccm.htm

Levin, G., and Green, A. R. *Implementing Program Management.* Boca Raton, FL: CRC Press/Auerbach, 2014.
Mandelbrot, B. B., and Hudson, R. L. *The (Mis)Behaviour of Markets: A Fractal View of Risk, Ruin, and Reward.* London, UK: Profile Books, Random House, 2008.
Marr, B. *What Are Key Performance Questions?* Management Case Study, The Advanced Performance Institute (www.ap-institute.com), 2010a.
Marr, B. *What Is a Modern Balanced Scorecard?* Management Case Study, The Advanced Performance Institute (www.ap-institute.com), 2010b.
Mayor of London and the London Underground. *Innovative Contractor Engagement.* SECBE, 2014. Provided to the author by Simon Addyman, and also available from: https://www.secbe.org.uk/content/panels/Report%20-%20Innovative%20Contractor%20Engagement%20Procurement%20Model%20-%20Bank%20Station%20Capacity%20Upgrade-6d5f2a.pdf downloaded 12 May 2015.
Mendelow, A. *Stakeholder Mapping,* Proceedings of the 2nd International Conference on Information Systems, Cambridge, MA, 1991. Cited by Johnson, G. and Scholes, K. in *Exploring Corporate Strategy.* Harlow, UK: *Financial Times*/Prentice Hall, 1998.
Murray-Webster, R., and Hillson, D. A. *Managing Group Risk Attitude.* Aldershot, UK: Gower, 2008.
Newbold, R. *Project Management in the Fast Lane: Applying the Theory of Constraints.* Boca Raton, FL: The CRC Press Series on Constraints Management, CRC Press, Taylor & Francis, 1998.
Office of Government Commerce (OGC). *An Introduction to PRINCE2: Managing and Directing Successful Projects.* Norwich, UK: The Stationery Office (TSO), 2009.
Office of Government Commerce (OGC). *Management of Risk: Guidance for Practitioners.* Norwich, UK: The Stationery Office (TSO), 2006.
Office of Government Commerce (OGC). *Managing Successful Programmes (MSP®).* Norwich, UK: The Stationary Office (TSO), 2011.
Ohno, Taiichi. *Toyota Production System—Beyond Large-Scale Production.* New York, NY: Productivity Press, 1988.
Piney, C. "A Matter of Life and Death—Applying Risk Management to Help Search and Rescue." 2014. Retrieved from http://www.pmi.org/learning/risk-management-help-search-rescue-1474
Piney, C. "Applying Utility Theory to Risk Management." *Project Management Journal* 34 (2003):26–31.
Piney, C. "Chapter 7." In *Program Management, a Life Cycle Approach,* edited by G. Levin. Boca Raton, FL: CRC Press, 2013.
Piney, C. *The Earned Benefit Method for Controlling Program Performance.* Proceedings of the PMI® EMEA Congress, 2011.
Porter, M. E. "What Is Strategy?" *Harvard Business Review* (November–December 1996). Available from https://hbr.org/1996/11/what-is-strategy
Project Management Institute. "Ethics, Standards, Accreditation: Special Report." In *Project Management Quarterly.* Newtown Square, PA: Project Management Institute, 1983.
Project Management Institute. *Implementing Organizational Project Management: A Practice Guide.* Newtown Square, PA: Project Management Institute, 2014.
Project Management Institute. *A Guide to the Project Management Body of Knowledge (PMBOK® Guide)—Fifth Edition.* Newtown Square, PA, USA: Project Management Institute, 2013a.
Project Management Institute. *Practice Standard for Earned Value Management.* Newtown Square, PA, USA: Program Management Institute, 2011a.
Project Management Institute. *Practice Standard for Project Estimating.* Newtown Square, PA: Project Management Institute, 2010.
Project Management Institute. *Practice Standard for Project Risk Management.* Newtown Square, PA: Project Management Institute, 2009.
Project Management Institute. *Practice Standard for Scheduling.* Newtown Square, PA: Project Management Institute, 2011b.
Project Management Institute. *The Project Management Body of Knowledge.* Newtown Square, PA, USA: Project Management Institute, 1987.
Project Management Institute. *The Pulse of the Profession®.* Newtown Square, PA, USA: Project Management Institute, 2016.
Project Management Institute. *The Standard for Portfolio Management.* Newtown Square, PA: Project Management Institute, 2006a.

Project Management Institute. *The Standard for Portfolio Management—Third Edition.* Newtown Square, PA: Project Management Institute, 2013b.

Project Management Institute. *The Standard for Program Management.* Newtown Square, PA: Project Management Institute, 2006b.

Project Management Institute. *The Standard for Program Management—Third Edition.* Newtown Square, PA: Project Management Institute, 2013c.

Santiago, F. *The BRM Tool*, 2012. Available online: https://www.youtube.com/watch?v=W-liDQt_WkE

Savage S. L. *The Flaw of Averages: Why We Underestimate Risk in the Face of Uncertainty.* Chichester, UK: John Wiley & Sons, 2009.

Schuyler, J. R. *Risk and Decision Analysis in Projects.* Newtown Square, PA: Project Management Institute, 2001.

Standards Australia/Standards New Zealand. *Risk Management Guidelines, Companion to AS/NZS 4360:2004.* Sydney, Australia and Wellington, New Zealand: Standards Australia/Standards New Zealand, 2004.

The Standish Group. *Chaos Report*, 2015. Available from https://www.standishgroup.com/store/services/chaos-report-2015-blue-pm2go-membership.html

Taleb, N. N. *The Black Swan: The Impact of the Highly Improbable.* New York, NY: Random House, 2008.

Thiry, M. *Program Management.* Surrey, UK: Gower Publishing Limited, 2015.

Walker, W. E., Harremoes, P., Rotmans, J., van der Sluijs, J. P., van Asselt, M. B. A., Janssen, P., and Krayer von Krauss, M. P. "Defining Uncertainty: A Conceptual Basis for Uncertainty Management in Model-Based Decision Support." *Integrated Assessment* 4, no. 1 (2003): 5–17.

Wideman, M. "Comparative Glossary of Project Management Terms." Retrieved from http://www.maxwideman.com/pmglossary/

Wideman, M. "Hierarchies." *Max's Project Management Wisdom* (website), 2015. http://www.maxwideman.com/musings/hierarchies2.htm

Wideman, M. "Issacons© (Issues and Considerations): Project Management Guidelines for Program Managers." *Max's Project Management Wisdom* (website), IAC #1439, 2006. http://www.maxwideman.com/issacons/1030.htm

Wideman, M. "Legacy." *Max's Project Management Wisdom* (website), 2007. http://www.maxwideman.com/musings/retire.htm

Index

C

D

E

J

K

L

M

N

O

P

Q

R

S

T

U

V

W